Handbook of Plasmonics

Handbook of Plasmonics

Edited by **Jonah Holmes**

New York

Published by NY Research Press,
23 West, 55th Street, Suite 816,
New York, NY 10019, USA
www.nyresearchpress.com

Handbook of Plasmonics
Edited by Jonah Holmes

International Standard Book Number: 978-1-63238-265-8 (Hardback)

Printed in the United States of America.

Contents

Preface

I am honored to present to you this unique book which encompasses the most up-to-date data in the field. I was extremely pleased to get this opportunity of editing the work of experts from across the globe. I have also written papers in this field and researched the various aspects revolving around the progress of the discipline. I have tried to unify my knowledge along with that of stalwarts from every corner of the world, to produce a text which not only benefits the readers but also facilitates the growth of the field.

This book examines the theory as well as the practical applications of plasmonics and serves as a compilation of the modern advancements and researches in the field. It also covers the technical issues related to the field. The information presented in this book covers topics under "emerging concepts with plasmonics" and "plasmonics applications". It will be beneficial to students, engineers and basic scientists.

Finally, I would like to thank all the contributing authors for their valuable time and contributions. This book would not have been possible without their efforts. I would also like to thank my friends and family for their constant support.

Editor

Emerging Concepts with Plasmonics

Application of Surface Plasmon Resonance Based on a Metal Nanoparticle

Amir Reza Sadrolhosseini, A. S. M. Noor and Mohd. Maarof Moksin

Additional information is available at the end of the chapter

1. Introduction

Several researchers have focused on biosensor for monitoring of biological interaction. Moreover, the detection of bimolecular is an extremely significant problem. Hence, the development of optical biosensors based on optical properties of noble metal nanoparticles using Surface Plasmon Resonance was considered.

Surface plasmon resonance (SPR) is a powerful technique to retrieve information on optical properties of biomaterial and nanomaterials. Biosensor based on SPR is a versatile technique for biological analysis applications. Essentially, SPR depends on the optical properties of metal layer [1] and enviromental changes so it is related to charge density oscillation at the interface between them [2]. Hence, biomolecular possess an extreme sensitivity to plasmon resonance and they remove the requirement for extrinsic biomolecular labeling [3]. One advantage of SPR is, the light beam never passes through the dielectric medium of interest and hence the effect of absorption of the light in the analyte can be ignored. Hence, the main potential of surface plasmon resonance is characterization of medium after the metal layer.

Sometimes polymer layer was coated on gold later for detection and improvement the sensor performance. So, the properties of nanostructures gave rised to interest in biosensing. Nano particles and Nano tube have improved the sensitivity, selectivity, and multiplexing capacity in biosensors. Different types of nanostructure-based biosensors including carbon nanotube, SnO_2 nanoparticles, nanowires, and nanogap biosensors have emerged..

Biomolecular interaction are determined and predicted via angular modulation sensor; hence, the out put of SPR biosensor sensor is angle shift (θ_{spr}), associated with the point of minimum reflected light intensity and is very sensitive to the changes in the dielectric constant of the medium. So, the sensitivity of modern SPR sensing systems based on the Kretschmann configuration is such that they are capable of detecting refractive indexes as high as approximately 5×10^{-7} refractive index units, which corresponds to a 1 pg/mm^2 surface coverage of biomolecules [4, 5].

Unfortunately, when sensing biomolecules of low molecular weight or in low concentration, the variation in SPR angle is very small and this causes difficulties in detecting the acquired signal [6]. Hence, several researchers have focused on enhancement of sensitivity of biomolecular sensor. Hence, the properties of metal nanoparticles (gold and silver) have contemplate as a result of the physical characteristics of surface-enhanced Raman scattering (SERS). Zhu at al., (1998) utilized colloidal gold nanoparticle films to enhance the the Raman scattering of the underlying molecules [7]. On the other hand the colloidal gold nanoparticles were used as a immobilized element on the thin gold film[8, 9]. Moreover, to improve the roughness of the film, the inclusion of colloidal Au nanoparticles was added to the gold film. SPR biosensors using colloidal Au-modified gold films has the large angle shift and increase in minimum reflectivity suggesting that SPR spectrum depend on size and volume fraction of the Au nanoparticles [10]. To improve the detection of SPR biosensors, He et al., (2000)[11] and; Lyon et al., (1998)[10] reported using biomolecule-tagged colloidal Au nanoparticles. In this case, the SPW damped strongly because the intraction of the biomolecules -colloidal Au were occouured with the immobilized bio-monolayer sensing. However, using this method to generate a biomolecular interaction signal involves complex procedures for the preparation of nano-sized colloidal Au particles and chemical immobilizations to tag the biomolecules. This process, eliminates the major advantage of SPR technology, namely that biomolecular labeling is not required [6]. On the methods for improving the the sensitivity of SPR biosensors was reported by Gestwicki et. al. [12]. They enhanced the sensitivity and response of biosensor based on absorption spectroscopy techniques [13, 14]. They shown the resonance angle can be improved by caused by the conformational changes of proteins and enzymes when they bind with analytes of low molecular weights and the response of sensor can be increased by adopting Au nanoshells or nanoclusters.

The Au nanocluster-embedded dielectric film is fabricated through a simple co-sputtering method utilizing a multi-target sputtering system. In the proposed method, the resolution of the SPR biosensor is enhanced by precisely regulating the size and volume fraction of the Au nanoclusters in order to control the surface plasmon effect [6].

Carbon nanotubes (CNTs)[15] have been used as both electrode and transducer components in biosensors becuse of their interesting electrochemical and electrical properties including thier flexibility for varying the electrical property (from metallic to semiconductor) and their, small diameters, high mechanical strength, and easy fabrication/integration. Wang [16] and Balasubramanian and Burghard [17], presented the design methodologies, performance characteristics, and potential applications of CNT biosensors, as well as the factors limiting their practical use. It is obvious that many challenges still do remain, especially regarding the sensitivity of the nanotube response to environmental effects, and also the chosen type of raw CNT material, and its functionalization.

The other nanostructured sensors is based on nano ceramic such as ZnO and SnO_2. For example, the SnO_2 sensor has demonstrated a potential for gas sensing applications. Liu et al. [18] developed a highly sensitive and nanostructured SnO_2 thin-film sensor for the detection of ethanol vapor. Its lower LOD (below 1 ppm) demonstrates an improvement over that of other works reported in the literature. Recently, Choi et al. [19] implemented a functionalized ZnO nanowire field effect transistor (FET) biosensor for the low level detection of biomolecular interactions. This biosensor was fabricated without any doping which showed an enhanced sensitivity compared to other nanowire biosensors.

Fundamentally, the base of the biosensor is the coupling of a ligand-receptor binding reaction [1] to a signal transducer. Moreover, angular, phase and polarization modulations are the various methods applicable for distinguishing the interaction of bimolecular with the receptor. Essentially, SPR is a quantum electromagnetic phenomenon that appears at the interface of the dielectric and the metal. Under certain conditions, the energy of the light beam is absorbed by collective excitation of the free electrons called surface Plasmon (SP), which lies between the interface of the dielectric and the metal. On the other hand, when the momentum of the photon matches that of the Plasmon, the resonance appears as an interface of two two media with dielectric constants of opposite signs, and the SP wave propagates along the interface. In accordance with the SP wave properties, the SPR is classified as propagating the SPR, the long-range SPR and the localized SPR.

In the case of propagating SPR (PSPR) [20], the amplitude of the SP wave is decaying exponentially normal to the interface, which the Plasmon propagates, for a distance of about one to ten microns. The interactions between the surface and molecules in the analyte lead to the SPR signal shift which can be observed in shift of the angle of resonance and the wavelength. A second type of Plasmon's wave is Long-range SPR (LRSPR) that occurs in thin metal films with low attenuation; hence, the traveling distance of the Plasmon wave is about a millimeter in the visible range.

The localized SPR (LSPR) [20, 21] is exited on a nano-scale. On the other hand, the LSPR appears in the metal structure as a lateral size that is less than half the wavelength of the excitation photon. Hence, the light beam will interacts with the nanoparticles, so as to leads to a locally Plasmon oscillation occurring around the particle with a specific frequency similar to LSPR. However, the LSPR is more sensitive to changes in the local dielectric analyte [20]. Typically, researchers will measure the changes in the local dielectric environment or analyte via making shifts in the LSPR wavelength measurements. Moreover a variant of angle-resolved sensing that is possible for LSPR is also possible, and LSPR can provide real-time kinetic data for the binding processes. In addition, the PSPR signal is more sensitive to changes in the bulk refractive index than are the LSPR signals, so when the LSPR and PSPR signals are examined the interaction with the molecular adsorption layer and, the response of the two methods becomes comparable in the short-range[20]. Consequently, both sensitivity and flexibility are the capability of the LSPR. Moreover, the decay length is shorter than PSPR; hence, the LSPR are offered for the sensor.

2. Optical properties of metal

Assume that the material such as metal consist of charges which can be set in motion by an oscillating electric field of light that is polarized in one direction. In accordance with Newton's second law of motion, displacement of the electron and damping factor which is a function of mass of electron and relaxation time, the polarization density (P_c) of the conduction electrons is given as

$$P_c = -nex_0(t) = -\frac{ne^2\tau}{m_e\omega} E \frac{\omega\tau - i}{\omega^2\tau^2 + 1} \tag{1}$$

where $x_0(t)$, m_e and τ are the displacement , mass of electron respectively and a decay or relaxation time [2]. Hence, the total polarization is

$$P = P_b + P_c \tag{2}$$

Wher P_b is the background polarization, and the susceptibility ($\chi = \dfrac{P}{E}$) therefore the dielectric function (ε) is

$$\varepsilon = 1 + 4\pi\chi \Rightarrow \varepsilon = 1 + 4\pi\frac{P_b + P_c}{E} \tag{3}$$

By using the equations (1-3) the dielectric function can be written as [2, 21],

$$\varepsilon = 1 + \frac{4\pi P_b}{E} - \frac{4\pi ne^2\tau}{m_e\omega} \cdot \frac{\omega\tau - i}{\omega^2\tau^2 + 1} \tag{4}$$

or

$$\varepsilon = \varepsilon' - i\varepsilon'' \tag{5}$$

In the case of high frequency of $\omega \to \infty$

$$\varepsilon_\infty = 1 + \frac{4\pi P_b}{E} \tag{6}$$

Then by using equations (4) and (6),

$$\varepsilon = \varepsilon_\infty - \frac{4\pi ne^2\tau^2}{m_e(\omega^2\tau^2 + 1)} + \frac{4\pi ne^2\tau}{\omega\, m_e(\omega^2\tau^2 + 1)} i \tag{7}$$

Two frequency regimes of interest in equation (7) are to be considered for studying the material response [2].

i.   Low frequency regime

For the case of the frequency ω is less than $1/\tau$, $\varepsilon'' = 4\pi ne^2/(\omega m_e)$ the real terms of equation (7) can be neglected [2]. In this case the media is absorbent and the absorption coefficient (α) is

$$\alpha = \frac{\omega}{c}\sqrt{2\varepsilon''} \ , \ \alpha = \frac{\omega}{c}\sqrt{\frac{8\pi n e^2 \tau}{m_e \omega}} \tag{8}$$

ii.   High frequency regime

Since the dielectric function is real in the high frequency regime, the imaginary part of Eq. (7) can be ignored. In this case the frequency ω is more then $1/\tau$, the dielectric function is therefore,

$$\varepsilon = \varepsilon_\infty - \frac{4\pi n e^2 \tau^2}{m_e (\omega^2 \tau^2 + 1)} \tag{9}$$

or,

$$\varepsilon = \varepsilon_\infty (1 - \frac{\omega_p^2}{\omega^2}) \tag{10}$$

Where, the plasma frequency (ω_p) is defined as

$$\omega_p = \frac{4\pi e^2 n}{\varepsilon_\infty m_e} \tag{11}$$

As can be seen from equation (10) the dielectric function vanishes at the plasma frequency. However, for $\omega > \omega_p$ the medium behaves like non-absorbing and the light beam traverses through the medium without attenuation. On the other hand when $\omega < \omega_p$, the medium is of high reflectivity behaviour. Therefore the plasma frequency is the frequency of collective oscillations of the electron gas (plasma). The electron density oscillates at the plasma frequency and plasma oscillations are longitudinal type excitation [2]. The quantum of the plasma oscillation is called *plasmon* which can be excited by X ray and light or by an inelastic electron scattering [22].

To obtain the conditions of excitation to result in resonance, let's consider the electrical field component in the x direction,

$$\vec{E} = \vec{E} e^{i(k_x x - \omega t)} e^{-\alpha z} \tag{12}$$

where α is the absorption coefficient and k_x is the wave number. The absorption coefficients and dielectric constants of two adjacent media is related by

$$\frac{\alpha_1}{\alpha_2} = -\frac{\varepsilon_1}{\varepsilon_2} \tag{13}$$

By using the transverse wave equation ($\nabla^2 \vec{E} = \frac{\varepsilon}{c^2} \frac{d^2 \vec{E}}{d^2 t}$) on the components of the electrical field, a dispersion relation for surface plasmon can be obtained as

$$k^2_{palasmon} = \frac{\omega}{c}\left[\frac{\varepsilon_1 \cdot \varepsilon_2}{\varepsilon_1 + \varepsilon_2}\right] \qquad (14)$$

The plot of this equation is shown in Figure 1a. In the figure the wave number of the surface plasmon wave is always larger than the photon wave number and hence the photon would not be able to excite surface plasmon. The dispersion curve of the photon and surface plasmon wave must cross each other as shown in Figure 1.b [2, 20] for surface plasmon resonance to occur when the momentum of incoming light is equal to momentum of the plasmons. This can be done by using prism coupling method as in the present work.

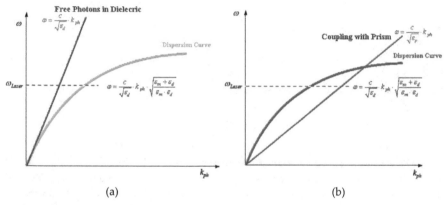

(a)   (b)

Figure 1. a) The wave number of photon is always less than the wave number of Plasmon at all frequencies and b) The wave number of photon may equal the wave number of number of plasmon.

3. Optical properties of the metal nano particles

The nanoparticles of interest are typically large enough to accurately apply classical electromagnetic theory to describe their interaction with light [23, 24]. However, they are also small enough to allow observation strong variations in the optical properties depending on the particle size, shape, and local environment. Because of the complexity of the systems being studied, efficient computational methods are the capable of treating large size materials are essential. In the past few years, several numerical methods have been developed to determine the optical properties of small particles, including the discrete dipole approximation (DDA) [23], T-matrix, and spectral representation methods (SR) [25]. We employed the DDA, which is a computational procedure suitable for studying the scattering and absorption of electromagnetic radiation by particles with sizes in the order or less of the wavelength than the incident light. DDA has been applied to a broad range of problems, including interstellar and interplanetary dust grains, ice crystals in the atmosphere, human blood cells, surface features of semiconductors, metal nanoparticles and their aggregates, and more. The DDA was first introduced by Purcell and Pennypacker [26] and has been subjected to several improvements, in particular those made by Draine et. al [27]. Below, we briefly describe the main characteristics of DDA and its numerical implementation, namely, the DDSCAT cod[1]. For a full description of DDA and DDSCAT [23], the reader can consult refs [24-28].

The main idea behind DDA is to approximate a scatter, in our case the nanoparticle, by large enough array of polarizable point dipoles. Once the location and polarizability of each dipole are specified, the calculation of the scattering and absorption efficiencies by the dipole array can be determined, depending on the accuracy of the mathematical algorithms and the capabilities of the computational hardware. Although the calculation of the radiated fields in DDA is, in theory and principle, also possible, that calculation is actually beyond the computational capabilities of any of the current systems.

Let us assume an array of N polarizable point dipoles located at $\{r_i\}$, $i = 1, 2, ..., N$, each one characterized by a polarizability α_i. The system is excited by a monochromatic incident plane wave $E_{inc}(r,t) = E_0 e^{ik.r - i\omega t}$ where r, t, ω, $k = \omega/c = 2\pi/\lambda$, c and λ are the position vector, time, the angular frequency, the wave vector, the speed of light and the wavelength of the incident light. Each dipole of the system is subjected to an electric field that can be split in two contributions: (i) the incident radiation field, plus (ii) the field radiated by all of the other induced dipoles. The sum of both fields is the so-called local field at each dipole and is given by

$$E_{i,loc} = E_{i,inc} + E_{i,dip} = E_0 e^{ik.r_i} - \sum_{i \neq j} A_{ij}.P_j \tag{15}$$

where P_i is the dipole moment of the ith element, and A_{ij} with $i \neq j$ is an interaction matrix with 3×3 matrixes as elements, such that [23]

$$\vec{A}_{ij}.\vec{P}_i = \frac{e^{ikr_{ij}}}{r_{ij}^3} \left\{ k^2 \vec{r}_{ij} \times (\vec{r}_{ij} \times \vec{P}_j) + \frac{(1 - ikr_{ij})}{r_{ij}^2} \left[r_{ij}^2 \vec{P}_j - 3\vec{r}_{ij}(\vec{r}_{ij}.\vec{P}_j) \right] \right\} \tag{16}$$

Here $r_{ij} = |\vec{r}_i - \vec{r}_j|$, and $\vec{r}_{ij} = \vec{r}_i - \vec{r}_j$, and we are using cgs units. Once we solve the 3N-coupled complex linear equations given by the relation

$$\vec{P}_i = \alpha_i.\vec{E}_{i,loc} \tag{17}$$

and determined each dipole moment $\vec{P}_i$, we can then find the extinction and absorption cross sections for a target, C_{ext} and C_{abs} in terms of the dipole moments as [23]

$$C_{ext} = \frac{4\pi k}{|\vec{E}_0|^2} \sum_{i=1}^{N} (\vec{E}_{i,inc}^* . \vec{P}_i) \tag{18}$$

---

[1] Discrete Dipole Scattering (DDSCAT) is a Fortran code for simulation and calculation of scattering and absorption of light by particles.

$$C_{abs} = \frac{4\pi k}{\left|\vec{E}_0\right|^2} \sum_{i=1}^{N} \left\{ \mathrm{Im}\left[\vec{P}_i.(\alpha_i^{-1})^\dagger P_i^\dagger \right] - \frac{2}{3}k^3 \left|P_i\right|^2 \right\} \tag{19}$$

where † means complex conjugate. The scattering cross section can be obtained using the following relation:

$$C_{ext} = C_{sca} + C_{abs} \tag{20}$$

Certain arbitrariness in the construction of the array of dipole points represents a solid target of a given geometry. For example, the geometry of the grid where the dipoles have to be located is not uniquely determined, and a cubic grid is thus usually chosen. Also, it is not obvious how many dipoles are actually required to adequately approximate the target or which choice is the best for the dipole polarization. If one chooses the separation between dipoles d such that $d \ll \lambda$, then one can assign the polarize for each particle i in a vacuum, using the lattice dispersion relation (LDR) polarization, α_i^{LDR} at a third order in k, given by [23, 27]

$$\alpha_i^{LDR} = \frac{\alpha_i^{CM}}{1 + \alpha_i^{CM}\left[b_1 + b_2\varepsilon_i + b_3 S\varepsilon_i\right]\left(k^2\big/d\right)} \tag{21}$$

where ε_i is the macroscopic dielectric function of the particle, α_i^{CM} is the polarizability given by the well-known Clausius- Mossotti relation

$$\alpha_i^{CM} = \left(\frac{d}{3}\right)^3 \frac{\varepsilon_i - 1}{\varepsilon_i + 2} \tag{22}$$

and S, b_1, b_2 and b_3 are coefficients of the expansion. Now the question is, how many dipoles do we need to mimic the continuum macroscopic particle with an array of discrete dipoles? The answer is not straightforward, because we have to consider the convergence of the physical quantities as a function of the dipole number. It has been found that $N \geq 10^4$ for an arbitrary geometry is a good starting number, as shown in the Appendix. However, we do have a matrix of $(3N)^2$ complex elements that require a large amount of computational effort. In this work, we have employed the code adapted by Draine and Flatau to solve the complex linear equations found in DDA. To solve the complex linear equations directly would require tremendous computer capabilities; however, one can use iterative techniques to compute the vector $\vec{P} \equiv \{\vec{P}_i\}$. In this case, each iteration involves the evaluation of matrix-vector products such as $\bar{A}.\vec{P}^{(n)}$, where n is the number of the iteration. The algorithm, named DDSCAT, locates the dipoles in a periodic cubic lattice, and then uses fast Fourier transform techniques to evaluate matrix-vector products such as $\bar{A}.\vec{P}$, which thus allows the whole computation of the final vector $\vec{P}$ for a large number of dipoles [28]. For a detailed description of DDA and DDSCAT code, the reader can look at refs [24-28].

In 2003, Sosa et. al. [23] defined the extinction, absorption and scattering efficiencies or coefficients, Q_{ext}, Q_{sca} and Q_{abs} as

$$Q_{ext} = \frac{C_{ext}}{A}, \; Q_{abs} = \frac{C_{abs}}{A}, \; Q_{sca} = \frac{C_{sca}}{A} \tag{23}$$

Where $A = \pi a_{eff}^2$ and a_{eff} is/are defined through the concept of an effective volume equal to $4\pi a_{eff}^3 / 3$.

In Figure 2 [23], we show Q_{ext}, Q_{sca} and Q_{abs} in dotted, dashed, and solid lines, respectively, as a function of the wavelength of the incident light, λ for nanometric-sized particles. The calculations were done for nanoparticles with $a_{eff} = 50nm$ and dielectric functions as measured on bulk silver and gold by Johnson and Christy [29]. The nanoparticles were represented or mimic by around 65000 point dipoles in order to have a good convergence on their optical properties, as discussed below. The number of these dipoles is quite large in comparison with the numbers used in previous studies on isolated and supported small metallic nanoparticles [30, 31] where, incidentally, only an extinction efficiency was reported. Our use of a large number of dipoles is in agreement with the results found in a previous study where we learned that even for small metallic nanoparticles with radii of about a few nanometers, one needed more than 12000 dipoles to achieve convergence on their optical properties [32]. A larger number of dipoles is necessary to achieve that convergence. In particular, we found that extinction efficiency converges very rapidly as a function of the number of dipoles. However, this is not the case for the absorption or scattering efficiencies where a very large number of dipoles were necessary to achieve that convergence. It is also well known that for small nanoparticles (<20 nm) the light absorption process dominates the extinction spectrum, whereas for very large nanoparticles (>100 nm), the light-scattering process dominates. As we show in this paper, to elucidate the optical properties of medium-size nanoparticles, it is indispensable to undertake an in-depth study of the scattering and absorption efficiencies, and not only of the extinction one, because both phenomena, namely, scattering and absorption, are the same order of magnitude. Figure 2 shows the optical efficiencies for a sphere with a radius of 50 nm. In the spectra, we can observe that at about 320 nm all the efficiencies have a local minimum that corresponds to the wavelength at which the dielectric functions of silver. Therefore, this feature of the spectra is inherent to the material properties and, as we observe below, that feature independent of the particle geometry. Below 320 nm, the absorption of light is mainly due to the intra-band electronic transitions of silver; therefore, this feature of the spectra should also be quite independent of the shape and size of the particles, as it is actually corroborated in all of the graphs shown below that corresponding to a silver particle. At about 350 nm, the spectrum of Q_{abs} shows a peak is related to the excitation of the surface Plasmon of the sphere; therefore, this feature is inherent to the geometry of the particle, although the position depends on its material properties. At larger wavelengths the spectrum of Q_{abs} shows specific features from 380 nm to about 500 nm that corresponds to plasmon excitations due to higher multi-polar charge distributions [23]. If we look only at the Q_{ext}, it is not possible to observe the same features because scattering effects hide them. The Q_{sca}

spectrum shows a broad structure from *320 nm* to *750 nm*, with a maximum being about *400 nm*, three times more intense than the maximum of Q_{abs}. The characteristics of the Q_{sca} spectrum are mainly due to the size of the particle. In a previous work [33], we found that this maximum is less pronounced as the radius of the sphere increases. Also, the Q_{sca} spectra of nanospheres decay slowly as the radius increases, which means that as the sphere becomes larger it scatters light at longer wavelengths, just as expected. The long tail in the extinction spectrum corresponds to Rayleigh scattering which also observed in the extinction spectra shown below.

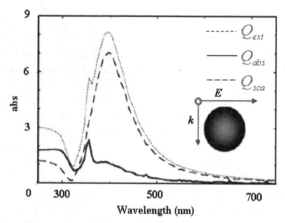

Figure 2. Optical coefficients for a silver nanosphere[23]

4. Theory

4.1. Surface plasmon resonance with prism coupling

We study SPR in the classical Kretschmann's configuration shown in Figure 3, where one side of the glass prism is coated thin gold or silver film in contact with the probed sample. Surface plasmons, which are collective oscillations of free electrons, are propagating along the interface between the sample and the metal film, and are thus sensitive to the optical properties of both the sample and the metal film. These plasmons can only be excited with TM-polarized light, which has an electric field component normal to the surface that generates the required surface charge distribution for the excitation [34].

The incident light is totally internally reflected at the interface and its intensity decreases with incident angle which is also called as coupling angle. By using the momentum and energy conservation laws [1, 35] the condition of resonance is obtained as follows:

$$k_{glass} \sin \theta_i = \frac{\omega}{c} n_{glass} \sin \theta_i = k_{plasmon} \tag{24}$$

where θ_i is the coupling angle.

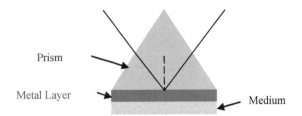

Figure 3. Kretschmann configuration

To excite the SPW the electrical and magnatisam fields are parallel and perpendicular in the incident plane which is defined by light propagation vector and the normal to the interface. Thus, in accordance with Figure 4 and the boundary conditions, the relationship between electrical and magnetism fields can be written as follows [2, 22]:

$$B_a = n_0\sqrt{\varepsilon_0\mu_0}(E_0 + E_{r1}) = n_1\sqrt{\varepsilon_0\mu_0}(E_{t1} + E_{i1}) \tag{25}$$

$$B_b = n_1\sqrt{\varepsilon_0\mu_0}(E_{i2} + E_{r2}) = n_2\sqrt{\varepsilon_0\mu_0}E_{t1} \tag{26}$$

$$E_a = (E_0 - E_{r1})\cos\theta_0 = (E_{t1} - E_{i1})\cos\theta_{t1} \tag{27}$$

$$E_b = (E_{i2} - E_{r2})\cos\theta_{t1} = E_{t2}os\theta_{t2} \tag{28}$$

The relationship between E_a, B_a and E_b, B_b are obtained by using phase change due to light passing through the different layers and $E_{i1} = E_{r2}e^{-i\delta}$, $E_{i2} = E_{t1}e^{-i\delta}$, $B_{i1} = B_{r2}e^{-i\delta}$, $B_{i2} = B_{t1}e^{-i\delta}$, $B = \dfrac{n}{c}E$ and $c = \dfrac{1}{\sqrt{\varepsilon_0\mu_0}}$ [2, 22], as follows [36]:

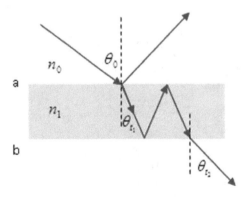

Figure 4. Reflection and refraction at the interface

$$E_a = (e^{-i\delta}\cos\delta)E_b - (i\frac{e^{-i\delta}}{\tau_1}\sin\delta)B_b \tag{29}$$

$$B_a = (-\tau_1 i e^{-i\delta} \sin\delta)E_b + (e^{-i\delta} \cos\delta)B_b \tag{30}$$

as τ_n is

$$\tau_N = \frac{n_N}{\cos\theta_{t_N}} \sqrt{\varepsilon_0 \mu_0} \tag{31}$$

Consequently, the layers cause the phase change of light and the matrix layers obtain from the coefficient of equations (30) and (31) as follows:

$$\begin{bmatrix} E_a \\ B_a \end{bmatrix} = M \begin{bmatrix} E_b \\ B_b \end{bmatrix}$$

where M is a matrix consist of

$$M = \begin{bmatrix} \cos\delta & -i\dfrac{\sin\delta}{\tau_1} \\ -i\tau_1\sin\delta & \cos\delta \end{bmatrix}, \tag{32}$$

where δ is the phase shift due to the beam passing through different layer given as

$$\delta = \frac{2\pi}{\lambda} d n_1 \cos\theta_{t_1} \tag{33}$$

where d is the thickness of layer. Considering more than one layer is involved for description and simulation of SPR, the matrix for N-layer can be achieved from the relation between tangential fields at the boundary of the first layer up to the N_{th} layer as follows [2, 36, 37]

$$\begin{bmatrix} E_a \\ B_a \end{bmatrix} = M_1 M_2 M_3 ..... M_N \begin{bmatrix} E_b \\ B_b \end{bmatrix} \tag{34}$$

Substituting,

$$M = M_1 M_2 M_3 .... M_N \tag{35}$$

and the characteristic matrix [38] is

$$M = \begin{bmatrix} m_{11} & m_{12} \\ m_{21} & m_{22} \end{bmatrix} \tag{36}$$

If one layer is deposited on the prism, the reflection and transmission coefficients are defined respectively as

$$r = \frac{E_{r_1}}{E_0} \tag{37}$$

and

$$\Gamma = \frac{E_{t_2}}{E_0} \tag{38}$$

and from equations (25, 26, 27, 28 and 36), it can be obtained that

$$\begin{bmatrix} (E_0 - E_{r_1})\cos\theta_0 \\ n_0\sqrt{\varepsilon_0\mu_0}(E_0 + E_{r_1}) \end{bmatrix} = \begin{bmatrix} m_{11} & m_{12} \\ m_{21} & m_{22} \end{bmatrix} \begin{bmatrix} E_{t_2}\cos\theta_{t_1} \\ n_2\sqrt{\varepsilon_0\mu_0}E_{t_2} \end{bmatrix} \tag{39}$$

By substituting equations (37 and 38) in equation (39) and after simplification, the reflection coefficient is as follows:

$$r = \frac{m_{21} + m_{22}\tau_2 - m_{11}\tau_0 - m_{12}\tau_2\tau_0}{m_{21} + m_{22}\tau_2 + m_{11}\tau_0 + m_{12}\tau_2\tau_0} \tag{40}$$

whereby the reflectivity is

$$R = rr^* \tag{41}$$

If there are three media of prism, n_0, metal layer, n_1 and air ($n_2 = 1$); and the thickness of metal layer, t, is less than wavelength ($\lambda = 632.8nm$) then the characteristic matrix is

$$M = \begin{pmatrix} 1 & -i(2\pi/\lambda)d\cos^2\theta_{t1} \\ -i(2\pi/\lambda)dn_1^2 & 1 \end{pmatrix} \tag{42}$$

According to equations (37, 40) and (41), the reflection coefficient is

$$r = \frac{\overbrace{\left(\tau_0 n_2\sqrt{\varepsilon_0\mu_0} - \cos\theta_{t_2}\right)}^{A} - i\alpha\overbrace{\left(n_1^2\cos\theta_{t_2}\tau_0 - \cos^2\theta_{t_1}\sqrt{\varepsilon_0\mu_0}\right)}^{B}}{\underbrace{\left(\tau_0 n_2\sqrt{\varepsilon_0\mu_0} + \cos\theta_{t_2}\right)}_{C} - i\alpha\underbrace{\left(n_1^2\cos\theta_{t_2}\tau_0 + \cos^2\theta_{t_1}\sqrt{\varepsilon_0\mu_0}\right)}_{D}} \tag{43}$$

Substituting α for $2\pi t/\lambda$, and after multiplying the numerator and denominator of equation (42) with the complex conjugate of the denominator, the reflection coefficient is derived as follows:

$$r = \frac{AC + \alpha^2 BD - i\alpha(BC - DA)}{C^2 + \alpha^2 D^2} \tag{44}$$

and the phase change [39] is

$$\tan\phi = \frac{\alpha(BC - DA)}{AC + \alpha^2 BD} \tag{45}$$

4.2. Theory of localized SPR

In a general and simple case, we can assume that the nanoparticle is as metallic sphere with the radius R. It is embedded in a liquid medium. So, the SP wave just can propagate less than the 100 nm or the size of particle; hence, becomes localized.

The electrical field outside the sphere is [20]:

$$E_{out}(x,y,z) = E_0\hat{z} - \left|\frac{\varepsilon_m - \varepsilon_d}{\varepsilon_m + 2\varepsilon_d}\right| \times a^3 E_0\left[\frac{\hat{z}}{r^3} - \frac{3z}{r^5}(x\hat{x} + y\hat{y} + z\hat{z})\right] \tag{46}$$

where E_0 is the applied field magnitude polarized in the z direction. The free electron clouds are dislocate by the electromagnetic wave field and produces uncompensated charges near the particle surface which produces corresponding opposing forces (as in Figure 5 [20]).

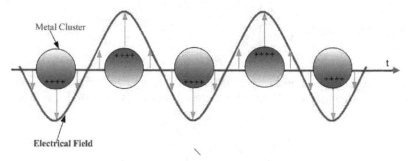

Figure 5. Collective Oscillations of free electrons due to applied electric field

The optical resonance related to these oscillations is called the localized SPR (LSPR) [20]. The origin of the term "surface" comes from the knowledge that the oscillations are caused by the polarization of the particle surface, and because the generated electric field is larger near the particle surface and decays farther away far from the surface, which makes that field similar to the evanescent field at a flat interface in the propagating SPR case. In the general case, the frequency of the collective oscillations does not coincide with the applied wave frequency and is determined by many factors, electron concentration, electron effective mass, the particle shape and size, interaction between the particles, and the influence of the environment. However, for an elementary description of the nanoparticles plasmon resonance it is sufficient to use the usual dipole approximation, and the Drude model. The electrostatic polarizability of particle α_0 [20] assigns the scattering and absorption of light by a small particle. The electrostatic polarizability of the particle can be calculated using the metal optical permittivity ε_m, the medium dielectric constant ε_d, and the particle geometrical dimensions [20, 40]:

$$\alpha_0 = a^3\left(\frac{\varepsilon_m - \varepsilon_d}{\varepsilon_m + 2\varepsilon_d}\right) \tag{47}$$

Moreover, static polarizability is [20]

$$\alpha = \frac{a\alpha_0}{a - (k^2 + 0.67ik^3 a)\alpha_0}$$ (48)

Where

$$k = \frac{2\pi\sqrt{\varepsilon_d}}{\lambda}$$

is the medium wave number. The fundamental assumption in the theoretical model is that of a spherical particle shape, which allows us to utilize extended the Mie theory. Furthermore, we can assume that the particles are distributed in a triangular two-dimensional array instead of in a disordered layer. This assumption allow us to perform efficient estimates of the inter particle coupling effects. The calculated quantity is the extinction cross-section, which is formally obtained from the integral [41]:

$$C_{ext} = \frac{1}{2I_0}\mathrm{Re}\int_A\left[E_i \times H_s^\dagger + E_s \times H_i^\dagger\right]\cdot\hat{n}dA$$ (49)

Here, ($E_i(r)$, $H_i(r)$) and ($E_s(r)$, $H_s(r)$ are the incident and total scattered fields, respectively, and I_0 is the incident irradiance. The integral is evaluated over an arbitrary surface A that encloses the entire nanoparticle system. In the case of an isolated spherical nanoparticle illuminated by a plane wave, the scattering cross-sections C_{ext} are calculated based on the Mie scattering theory [42]. From Eqs.(47) and (49) and badi the scattering cross-sections is[20]

$$C_{ext} \cong 4\pi k\,\mathrm{Im}(\alpha)$$ (50)

In the static case the extinction cross-section will be [43] explained by Eq (50). One can see from the latter expression that the extinction spectrum has a strong resonance when:

$$\varepsilon_m(\omega = \omega_{res}) = -2\varepsilon_d$$ (51)

The permittivity of bulky metal is explained by the Drude model; hence, we utilize it to obtain the resonance frequency:

$$\varepsilon_m(\omega) = \varepsilon_{ib} - \frac{\omega_p^2}{\omega(\omega + i\gamma_b)}$$ (52)

where ε_{ib} is the contribution of the inter band electronic transitions and γ_b is the volume decay constant. The metal permittivity is a main parameter for determining the LSPR frequency. So, we consider the exact Drude model. From Eq. (51) and Eq. (52) making the assumption that the significant contribution to the extinction spectrum is caused by absorption, the effective absorption cross-section can be expressed by the resonance frequency as follows

$$C_{abs} = \frac{12\pi k}{a^3} \frac{\varepsilon_d \operatorname{Im}(\varepsilon_m)}{|\varepsilon_m - \varepsilon_d|^2} \times |\alpha| = \pi a^2 \frac{12 \cdot k \cdot a \cdot \varepsilon_d}{(2\varepsilon_d + \varepsilon_{ib}^2)} \times \frac{\omega_p^2 (\gamma_b/\omega)(\omega + \gamma_b)^2}{(\omega^2 + \gamma_b^2 - \omega_{res}^4)^2 + \omega_{res}^4 \times (\gamma_b^2/\omega^2)} \tag{53}$$

The resonance frequency and resonance wavelength are given by [20]

$$\omega_{res} = \frac{\omega_p}{\sqrt{\varepsilon_{ib} + 2\varepsilon_d}} \quad \lambda_{res} = \lambda\sqrt{\varepsilon_{ib} + 2\varepsilon_d} \tag{54}$$

where

$$\omega_{res}, \ \lambda_{res} \text{ and } \lambda_p = \frac{2\pi c}{\omega_p}$$

are the resonance frequency, resonance wavelength and plasma wavelength respectively, demonstrated as [44, 45] that is near the resonance, and sometimes for comparison between the metallic bulk permittivity and the metallic nano particle permittivity, the absorption spectrum can be approximately reduced to a Lorenzian profile.

The wavelength modulation is the common method used for LSPR sensor. So, it is the shift in the maximum (or minimum) of the spectrum curve that is monitored as a function of changes in the local dielectric environment that is caused by analyte absorption. The mechanism of this method is described in Eq. (48), and it has been demonstrated for variation of refractive index or the length of a molecular which was adsorbed to receptor. In this equation, the sensitivity to the bulk refractive index ($S_{\lambda \to bulk}$) plays an important role in determining the total response of the LSPR. This sensitivity depends on various factors such as the size and shape of the nanoparticles. In the following section, we discuss the effect of the shape and design of the particle on the sensing process. Apparently, the resonance position is not dependent on the particle size as can be seen by Eq. (54) for a sphere and Eq. (50) for an arbitrary smooth particle and only variation in the intensity of the SPR band with particle size is observed. However, experimentally the SPR bandwidth as well as the band position is observed depending on the nanoparticle size. In the modification to the Mie theory, Drude and Sommerfeld considered that the relative permittivity of the nanoparticle depends on the size $\varepsilon(\omega, R)$ in lieu of $\varepsilon(\omega)$. The real and the imaginary parts of the modified relative permittivity are given by [46, 47]:

$$\varepsilon'(\omega) = \varepsilon_{ib} - \frac{\omega_p^2}{\omega^2 - \omega_d^2} \quad \text{and} \quad \varepsilon''(\omega) = \frac{\omega_p^2 \times \omega_d}{\omega(\omega^2 + \omega_d^2)} \tag{55}$$

Where ε_{ib} is the high-frequency limit of $\varepsilon(\omega)$, arising from the response of the core electrons (electrons in completely filled bands); ω_d is the relaxation or damping frequency that represents collisions of electrons with the lattice (phonons) and defects; and ω_d is the bulk plasmon frequency. Dependence of the relative permittivity on the particle radius R

was introduced by assuming that the particle size is smaller than the mean free path of the conduction electrons. The damping frequency ω_d is related to the mean free path of the conduction electrons in the bulk metal R_{bulk} and the Fermi velocity of electrons υ_f by:

$$\omega_d = \frac{\upsilon_f}{R_{bulk}} \tag{56}$$

When R becomes smaller than the mean free path, surface scattering is dominant. Such surface scattering results in the peak broadening and induces a $1/R$ dependence of the SPR bandwidth. In this case, the mean free path R_{eff} becomes size-dependent according to:

$$\frac{1}{R_{eff}} = \frac{1}{R} + \frac{1}{R_{bulk}} \tag{57}$$

One can see from the above equations that the metal particle's shape dictates the spectral signature of its plasmon resonance; the ability to change this parameter and study the effect on LSPR response is a very important experimental challenge. The development of increasingly sophisticated lithographic and chemical methods now allows for the routine production of a wide variety of complex NPs and their assemblies [50].

4.3. Metal cluster

The intraction of light with the metal cluster is the atractive aspect in nanotechnology. The electrons between metal atoms in small clusters are localized, and the plasmon resonance is a size dependent phenomenon.

The absorption and scattering cross section C_{abs} and C_{sca} are the common parameter to express the optical properties. Hence, absorption and scattering particles are not purely, so extinction cross section should be considered, which is the sum of both C_{abs} and C_{sca} or and $C_{ext} = C_{abs} + C_{sca}$. In the quasi-static the metal cluster is a spherical form with the radius is smaller than the wavelength ($R \ll \lambda$). On the other hand, extinction cross section may be written in other form as follows:

$$C_{ext}(\omega) = 9 \frac{\omega}{c} \varepsilon_m^{3/2} V_0 \frac{\varepsilon_2(\omega)}{\left[\varepsilon_1(\omega) + 2\varepsilon_m\right]^2 + \varepsilon_2^2(\omega)} \tag{58}$$

Where ω/c, V_0, ε_m and $\varepsilon(\omega)$ are the wave vector, cluster volume, dielectric function of the embedding medium and dielectric function of the particle metal. The extinction cross section is due to dipolar absorption.

A multidipolar contribution, quadrupole extinction and quadrupole scattering are conquered in the scattering cross section in this region. Consequantly, the position and shape of the plasmon resonance will depend on the size, the dielectric functions of the metal clusters, the dielectric function of the surrounding medium and the shape of the particles, its aggregations and, interactions with the substrate etc..

4.4. The nanoparticle in liquid matrix

Today, any investigation of nano-fluid is considerable. Hence, in this instance, the complex refractive index of liquids containing isotropic nanoparticles and optical the response of medium are investigated. The diameter of the particles is much smaller than the wavelength of the incident light. The response of a medium to an external electromagnetic field cannot be completely described using macroscopic fields as local field effects play an important role. This pont was first demonstrated by Lorentz who showed an impressive accuracy when predicting the dielectric function of the medium [49]. The external field induces a collection of dipole moments [50] and the local field factor relates the macroscopic fields to the local ones. These local field factors are important in the derivation of various effective medium theories (EMTs) considered next here. Regardless of the composition of the nanocomposite, the effective dielectric function can always be estimated by Wiener [51]

$$\varepsilon_{eff}(\omega) \le \sum_{i=1}^{N} f_i \varepsilon_i(\omega) \ , \ \frac{1}{\varepsilon_{eff}(\omega)} \le \sum_{i=1}^{N} \frac{f_i}{\varepsilon_i(\omega)} \tag{59}$$

where the summation is more than the number of constituents N, and their corresponding volume fractions f_i and complex dielectric functions $\varepsilon_i(\omega)$. These limits correspond to capacitors that are connected in parallel or as a series *i.e.* the Wiener bounds give upper and lower limits for the effective complex dielectric function for a certain volume fraction irrespective of the geometry of the nanostructure. Even tighter limits, such as Hashin–Shtrikman and Bergman–Milton, can be utilized in the estimation of the effective dielectric function for two-phase nanocomposites with known volume fractions of the constituents [51–54]. For a more detailed description of the different models see, *e.g.*, a review article by Hale [55].

The Maxwell Garnet (MG) EMT can be used to describe the effective dielectric function of a solution containing a small number of spherical nanoparticles. This dilute limit must be assumed as the MG model itself does not include interaction between nanoparticles or agglomeration. In such a case the effective dielectric function of a medium is given by

$$\frac{\varepsilon_{eff}(\omega) - \varepsilon_h(\omega)}{\varepsilon_{eff}(\omega) + 2\varepsilon_h(\omega)} = f_i \frac{\varepsilon_i(\omega) - \varepsilon_h(\omega)}{\varepsilon_i(\omega) + 2\varepsilon_h(\omega)} \tag{60}$$

where $\varepsilon_i(\omega)$ and $\varepsilon_h(\omega)$ are the frequency-dependent complex dielectric functions of the inclusions and the host, respectively. Naturally the corresponding volume fractions f_i and satisfy $f_i + f_h = 1$. The predicted optical properties of an MG medium are in fine agreement with the measured ones/properties for relatively low volume fractions of the inclusions.

Mathematically this low volume fraction limit of the MG formalism arises from the asymmetric equation for the effective dielectric function since the effective dielectric function would become dependent on the choice of the host material for a two-phase MG system with comparable volume fractions. The Bruggeman effective medium theory (BMT) [56] is based on the assumption that the inclusions are embedded in the effective medium itself, which is invariant if the constituents are replaced by each other. The Bruggeman formalism has been extended to inclusions with non-spherical shape by introducing a geometric (or depolarization) factor g [57], which depends on the shape of the inclusions. This generalized BMT is

$$f_h \frac{\varepsilon_h(\omega) - \varepsilon_{eff}(\omega)}{\varepsilon_h(\omega) + g[\varepsilon_h(\omega) - \varepsilon_{eff}(\omega)]} + f_i \frac{\varepsilon_i(\omega) - \varepsilon_{eff}(\omega)}{\varepsilon_i(\omega) + g[\varepsilon_i(\omega) - \varepsilon_{eff}(\omega)]} = 0 \tag{61}$$

For example, the classical Bruggeman formula is obtained with a value of $g = 1/3$ which corresponds to spherical inclusions whereas for two-dimensional circular inclusions the value of $g = 1/2$. Thus by extracting the effective dielectric function of a two-phase nanocomposite with known volume fractions and dielectric functions of the constituents, we can solve 10 for g and then obtain the geometrical shape of the investigated particles.

The quadratic Eq. (61) has two solutions for the complex effective dielectric function but only the positive branch of the square root is physically reasonable and is given by

$$\varepsilon_{eff}(\omega) = \frac{-c(\omega) + [c^2(\omega) + 4g(1-g)\varepsilon_h(\omega)\varepsilon_i(\omega)]^{1/2}}{2(1-g)} \tag{62}$$

Where

$$c(\omega) = 3(f_h - 1)\varepsilon_h(\omega) + (3f_i - 1)\varepsilon_i(\omega) \tag{63}$$

The frequency-dependent effective refractive index $n_{eff}(\omega)$ and the effective extinction coefficient $k_{eff}(\omega)$ keff(ω) of a nanoliquid are connected to effective dielectric function $\varepsilon_{eff}(\omega)$ as follows:

$$n_{eff}(\omega) = \frac{1}{\sqrt{2}}[\text{Re}\{\varepsilon_{eff}\} + (\text{Re}\{\varepsilon_{eff}\}^2 + \text{Im}\{\varepsilon_{eff}\}^2)^{1/2}]^{1/2} \quad k_{eff}(\omega) = \frac{\text{Im}\{\varepsilon_{eff}\}}{2n_{eff}} \tag{64}$$

We show in Figure 4.4 the effective dielectric functions for MG and BR type curves with geometric factors $g = 1/3$ and $g = 1/2$, which are inside the Wiener bounds as expected.

Further, at low volume fractions these three curves are almost overlap, whereas there is a rather larger difference for larger volume fractions. Hence it is important to know how to model the medium with nanoparticles to gain an accurate retrieval of the complex refractive index of an individual nanoparticle. Before closing this section we want to mention that the MG model can be generalized to hold for ellipsoids [58] by using a geometric factor such as that for BR [59]. The calculation of this geometric factor g for ellipsoids can be found from [60].

In the Rayleigh limit $r_0 \ll \lambda$, the optical intensity at SPR resonance decays approximately as $I \propto \left(\frac{r_0}{r}\right)^6$ where r is the distance from the center of the metal particle. We can use this dependence in order to define an effective refractive index of the medium surrounding the metal core according to:

$$n_{eff} = \frac{n_2 \int\limits_{r_0}^{r_0+d} 4\pi r^2 \left(\frac{r_0}{r}\right)^6 \partial r + n_1 \int\limits_{r_0+d}^{\infty} 4\pi r^2 \left(\frac{r_0}{r}\right)^6 \partial r}{\int\limits_{r_0}^{\infty} 4\pi r^2 \left(\frac{r_0}{r}\right)^6 \partial r} = n_2 - \frac{n_2 - n_1}{\left(1 + \left(\frac{d}{r_0}\right)^3\right)} \qquad (65)$$

5. SPR sensor parameters

The main performance parameters of SPR sensors for nano application are sensitivity, linearity, resolution, accuracy, reproducibility and limitation of detection.

Sensitivity

Sensor sensitivity is the ratio of the change in the sensor output to the change in the refractive index [1].

$$S = \frac{\partial Y}{\partial n} \qquad (66)$$

S, Y and n are the sensitivity, sensor output and refractive index of the probe medium respectively.

Sensor linearity

Sensor linearity is related to linear relationship of output sensor to quantity which should be measured.

Resolution and Reproducibility

The resolution of a sensor is the smallest change in the measured quantity which produces a detectable change in the sensor output. The resolution of SPR sensor usually refers to a bulk refractive index resolution. In SPR biosensor, an equivalent of this term is the limit of detection. Sensor accuracy describes the closeness of agreement between the measured quantity and the true value of the quantity such as refractive index or concentration of the analyte. It is usually expressed in absolute terms or as a percentage of the error/output ratio.

Reproducibility is the ability of a sensor to provide the same output when measuring the same value of the quantity measurement such as concentration or refractive index under the same operating condition over a period of time. It is expressed as the percentage of full range. The limitation of detection is an important parameter for evaluation of sensor. It is related to minimum detection of concentration of analyte. The minimum variation of quantity which measured with the sensor is related to minimum concentration value which is expressed the limitation detection.

6. Simulation of SPR Curve for various refractive index

Figure 6 shows the simulation of SPR curve for variation of the refractive index of analyte. The refractive index of prism and gold layer are 1.83 and .235+3.31i respectively. The wavelength of incident light was 632.8 nm (He-Ne Laser). When the incident angle was changed from 30º to 90º the refractive index of analyte shifted from 1.1 up to 1.6. Figure 7 shows the resonance angle shifted from 39.76º to 80.24º. Moreover, in accordance with Figures 8 and 9, the changes in refractive index are 0.01 and 0.001 respectively. As a result, the resonance angle shifts for small variation of refractive index of analyte are 10.944 and 0.08 respectively [61].

Figure 10 shows the variation of reflectivity with refractive index of analyte. Consequently, the reflectivity increased with increasing refractive index of analyte.

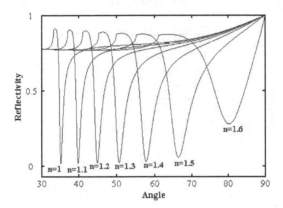

Figure 6. Simulation of SPR curve for various refractive index of analyte from 1.1 to 1.6[61]

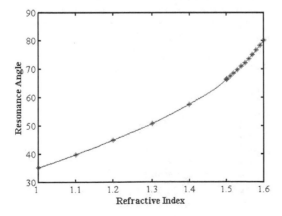

Figure 7. The variation of resonance angle with refractive index of analyte[61]

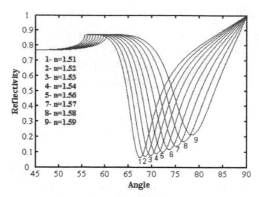

Figure 8. Simulation of SPR curve for various refractive index of analyte from 1.51 to 1.59[60]

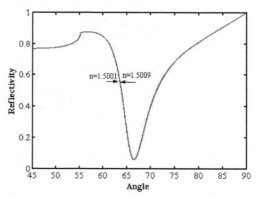

Figure 9. Simulation of SPR curve for variation of refractive index of analyte for 1.5001 and 1.5009[61]

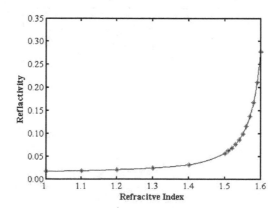

Figure 10. Reflectivity with refractive index of analyte[61]

If the nano-fluid is considered, the refractive index of analyte has an imaginary part which is changed from 0.001 to 0.2, and the resonance angle and reflectivity are shifted from 52.34º to 54.628º and .0252 to 0.6587 respectively, that are depicted in figure 11.

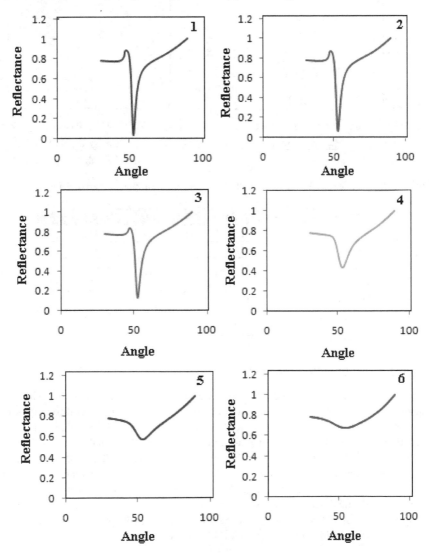

Figure 11. SPR signals related to simulation of nano fluid

The pertinent parameters were sorted in Table 1. Figure 12a and 12b show the variation of resonance angle and reflectivity versus the imaginary part.

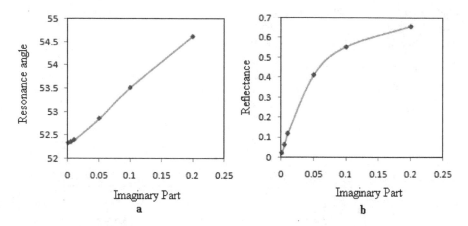

Figure 12. Variation of resonance angle and reflectance versus the imaginary part of liquid

Refractive index	Resonance angle	Reflectance
1.331+0.001i	52.34	0.0252
1.331+0.005i	52.364	0.0663
1.331+0.01i	52.405	0.1221
1.331+0.05i	52.864	0.4140
1.331+0.1i	53.523	0.5546
1.331+0.2i	54.628	0.6587

Table 1. Pertinent parameters for simulation of nanofluid

7. Application of SPR nano-metal

One of the most attractive aspects of collective excitations of Plasmon is their use to concentrate light in sub wavelength structures and enhance transmission through periodic arrays of sub wavelength holes in optically thick metal films. SPs are tightly bound to metal–dielectric interfaces penetrating around 10 nm into the metal (the so-called skin depth) and typically more than 100 nm into the dielectric (depending on the wavelength). SPs at the optical wavelengths concentrate light in a region that is considerably smaller than their wavelength, a feature that suggests the possibility of using surface – plasmon polaritons for the fabrication of nanoscale photonic circuits operating at optical frequencies. In addition the possibility of manufacturing such nano beds in planar form makes them remarkably important for biosensing towards biochip application.

Aslan et. al., reported on Nanogold-plasmon-resonance-based glucose sensing in 2004 [62]. Their sensor based on the aggregation and disassociation of 20-nm gold particles. The plasmon absorption was shifted due to interaction of nanoparticle and glucose and the response range is $\mu M \rightarrow mM$ glucose levels in many biological sample.

Application of Au nanocluster-embedded film was reported by Hu, et. al. in 2004 [6]. The used the mentioned dielectric film (SiO2) in biosensor and prepared a novel ultrahigh-resolution surface plasmon resonance biosensor. The limitation of sensor was reported 0.1 pg/mm^2 and this sensor can detect the interaction of small molecules in low concentrations. The sensitivity of sensor depend on the size and volume fraction of the embedded Au nanoclusters [6].

Another application of gold nanoparticle in SPR sensor was reported by Chau et al in 2006 [63]. A novel localized surface plasmon resonance based on fiber-optic was used for sensing Ni^{2+} ion and label-free detection of streptavidin and staphylococcal enterotoxin B at the picomolar level using self-assembled gold nanoparticles [63].

In order to develop the fully integrated portable surface plasmon resonance (SPR) system for detection of explosives, the amplification strategy of SPR signal was investigated. Indirect competitive inhibition method allowed the middle-sized SPR sensor to detect trinitrotoluene (TNT) at ppt level [64]. However, this enhanced SPR signal was not high enough to detect TNT at ppt level by a miniaturized SPR sensor. Therefore, localized surface plasmon resonance (LSPR) effect using Au nanoparticle as further signal amplification approach was used. The amplification method of indirect competitive inhibition and LSPR were combined together for fabrication of the immunosurface using Au nanoparticle. TNT detectable range of this immune surface was from 10 ppt (10 pg/ml) to 100 ppb (100 ng/ml), which was almost comparable to that without Au nanoparticle. The observed resonance angle change due to binding monoclonal TNT antibody (M-TNT Ab) with the immune surface modified with Au nanoparticlewas amplified to four times higher than that in absence of Au nanoparticle [64].

LSPR sensors and silver nanoparticle were applied with various thiolate self-assembled monolayers (SAM) to provide chemical selectivity for detection of volatile organic compounds (VOCs) [65]. LSPR spectrums of silver nanoparticles were shifted due to binding and for detection should be measured as the response signal. The limitation of sensor detection was as low as 18–30ppm for heptanone, depending on the surface modification of Ag nanoparticles [65].

Moreover for colorimetric hydrogen peroxide, the LSPR of poly(vinylalcohol) capped silver nanoparticles was used [66]. The silver nanoparticles are directly synthesized in the PVA matrix treatment in aqueous medium. No other reagent is used. . The silver nanoparticles have the catalytic ability for the decomposition of hydrogen peroxide; then the decomposition of hydrogen peroxide induces the degradation of silver nanoparticles. Hence, a remarkable change in the localized surface plasmon resonance absorbance strength could be observed. As a result, the yellow color of the silver nanoparticle–polymer solution was gradually changed to transparent colour. Furthermore, when this transparent solution was subjected to thermal treatment, it became again yellow and the UV–vis spectroscopy confirmed that nanoparticles were again formed, suggesting the renewability of this sensor. The determination of reactive oxygen species such as hydrogen peroxide has possibilities for applying to medical and environmental applications [66].

Salmonella was detected using LSPR and Au nanoparticles [67]. In this case, the small contact area between the nanoparticle and the bacteria was the cause of the short range interaction of the local electric field and bacteria. Hence, the SPR signals were shifted to higher value that can be explained by the Mie theory and effective medium theory. So, this is alternative methods for detection of Salmonella [67].

Author details

Amir Reza Sadrolhosseini
*Center of Excellence for Wireless and Photonics Networks (WiPNet), Faculty of Engineering
Universiti Putra Malaysia, UPM Serdang, Malaysia*

A. S. M. Noor
*Center of Excellence for Wireless and Photonics Networks (WiPNet), Faculty of Engineering
Universiti Putra Malaysia, UPM Serdang, Malaysia
Department of Computer and Communication Systems Engineering, Faculty of Engineering,
Universiti Putra Malaysia, UPM Serdang, Malaysia*

Mohd. Maarof Moksin
Department of Physics, Faculty of Science, Universiti Putra Malaysia, Serdang, Malaysia

8. References

[1] Homola J. Surface plasmon resonance based sensors. Verlag Berlin Heidelberg: Springer; 2006.

[2] Peyghambarian N, Koch S W, Mysyrowicz A. Introduction to Semiconductor Optics. New Jersey: Prentice Hall; 1993.

[3] Liedberg, B., Nylander, C., Lundstrom, I. 1983. Surface plasma resonance for gas detection and biosensing. Sensors and Actuators B 1983; 4 299.

[4] Stenberg E, Persson B, Roos H, Urbaniczky, C. Quantitative determination of surface concentration of proteins with surface plasmon resonance using radiolabeled protein. J. Colloid. Interf. Sci. 1991; 143 513–526.

[5] Homola J, Yee S, Gauglitz G. Surface plasmon resonance sensors: review. Sens. Actuat. B 1999; 54 3–15.

[6] Hu W, Chenb S-J, Huangc K-T, Hsud J H, Chend W Y, Changa G L, Lai K-A. A novel ultrahigh-resolution surface plasmon resonance biosensor with an Au nanocluster-mbedded dielectric film. Biosensors and Bioelectronics 2004;19 1465–1471.

[7] Zhu T, Zhang X, Wang J, Fu X, Liu Z . Assembling colloidal Au nanoparticles with functionalized self-assembled monolayers. Thin Solid Films 1998; 327–329 595–598.

[8] Gradar K C, Brown K R, Keating C D, Stranick S J, Tang S-L, Natan M J. Nanoscale characterization of gold colloid monolayers: a comparison of four techniques. Anal. Chem. 1997;69 471–477.

[9] Brown K R, Lyon L A, Fox A P, Reiss B D, Natan M J. Hydroxylamine seeding of colloidal Au nanoparticles. Part 3. Controlled formation of conductive Au films. Chem. Mater. 2000; 12 314–323.

[10] Lyon L A, Musick M D, Natan M J. Colloidal Au-enhanced surface plasmon resonance immunosensing. Anal. Chem. 1998; 70 5177–5183.

[11] He L, Musick M D, Nicewarner S R, Salinas F G, Benkovic S J, Natan M J, Keating C D. Colloidal Au-enhanced surface Plasmon resonance for ultrasensitive detection of DNA hybridization. J. Am. Chem. Soc. 2000; 122 9071–9077.

[12] Gestwicki J E, Hsieh H V, Pitner J B. Using receptor conformational change to detect low molecular weight analytes by surface plasmon resonance. Anal. Chem. 2001; 73 5732–5737.

[13] Sun Y, Xia Y. Increased sensitivity of surface plasmon resonance of gold nanoshells compared to that of gold solid colloids in response to environmental changes. Anal. Chem. 2002; 74 5297–5305.

[14] Mayer C, Stich N, Schalkhammer T, Bauer G. Slide-format proteomic biochips based on surface-enhanced nanocluster-resonance. Fresenius J. Anal. Chem. 2001; 371 238–245.

[15] Kouzani A Z, Dai X J, Michalski W P. Investigation of the effect of design parameters on sensitivity of surface plasmon resonance biosensors. Biomedical Signal Processing and Control - Biomed Signal Process Control 2011 6(2) 147-156, 2011.

[16] Wang J. Carbon-nanotube based electrochemical biosensors: a review. Electroanalysis 2005; 17 7–14.

[17] Balasubramanian K, Burghard M. Biosensors based on carbon nanotubes, Analytical and Bioanalytical Chemistry 2006; 385 452–468.

[18] Liu Y, Koep E, Liu M. A highly sensitive and fast-responding SnO2 sensor fabricated by combustion chemical vapor deposition. Chemistry of Materials 2005; 17 3997–4000.

[19] Choi A, Kim K, Jung H-I, Lee S Y. ZnO nanowire biosensors for detection of biomolecular interactions in enhancement mode. Sensors and Actuators B: Chemical 2010; 148 577–582.

[20] Zalevsky Z, Abdulhalim I. Integrated nanophotonics devices Elsevier Science 2010.

[21] Willets K A, Van Duyne R P. Localized Surface Plasmon Resonance Spectroscopy and Sensing. Annu. Rev. Phys. Chem. 2007; 58:267–97.

[22] Kittle C. Solid State Physics New York: Wiley; 1996.

[23] Sosa I O, Noguez C, Barrera R G. Optical Properties of Metal Nanoparticles with Arbitrary Shapes. J. Phys. Chem. B 2003; 107 6269-6275.

[24] Kelly K L, Lazarides A A, Schatz G C. Computational electromagnetics of metal nanoparticles and their aggregates. Computing in Science and Engineering, 2001; 3(4) 67.

[25] Mishchenko M I, Hovenier J W, Travis L D. Light Scattering by Nonspherical Particles. San Diego: Academic Press; 2000.

[26] Toma H E, Zamarion V M, Toma S H, Araki K. The coordination chemistry at gold nanoparticles J. Braz. Chem. Soc. 2010; 21(7).

[27] Draine B T. Astrophys. J. 1998; 333 848.

[28] Draine B T, Flatau P J. Program DDSCAT; University of California at San Diego: San Diego, CA.

[29] Johnson P B, Christy, R. W. Phys. ReV. B 1972; 6 4370.

[30] Yang W-H, Schatz G C, van Duyne R P. Discrete dipole approximation for small particles with arbitrary shapes. J. Chem. Phys. 1995; 103 869.

[31] Jensen T R, Duval M L, Kelly K L, Lazarides A A, Schatz G C, Van Duyne, R P. Nanosphere lithography: Effect of the external dielectric medium on the surface plasmon resonance spectrum of a periodic array of silver nanoparticles, Journal of Physical Chemistry B, 1999; 103(45) 9846 – 9853.

[32] Noguez C, Sosa I, Barrera R G. Optical characterization of isolated nanoparticles with arbitrary shapes MRS Proc. 2002; 704 275.

[33] Noguez C, Sosa I, Barrera R G. MRS Proc. 2002, 704, 275; Singh, R. K., Partch, R., Muhammed, M., Senna, M., Hofmann, H., Eds.

[34] Fontana E. Thickness Optimization of Metal Films for the Development of Surface Plasmon Based Sensors for non-absorbing media. Journal of Applied optics 2006.

[35] Schasfoort, R. and Tudos, A. 2008, Handbook of Surface Plasmon Resonance. RSC Publishing, Cambridge, U.K.

[36] Sharma K. 2006. Optics. Academic Press, California, 301-316.

[37] Beketov G V, Shirshov Y M, Shynkarenko O V, Chegel V I. Surface plasmon resonance spectroscopy: prospects of superstrate refractive index variation for separate extraction of molecular layer parameters. Sensors and Actuators B. 1998; 48 432.

[38] Sadrolhosseini A R, Moksin M M, Yunus W M M, Talib Z A, Abdi M M. Surface Plasmon Resonance Detection of Copper Corrosion of Biodiesel Using Polypyrrole-Chitosan Layer Sensor. Optical review 2011 18(4) 331-337.

[39] Yuan X-C, Ong B, Tan Y, Zhang D, Irawan R, Tjin S. Sensitivity-stability-optimized surface plasmon resonance sensing with double metal layers. Journal of optics 2006; 959 963-8.

[40] Stewart ME, Anderton CR, Thompson LB, Maria J, Gray SK, Rogers JA . Nanostructured plasmonic sensors. Chem Rev 2008;108: 494–521.

[41] Xu H, Kall M. Modeling the optical response of nanoparticle- based surface Plasmon resonance sensors. Sensors and Actuators B 2002; 87 244-249.

[42] Mie G. Beitrage zer optik truber meiden speziell kolloidaler metallosungen. Ann. Phys. (Leipzig) 1908; 25 377–445.

[43] Gupta S, Huda S, Kilpatrick PK, Velev OD. Characterization and optimization of gold nanoparticle-based silver-enhanced immu- noassays. Anal Chem 2007; 79 3810–20.

[44] Baptista P, Pereira E, Eaton P, Doria G, Miranda A, Gomes I. Gold nanoparticles for the development of clinical diagnosis methods. Anal Bioanal Chem 2008; 391 943–50.

[45] Luo PG, Stutzenberger FJ. Nanotechnology in the detection and control of microorganisms. Adv Appl Microbiol 2008;63 145–81.

[46] Bruzzone S, Arrighini GP, Guidotti C. Some spectroscopic proper- ties of gold nanorods according to a schematic quantum model founded on the dielectric behavior of the electron-gas confined in a box. Chem Phys 2003; 291 125–40.

[47] Prodan E, Radloff C, Halas NJ, Nordlander P. A hybridization model for the plasmon response of complex nanostructures. Science 2003; 302 419–22.

[48] Wang H, Brandl DW, Nordlander P, Halas NJ. Plasmonic nanos- tructures: artificial molecules. Acc Chem Res 2007; 40 53–62.

[49] Lorentz H A. -ber die beziehungzwischen der fortpflanzungsgeschwindigkeit des lichtes der körperdichte. Ann. Phys. 1880; 9 641–665.

[50] Sutherland R L. Handbook of Nonlinear Optics New York: Marcel Dekker; 1996.

[51] Hashin Z, Shtrikman S. A variational approach to the theory of the effective magnetic permeability of multiphase materials. J. Appl. Phys. 1963; 33 3125–3131.

[52] Milton G W. Bounds on the complex dielectric constant of a composite material. Appl. Phys. Lett. 1980; 37 300–302.

[53] Bergman D J. Exactly solvable microscopic geometries and rigorous bounds for the complex dielectric constant of a two-component composite material. Phys. Rev. Lett. 1980; 44 1285–1287.

[54] Aspnes D E. Bounds to average internal fields in two-component composites. Phys. Rev. Lett. 1982; 48 1629–1632.

[55] Hale D K. The physical properties of composite materials. J. Mater. Sci. 1976; 11 2105–2141.

[56] Sipe J E, Boyd R W. Nonlinear susceptibility of composite optical materials in the Maxwell Garnett model. Phys. Rev. A 1992; 46 1614–1629.

[57] Zeng X C, Bergman D J, Hui P M, Stroud D. Effective-medium theory for weakly nonlinear composites. Phys. Rev. B 1988; 38 10970–10973.

[58] Cohen R W, Cody G D, Coutts M D, Abeles B. Optical properties of granular silver and gold films. Phys. Rev. B 1973; 8 3689–3701.

[59] Lagarkov A N, Sarychev A K. Electromagnetic properties of composites containing elongated conducting inclusions. Phys. Rev. B 1996; 53 6318–6336.

[60] Grosse P, Offermann V. Analysis of reflectance data using the Kramers–Kronig relations. Appl. Phys. A 1991; 52 138–144.

[61] Sadrolhosseini A R. Surface Plasmon Resonance Characterization Of Biodiesel. PhD thesis. Universiti Putra Malaysia; 2011.

[62] Kadir Aslan K, Lakowicz J R, Geddes C D. Nanogold-Plasmon-Resonance-based glucose sensing. Analytical Biochemistry 2004; 330 145-155.

[63] Chau L K, Lin Y F, Cheng S F, Lin T J. Fiber-optic chemical and biochemical probes based on localized surface plasmon resonance. Sensors and Actuators B 2006; 113 100–105.

[64] Kawaguchi T, Shankaran D R, Kim S J, Matsumoto K, Toko K, Miura N. Surface plasmon resonance immune sensor using Au nanoparticle for detection of TNT. Sensors and Actuators B 2008; 133 467–472.

[65] Chena Y Q, Lub C J. Surface modification on silver nanoparticles for enhancing vapor selectivity of localized surface plasmon resonance sensors. Sensors and Actuators B 2009; 135 492–498.

[66] Filippo E, Serra A, Manno D. Poly(vinyl alcohol) capped silver nanoparticles as localized surface Plasmon resonance-based hydrogen peroxide sensor. Sensors and Actuators B 2009 138 625–630.

[67] Fua J, Park B, Zhaoa Y. Limitation of a localized surface plasmon resonance sensor for Salmonella detection. Sensors and Actuators B 2009; 141 276–283.

Chapter 2

Resonant Excitation of Plasmons in Bi-Gratings

Taikei Suyama, Akira Matsushima, Yoichi Okuno and Toyonori Matsuda

Additional information is available at the end of the chapter

1. Introduction

Metal gratings have an interesting property known as resonance absorption in the optics region [Raeter 1982], which causes partial or total absorption of incident light energy. This absorption is associated with the resonant excitation of plasmons on a grating surface; incident light couples with surface plasmons via an evanescent spectral order generated by the grating [Nevièr 1982]. Resonance absorption in metal film gratings has been the subject of many theoretical [Nevièr 1982] and experimental investigations focused on various applications including chemical sensing [DeGrandpre 1990, Zoran 2009], surface enhanced phenomena such as Raman scattering [Nemetz 1994], and photonic bandgaps [Barnes 1995, Tan 1998].

A thin-film metal grating, which is a corrugated thin metal film, also results in absorption similar to that observed for thick gratings [Inagaki, Motosuga 1985, Chen 2008, Bryan-Brown 1991, Davis 2009]. Absorption in thin-film metal gratings, however, is much more complicated than in thick gratings because of the existence of coupled plasmon modes in addition to those observed in thick gratings. If the metal film is sufficiently thick, single-interface surface plasmons (SISPs) alone are excited [Raeter 1977, Okuno 2006, Suyama 2009]. However, if the film is sufficiently thin, simultaneous excitation of surface plasmons occurs on both surfaces; these plasmons interfere with each other and produce two coupled plasmon modes, short-range surface plasmons (SRSPs) and long-range surface plasmons (LRSPs) [Chen 1988, Hibbins 2006].

Most previous studies on resonance absorption have mainly dealt with metal gratings whose surfaces are periodic in one direction. Metal bi-gratings, which are periodic in two directions, also yield plasmon resonance absorption, similar to singly periodic gratings [Glass 1982, Glass 1983, Inagaki, Goudonnet 1985, Harris 1996]. In this work, we therefore investigated coupled plasmon modes excited in multilayered bi-gratings [Matsuda 1993, Matsuda 1996, Suyama 2010]. We anticipated interesting behavior in the resonance phenomenon due to the presence of the double periodicity. Further, in view of the fact that layered gratings are interesting structures for optical device applications, we investigated a multilayered bi-gratings, which is a stack of thin-film bi-gratings made of a dielectric or metal. This paper is structured as follows. After formulating the problem in Section **2**, we briefly describe a method for obtaining a solution in Section **3**. Focusing our attention on the resonant excitation of plasmon modes, we then show the computational results in Section **4**, before presenting the conclusions of this study.

2. Formulation of the problem

Here, we formulate the problem of diffraction from multilayered bi-gratings when an electromagnetic plane wave is incident on it. The time-dependent factor, $\exp(-i\omega t)$, is suppressed throughout this paper as customary.

2.1. Geometry of the gratings

Figure 1 shows the schematic representation of multilayered sinusoidal gratings with double periodicity. The grating, with $L-1$ laminated grating layers, has a period d in both the X- and Y-directions. The semi-infinite regions corresponding to the medium above the

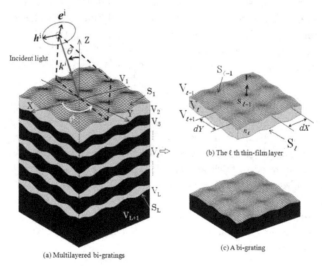

(a) Multilayered bi-gratings

(b) The ℓ th thin-film layer

(c) A bi-grating

Figure 1. Schematic representation of bi-gratings. (a) Multilayered bi-gratings; (b) The l th thin-film bi-grating; (c) A bi-grating

grating and the substrate are denoted by V_1 and V_{L+1}, respectively. The individual layers in the grating, beginning from the upper layer (light-incidence side), are denoted by $V_\ell (\ell = 2,3,...,L)$. All of the regions $V_\ell (\ell = 1,2,...,L+1)$ are filled with isotropic and homogeneous media with refractive indices n_ℓ, and the permeability of each region is equal to that of vacuum, μ_0. The interface between V_ℓ and $V_{\ell+1}$ is denoted by S_ℓ ($\ell = 1, 2, \ldots, L$). The profile of S_ℓ is sinusoidal and is given by

$$z = \eta_\ell(x, y) = \frac{h_\ell}{4}\left\{\sin\left(\frac{2\pi x}{d} + p_\ell\right) + \sin\left(\frac{2\pi y}{d} + p_\ell\right)\right\} - \sum_{i=1}^{\ell-1} e_i \tag{1}$$

Here, p_ℓ represents the phase of S_ℓ in each direction, and e_ℓ denotes the average distance between S_ℓ and S_{l+1}. The value h_ℓ is regarded as the groove depth of the boundary S_ℓ.

It should be noted that we can easily generalize the shape of S_ℓ by distinguishing between d and p_ℓ in the X-and Y-directions, by writing them as d_x, d_y, $p_{x\ell}$ and $p_{y\ell}$, respectively. A singly periodic grating is a special case where $d_y \to \infty$ in the doubly periodic case. In the present paper, we concentrate our attention on describing the analysis only for the doubly periodic case, since it reduces to the singly periodic case through a simple procedure.

2.2. Incident wave

The electric and magnetic fields of an incident wave are given by

$$\begin{pmatrix} E^i \\ H^i \end{pmatrix}(P) = \begin{pmatrix} e^i \\ h^i \end{pmatrix} \exp\left(ik^i \cdot P\right) \tag{2}$$

with

$$h^i = \left(1/\omega\mu_0\right) k^i \times e^i \tag{3}$$

Here, P is the position vector for an observation point $P(X, Y, Z)$, and k^i is the wave vector of the incident wave defined by

$$k^i = \left[\alpha, \beta, -\gamma\right]^{\mathrm{T}} \tag{4}$$

where $\alpha = n_0 k \sin\theta\cos\varphi$, $\beta = n_0 k \sin\theta\sin\varphi$, and $\gamma = n_0 k\cos\theta$. The symbol k ($= 2\pi/\lambda$) is the wave number in vacuum and λ is the wavelength of the incident wave. We define θ and φ as polar and azimuth angles, respectively, as shown in Fig. 1(a), and the superscript "T" denotes a transposition.

The amplitude of the incident electric field can be decomposed into TE and TM modes [Chen 1973] (with respect to the Z-axis) and is written as

$$e^i = \cos\delta\, e^{\mathrm{TE}} + \sin\delta\, e^{\mathrm{TM}} \tag{5}$$

$$e^{\mathrm{TE}} = \left[\sin\varphi, -\cos\varphi, 0\right]^{\mathrm{T}} \tag{6}$$

$$e^{\mathrm{TM}} = \left[\cos\theta\cos\varphi, \cos\theta\sin\varphi, \sin\theta\right]^{\mathrm{T}} \tag{7}$$

Here, the superscript TE (TM) indicates the absence of a Z component of the electric (magnetic) field.

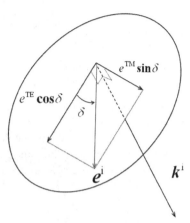

Figure 2. Definition of a polarization angle.

The symbol δ is the polarization angle between e^{TE} and e^i as shown in Fig. 2; for $\delta = 0°$ ($\delta = 90°$), this represents TE-mode (TM-mode) incidence.

2.3. Diffracted wave

We denote the diffracted fields as $E_\ell(P)$ and $H_\ell(P)$ in the region $V_\ell(\ell = 1,2,...L,L+1)$. These satisfy the following conditions:

(C1) Helmholtz equation:

$$\left(\nabla^2 + n_\ell^2 k_\ell^2\right)\binom{E_\ell}{H_\ell}(P) = 0 \quad (\ell = 1,2,...L,L+1) \tag{8}$$

(C2) Radiation conditions:

E_1 and H_1 propagate or attenuate in the positive Z-direction.

E_{L+1} and E_{L+1} propagate or attenuate in the negative Z-direction.

(C3) Periodicity conditions:

$$f(X+d,Y,Z) = \exp(i\alpha d)f(X,Y,Z) \tag{9}$$

$$f(X,Y+d,Z) = \exp(i\beta d)f(X,Y,Z) \tag{10}$$

Here, f denotes any component of $E_\ell(P)$ or $H_\ell(P)$.

(C4) Boundary condition ($0 < X < d$; $0 < Y < d$; $\ell = 1,2,...,L$):

$$v \times \left[\begin{pmatrix} E_{\ell-1} \\ H_{\ell-1} \end{pmatrix} + \delta_{\ell 1} \begin{pmatrix} E^i \\ H^i \end{pmatrix} - \begin{pmatrix} E_\ell \\ H_\ell \end{pmatrix} \right] = 0 \quad \left(\text{on } S_\ell \right) \tag{11}$$

where $\delta_{\ell 1}$ is the Kronecker delta, and v_ℓ denotes the unit vector normal to the surface S_ℓ, which is given by

$$v_\ell = \frac{\left[-\dfrac{\partial \eta_\ell (X,Y)}{\partial X}, -\dfrac{\partial \eta_\ell (X,Y)}{\partial Y}, 1 \right]}{\sqrt{\left(\dfrac{\partial \eta_\ell (X,Y)}{\partial X} \right)^2 + \left(\dfrac{\partial \eta_\ell (X,Y)}{\partial Y} \right)^2 + 1}} \tag{12}$$

3. Mode-matching method

We next explain the mode-matching method [Yasuura 1965, Yasuura 1971, Okuno 1990] for determining the diffracted field produced by the multilayered bi-grating. We introduce vector modal functions in the region $V_\ell (\ell = 1, \ldots, L+1)$ to express the diffracted field in each individual region. To construct the wave functions $E_{\ell N}^d (P)$ and $H_{\ell N}^d (P)$, we define the electric modal function $\phi_{\ell mn}^{TE,TM\pm} (P)$ and the magnetic modal function $\psi_{\ell mn}^{TE,TM\pm} (P)$ as

$$\varphi_{\ell mn}^{TE,TM\pm} (P) = e_{\ell mn}^{TE,TM\pm} \phi_{\ell mn}^{\pm} (P) \tag{13}$$

$$e_{\ell mn}^{TE\pm} = \frac{k_{\ell mn}^{\pm} \times i_Z}{\left| k_{\ell mn}^{\pm} \times i_Z \right|}, \, e_{\ell mn}^{TM\pm} = \frac{e_{\ell mn}^{TE\pm} \times k_{\ell mn}^{\pm}}{\left| e_{\ell mn}^{TE\pm} \times k_{\ell mn}^{\pm} \right|} \tag{14}$$

$$\psi_{\ell mn}^{TE,TM\pm} (P) = \frac{1}{\omega \mu_0} k_{\ell mn}^{\pm} \times \varphi_{\ell mn}^{TE,TM\pm} (P) \tag{15}$$

where $\phi_{\ell mn}^{\pm} (P)$ is the solution of the Helmholtz equation satisfying the periodic condition in the region $V_\ell (\ell = 1, .2, .., L+1)$. It is written as

$$\varphi_{\ell mn}^{\pm} (P) = \exp\left(ik_{\ell mn}^{\pm} \cdot P \right) \quad (m,n = 0, \pm 1, \pm 2, \ldots) \tag{16}$$

where the positive and negative signs match on either side of the equation, and $k_{\ell mn}^{\pm}$ is the wave vector of the (m,n)th order diffracted wave given by

$$k_{\ell mn}^{\pm} = \left[\alpha_m, \beta_n, \pm \gamma_{\ell mn} \right]^T \tag{17}$$

$$\alpha_m = \alpha + \frac{2m\pi}{d}, \, \beta_n = \beta + \frac{2n\pi}{d} \tag{18}$$

$$\gamma_{\ell mn} = \left(n_\ell^2 k^2 - \alpha_m^2 - \beta_n^2\right)^{1/2}, \ \mathrm{Re}\left(\gamma_{\ell mn}\right) \geq 0 \ \text{and} \ \mathrm{Im}\left(\gamma_{\ell mn}\right) \geq 0 \tag{19}$$

Note that the superscripts + and - represent upwardly and downwardly propagating waves in the positive and negative Z-direction, respectively.

In terms of the linear combinations of the vector modal functions, we form approximate solutions for the diffracted electric and magnetic fields in V_ℓ :

$$\begin{aligned}
\begin{pmatrix} E_{\ell N}^d \\ H_{\ell N}^d \end{pmatrix}(P) &= \sum_{m,n=-N}^{N} A_{\ell mn}^{TE+}(N) \begin{pmatrix} \varphi_{\ell mn}^{TE+} \\ \psi_{\ell mn}^{TE+} \end{pmatrix}(P) \\
&+ \sum_{m,n=-N}^{N} A_{\ell mn}^{TM+}(N) \begin{pmatrix} \varphi_{\ell mn}^{TM+} \\ \psi_{\ell mn}^{TM+} \end{pmatrix}(P) \\
&+ \sum_{m,n=-N}^{N} A_{\ell mn}^{TE-}(N) \begin{pmatrix} \varphi_{\ell mn}^{TE-} \\ \psi_{\ell mn}^{TE-} \end{pmatrix}(P) \\
&+ \sum_{m,n=-N}^{N} A_{\ell mn}^{TM-}(N) \begin{pmatrix} \varphi_{\ell mn}^{TM-} \\ \psi_{\ell mn}^{TM-} \end{pmatrix}(P) \\
&\qquad\qquad (\ell = 1,2,\dots,L+1)
\end{aligned} \tag{20}$$

with

$$\psi_{\ell mn}^{TE, TM\,\pm}(P) = \frac{1}{\omega\mu_0} k_{\ell mn}^\pm \times \varphi_{\ell mn}^{TE, TM\,\pm}. \tag{21}$$

Here, $A_{1mn}^{TE-}(N) = A_{1mn}^{TM-}(N) = 0$ and $A_{L+1mn}^{TE+}(N) = A_{L+1mn}^{TM+}(N) = 0$ because of the radiation conditions (C3) stated in **Section 2.3**.

The approximate solutions $E_{\ell N}^d(P)$ and $H_{\ell N}^d(P)$ already satisfy the Helmholtz equation (C1), the periodicity conditions (C2), and the radiation conditions (C3). The unknown coefficients $A_{\ell mn}^{TE\pm}(N)$ and $A_{\ell mn}^{TM\pm}(N)$ are therefore determined such that the solutions approximately satisfy the boundary conditions (C4). In the mode-matching method [Yasuura 1996, Yasuura 1971, Okuno 1990], the least-squares method is employed to fit the solution to the boundary conditions [Hugonin 1981]. That is, we find coefficients that minimize the weighted mean-square error by

$$\begin{aligned}
I_N &= \int_{S_1} \left| \nu \times \left[E_{1N}^d + E^i - E_{2N}^d \right](s_1) \right|^2 ds \\
&+ |\Gamma_1|^2 \int_{S_1} \left| \nu \times \left[H_{1N}^d + H^i - H_{2N}^d \right](s_1) \right|^2 ds \\
&+ \sum_{\ell=2}^{L} \left\{ \int_{S_\ell} \left| \nu \times \left[E_{\ell N}^d - E_{\ell+1N}^d \right](s_\ell) \right|^2 ds + \right. \\
&\qquad \left. |\Gamma_\ell|^2 \int_{S_\ell} \left| \nu \times \left[H_{\ell N}^d - H_{\ell+1N}^d \right](s_\ell)^2 \right| ds \right\}
\end{aligned} \tag{22}$$

Here, S'_ℓ is a one-period cell of the interface S_ℓ, and Γ_ℓ is the intrinsic impedance of the medium in V_ℓ.

To solve the least-squares problem on a computer, we first discretize the weighted mean-square error I_N by applying a two-dimensional trapezoidal rule where the number of divisions in the x-and y-direction is chosen as $2(2N + 1)$ [Yasuura 1965, Yasuura 1971, Okuno 1990]. We then solve the discretized least-squares problem by the QR decomposition method. Computational implementation of the least-squares problem is detailed in the literature [Lawson 1974, Matsuda 1966, Suyama 2008].

4. Numerical results

Here we show some numerical results obtained by the method described in the preceding section. After making necessary preparation, we show the results for three the bi-grating, thin-film bi-grating and multilayered thin-film bi-gratings cases.

4.1. Preparation

It is known that the solutions obtained by mode-matching method [Yasuura 1965, Yasuura 1971, Okuno 1990] have proof of convergence. We, therefore, can employ the coefficient $A^{TM,TE\pm}_{\ell mn}(N)$ with sufficiently large N for which the coefficients are stable in evaluating diffracted fields.

The power reflection and transmission coefficient of the (m, n) order propagating mode in V_1 and V_{L+1} are given by

$$\rho_{mn} = \rho^{TE}_{mn} + \rho^{TM}_{mn} \ [\text{Re}(\gamma_{1mn}) \geq 0], \ \rho^{TE}_{mn} = \frac{\gamma_{1mn}}{\gamma} \left| A^{TE+}_{1mn} \right|^2, \ \rho^{TM}_{mn} = \frac{\gamma_{1mn}}{\gamma} \left| A^{TM+}_{1mn} \right|^2 \tag{23}$$

$$\tau_{mn} = \tau^{TE}_{mn} + \tau^{TM}_{mn} \ [\text{Re}(\gamma_{L+1mn}) \geq 0], \ \tau^{TE}_{mn} = \frac{\gamma_{L+1mn}}{\gamma} \left| A^{TE-}_{L+1mn} \right|^2, \ \tau^{TM}_{mn} = \frac{\gamma_{L+1mn}}{\gamma} \left| A^{TM-}_{L+1mn} \right|^2 \tag{24}$$

The coefficient defined above is the power carried away by propagating diffraction orders normalized by the incident power. We calculate the total diffraction efficiency $\rho^{\text{total}} = \sum' \rho_{mn}$ where $\sum'$ denotes a summation over the propagating orders.

Although it is known that the solutions obtained by mode-matching modal expansion method [Yasuura 1965, Yasuura 1971, Okuno 1990] have proof of convergence for problems of diffraction by gratings, we compare our results with other existing theoretical [Glass 1983] and experimental results [Inagaki, Goudonnet 1985] on plasmon resonance absorption in bi-gratings to show the validity of the present method. Figure 3 shows the reflectivity curves calculated by the present method and those from Rayleigh's method [Glass 1983] for three sinusoidal silver bi-gratings with different corrugation amplitudes. As confirmed in Fig. 3, the reflectivity curves from the present method are coincident with those of the

Rayleigh's method [Glass 1983]. Next, we make comparison with the experimental results [Inagaki, Goudonnet 1985] in which the resonance angle θ_d, i.e., the polar angle at which the dip of reflectivity occurs, is observed near 55° for a sinusoidal silver bi-grating with $h = 0.048$ μm, $d = 2.186$ μm, $\lambda = 0.633$ μm, and $\phi = 45°$. The resonance angle calculated from the present metod for these parameters is $\theta_d = 54.1°$, which is close to the experimental data. These examples show that the present method gives reliable results for the analysis of plasmon-resonance absorption in metal bi-gratings.

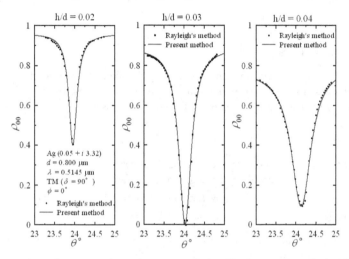

Figure 3. Comparison of resonance absorption curves calculated by the present method with other existing theoretical results. Solid curves show our results, and dotted curves are taken from Figure 2. of Ref. [Glass 1983].

In the numerical examples presented here, we deal with a shallow sinusoidal silver bi-grating with height $h = 0.030$ μm and period $d = 0.556$ μm. The wavelength of an incident light is chosen as $\lambda = 0.650$ μm. We take $n_2 = 0.07+i4.20$ as the refractive index of silver at this wavelength [Hass 1963].

It should be noted, however, that the index of a metal film depends not only on the wavelength but also on the thickness of the film, in particular when the film is extremely thin it may take unusual values if circumstances require. When dealing with a thin metal structure, hence, we should be careful in using the index value given in the literature. As for the value taken in our computation, we assume that $n_2 = 0.07+i4.20$ is available even for the case of $e/d = 0.02$. This is because a similar assumption was supported by experimental data in a problem of diffraction by an aluminum grating with a thin gold over-coating.

4.2. A bi-grating case

Using the numerical algorithm stated in the previous section, we first investigate the absorption in a metal bi-gratings by $L = 1$ as shown in Fig. 1(c). The semi-infinite regions corresponding to the medium above the grating and the substrate are denoted by V_1 and V_2, respectively. V_1 is vacuum (V) with a relative refractive index $n_1=1$ and V_2 consists of a lossy metal characterized by a complex refractive index n_2.

4.2.1. Diffraction efficiency

Figure 4 shows the total diffraction efficiency of a sinusoidal silver bi-grating as functions of a polar angle θ when the azimuthal angle $\phi = 30°$ is fixed. In the efficiency curves we observe four dips which occur at the same angles of incidence for both TE and TM polarized incident light. In this subsection, we demonstrate that the dips are associated with absorption that is caused by the coupling of surface plasmons with an evanescent mode diffracted by a sinusoidal bi-grating. For convenience, the four dips in Fig. 4 are labeled as A, B, C, and D.

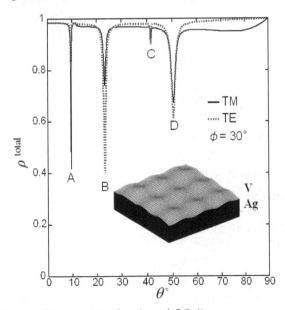

Figure 4. Total diffraction efficiencies ρ^{total} as functions of θ ($L=1$).

4.2.2. Expansion coefficients

In Fig. 5 we plot the expansion coefficients of the (0, -1)st-order and (-1, 0)th-order TM vector modal function, which are two evanescent modes, as a functions of θ under the same parameters as in Fig. 4. Solid curves in Fig. 5 represent the real part of the expansion coefficient, and dashed curves for the imaginary part. We observe a resonance curve of the expansion coefficient $A_{1(-1,0)}^{TM}$ at the angles of incidence $\theta = 9.5°$, i.e., dip A, and $A_{1(0,-1)}^{TM}$ at the angles of incidence $\theta = 23.3°$, $41.5°$ and $49.5°$, i.e., dips B, C, and D in Fig. 4 for both TE and TM incidence.

This implies that the TM component of the (0, -1)st-order and (-1, 0)th-order evanescent mode couples with surface plasmons at dips B and D. We can similarly confirm that dips A and C are associated with the coupling of the TM component of the (-1, 0)th- and (-1, -1)st-order evanescent mode, respectively, with surface plasmons, although we do not include any numerical example here.

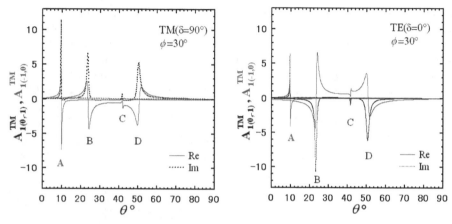

Figure 5. $A_{1(0,-1)}^{TM}$ and $A_{1(-1,0)}^{TM}$ as functions of θ for both TM- (a) and TE- (b) polarized incident light.

4.2.3. Field distributions and energy flows

In order to investigate the resonant excitation of surface plasmons, we study field distributions and energy flows in the vicinity of the grating surface when the absorption occurs. Here we consider the case of the dip B at which the TM component of the (0, -1)st-order evanescent mode couples with surface plasmons. We calculate the electric field of the TM component of the (0, -1)st-order evanescent mode $E_{\ell 0-1}^{TM} = A_{\ell 0-1}^{TM}(N)\ \varphi_{\ell 0-1}^{TM}(\ell=1,2)$ and the total electric field E^{t}. The magnitude of these fields along the Z-axis where $Y = (\beta_{-1}/\alpha_0)*X$ is plotted in Fig. 6(a). We observe in this figure that very strong electric fields are induced at the grating metal surface and the fields exponentially decay away from the surface.

Next we show the energy flows **S** that are the real part of Poynting's vectors for the total field. The X and Y components of the energy flows **S** are plotted as the vector (S_X, S_Y) in Fig. 6(b). The energy flows are calculated over the region close to the grating surface:

$$\{P(x,y,z) : 0 \leq x \leq d, 0 \leq y \leq d,$$
$$z = (h/4)[\sin(2\pi x/d) + \sin(2\pi y/d)] + 0.01d\}. \tag{25}$$

The energy flow at a point P is given by Re[$S(P)$], where $S(P)=(1/2)\ E^{t}(P) \times \overline{H^{t}}(P)$ stands for Poynting's vector, E^{t} and H^{t} denote total fields, and the over-bar means complex conjugate.

We calculate the energy flow at each point located densely near the grating surface and show the results in Fig. 6(b).

Figure 6(b) shows that the energy of electromagnetic fields in the vicinity of the grating surface flows uniformly in the direction that the (0, -1)st-order evanescent mode travels in the XY plane. We thus confirm that surface plasmons are excited on the grating surface through the coupling of the TM component of an evanescent mode.

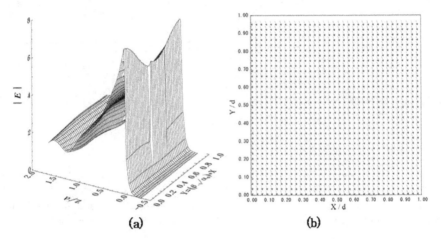

(a)   **(b)**

Figure 6. Field distribution (a) and Energy flow (b) for the total field when plasmon resonance absorption occurs at $\phi = 30°$, $\theta = 23.3°$.

4.2.4. Polarization conversion through plasmon resonance absorption

Diffracted fields from a sinusoidal metal bi-grating have both TE and TM component for an arbitrary polarized incident light. We therefore observe polarization conversion that a TM (or TE) component of the incident light is converted into a TE (or TM) component of the reflected light. It has been pointed out [Chen 1973, Inagaki, Goudonnet 1985] that the polarization conversion [Elston 1991, Matsuda 1999, Suyama 2007] is strongly enhanced when the plasmon-resonance absorption occurs in a sinusoidal metal bi-grating. Our study confirms the enhancement of polarization conversion through plasmon-resonance absorption. In Fig. 7, the TE and TM component of the diffraction efficiency ρ_{00} ($\rho_{00} = \rho_{00}^{TE}$ + ρ_{00}^{TM}) of Fig. 4 are shown for the case of the TM incidence. The TM component ρ_{00}^{TM} is decreased at the position of the plasmon-resonance absorption, but the TE component ρ_{00}^{TE} is contrary increased there. That is, the resonant excitation of surface plasmons causes the enhancement of polarization conversion. On the other hand, in the case of the TE incidence the conversion from a TE to a TM component occurs through the plasmon-resonance absorption. It should be noted that the conversion efficiency depends on the azimuthal angle φ and the depth of the grating surface h.

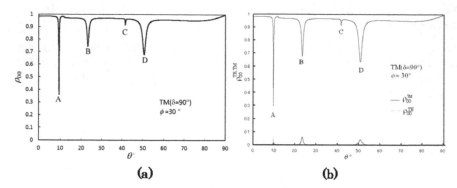

Figure 7. Diffraction efficiencies of ρ_{00} (a) ρ_{00}^{TE} and ρ_{00}^{TM} (b) as functions of θ ; parameters are the same as in Figure 4.

4.2.5. Prediction of resonance angles

We seek to determine a complex incidence angle θ_c for which total or partial absorption occurs, i.e., ρ_{total} $(=\rho_{00})$ takes a minimum. This angle relates to the propagation constant of the surface plasmon on the corrugated surface. Here, we denote by $\hat{\alpha}_{SP}$ and $\hat{\beta}_{SP}$ the X and Y components of the surface plasmon wave vector normalized by the wave number k_1:

$$\hat{\alpha}_{SP} = \sin\theta_c \cos\phi + m\lambda / d,$$
$$\hat{\beta}_{SP} = \sin\theta_c \sin\phi + n\lambda / d. \tag{26}$$

In reality, we cannot realize a complex incidence angle θ_c. We can, however, estimate the real angle of incidence at which the absorption occurs by taking the real part of Eq. (26).

If the wavevector of the surface plasmon ($\hat{\alpha}_{SP}, \hat{\beta}_{SP}$) is obtained, we can estimate the resonance angle θ_{SP} for each azimuthal angle φ from the phase-matching condition for coupling of a surface plasmon wave with the (m, n)th-order evanescent mode:

$$\text{Re}\{\hat{\alpha}_{SP}\} = \sin\theta_{SP}\cos\varphi + m\lambda / d,$$
$$\text{Re}\{\hat{\beta}_{SP}\} = \sin\theta_{SP}\sin\varphi + n\lambda / d. \tag{27}$$

We solve the homogenous problem [Nevièr 1982] for a sinusoidal metal bi-grating by present method and then obtain the surface-plasmon wave vector. Table 1 shows the propagation constants of the surface plasmon and the resonance angles θ_{SP}. The data demonstrate that the estimated resonance angle θ_{SP} agrees with θ_d which is the absorption peak in Fig. 4. Figure 8 shows the estimated resonance angle θ_{SP} as a function of the azimuthal angle φ for the sinusoidal bi-grating considered in Fig. 4. Note that points A, B, C, and D in Fig. 8 are results obtained from the absorption peak of total-efficiency curves in Fig. 4. From this figure, we can find the resonance angle for each azimuthal angle.

Dip	Mode	Re($\hat{\alpha}_{SP}$)	Im($\hat{\alpha}_{SP}$)	Re($\hat{\beta}_{SP}$)	Im($\hat{\beta}_{SP}$)	θ°_{SP}	θ°_d
A	(-1, 0)	-1.026277	-0.000804	0.087099	-0.001279	9.49	9.5
B	(0, -1)	0.343795	-0.004910	-0.971021	-0.002498	23.39	23.3
C	(0, -1)	-0.595897	-0.001282	-0.837168	-0.000549	41.44	41.5
D	(0, -1)	0.668265	-0.004365	-0.778336	-0.005224	50.50	50.5
E	(0, -1)	0.149639	-0.000679	-1.019425	-0.000679	12.22	12.2
	(-1, 0)	-1.019426	-0.000679	0.149639	-0.000679	12.22	

[a]$h = 0.03\,\mu m$, $d = 0.556\,\mu m$, $\lambda = 0.650\,\mu m$

Table 1. Propagation Constants and Estimated Resonance Angles[a]

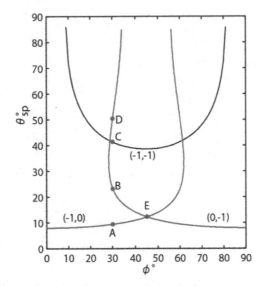

Figure 8. Resonance angles θ_{SP} as functions of azimuthal angle ϕ.

4.2.6. Simultaneous resonance absorption

From Fig. 8, it is predicted the angle of incidence at which the simultaneous resonance absorption occurs from the position of the intersection of the (-1,0) and (0,-1) curve. At the intersection E, the (0,-1)st- and (-1,0)th-order evanescent modes couple simultaneously with two surface-plasmon waves at the same angle of incidence. Thus, two surface-plasmon waves are excited simultaneously in directions symmetric with respect to the plane of incidence and interact with each other. The interference of the surface-plasmon waves causes the standing wave in the vicinity of the grating surface. This is confirmed from Figs. 9, the strong fields along the Z-axis where Y=X, and where the X and Y components of Poynting's vectors **S** on a surface 0.01d above the one-unit cell of the grating surface are plotted as the vector (S_X, S_Y). We further observe in Fig. 10 that the simultaneous excitation of the surface plasmons waves causes the strong absorption for both TE- and TM-polarized incident light.

It has been reported [Barnes 1995, Ritchie 1968] that surface-plasmon band gaps exist at the angles of incidence at which simultaneous excitation of plasmon waves occurs, and that the appearance of the band gaps depends strongly on the surface profile. Hence, there is a possibility that a band gap will be observed at the point E in Figs. 9 and 10 provided that the grating profile is appropriately chosen, because two plasmon modes are excited at that point.

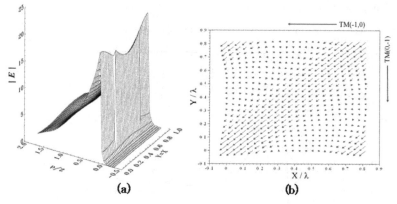

(a)　　　　　　　　　　　　　　　　**(b)**

Figure 9. Field distributions and Energy flows for the total field when plasmon resonance absorption occurs.

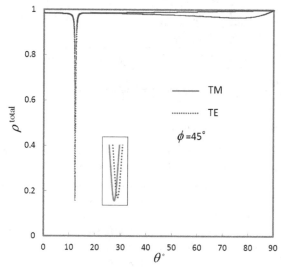

Figure 10. Total diffraction efficiencies ρ total as functions of θ.

4.3. A Thin-film Bi-grating case

As a numerical example, we consider a sinusoidal silver (Ag) film bi-grating having a common period as shown in Fig. 1(b). The values of the parameters are the same as those in Fig. 4 except for the thickness of the silver film. Using the present algorithm, we calculated the diffraction efficiencies and field distributions to clarify the properties of the coupled surface plasmon modes.

4.3.1. Diffraction efficiency

First, we consider a sinusoidal silver-film bi-grating. The bi-grating is denoted by $L = 2$ (V/Ag/V). Figure 11 shows the (0,0)th order power reflection ρ_{00} in V_1 (Vacuum) (a) and the transmission coefficient τ_{00} in V_3 (Vacuum) (b) as functions of the incident angle θ for two values of e/d when the azimuth angle $\phi = 0°$ is fixed; e is the thickness of the silver film. We observe partial absorption of the incident light as dips in the efficiency curves in Fig. 11(a), in addition to the constant absorption corresponding to the reflectivity of silver. We assume that the dips are caused by resonant excitation of surface plasmons. If this is the case, each of the dips can be related to one of the three types of plasmon modes: a SISP that is observed as a single dip at $\theta = 8.0°$ on the $e/d = 0.4$ curve, and a SRSP and LRSP corresponding to the dips in the $e/d = 0.08$ curve at $\theta = 6.54°$ and $8.8°$, respectively.

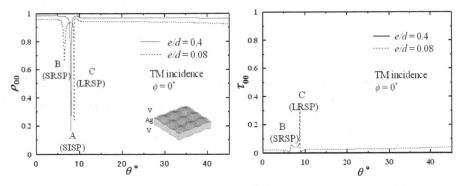

Figure 11. The (0, 0)-th order diffraction efficiencies ρ_{00} (a) and τ_{00} (b) as functions of θ for two values of e/d (L=2).

When the grating is thick ($e/d = 0.4$), the power can be seen in V_1 alone and no transmitted power exists in V_3 in Fig. 11. Increasing θ from $0°$, we first observe the dip at $\theta = 8.0°$ corresponding to absorption in Fig. 11(a). If the grating is relatively thin ($e/d = 0.08$), the power exists in both V_1 and V_3. Although the power in V_3 is generally small, it becomes large at the incidence angles for which absorption was observed in Fig. 11. This suggests that coupled oscillations occur on the upper and lower surfaces of the grating.

4.3.2. Expansion coefficients

We examined the same phenomena observing the modal expansion coefficients in $V_l (l = 1,3)$.
Figures 12 and 13 illustrate the (-1,0)th-order coefficients $A_{1(-1,0)}^{TM+}$ for $e/d = 0.4$ and 0.08,
respectively. We observe, in Fig. 12(a) ($e/d = 0.4$), the resonance characteristics (enhancement
and rapid change in phase) of the coefficient $A_{1(-1,0)}^{TM+}$ near $\theta = 8.0°$. The coefficient $A_{3(-1,0)}^{TM-} = 0$
remains unchanged, as seen in Fig. 12(b). This means that the incident wave illuminating the
grating at this angle causes coupling between the (-1,0)th-order evanescent mode and some
oscillation excited on the upper surface of the grating. The oscillation exists locally in the
vicinity of the illuminated surface and hence does not have any influence on the field in V_3 .

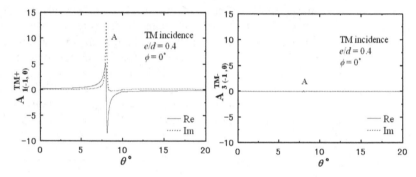

Figure 12. The (-1, 0)th order modal coefficients $A_{1(-1,0)}^{TM+}$ (a) and $A_{3(-1,0)}^{TM-}$ (b) as functions of θ at
$e/d = 0.4$.

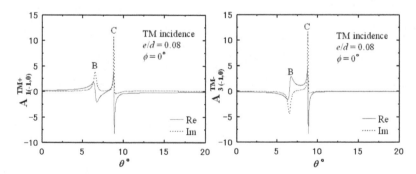

Figure 13. The (-1, 0)th order modal coefficients $A_{1(-1,0)}^{TM+}$ (a) and $A_{3(-1,0)}^{TM-}$ (b) as functions of θ at
$e/d = 0.08$.

In Fig. 13 ($e/d = 0.08$), we find the resonance characteristics in both $A_{1(-1,0)}^{TM+}$ and $A_{3(-1,0)}^{TM-}$. In
addition, they appear around two incidence angles: $\theta = 6.54°$ and $\theta = 8.8°$. This means that

the oscillation in the vicinity of the upper surface causes another oscillation on the lower surface at this thickness. The oscillations interfere with each other and result in two coupled oscillating modes: the SRSP and LRSP. This means that the TM component of the (-1,0)th order evanescent mode couples with the surface plasmons simultaneously excited on the upper and lower surface of the film grating. The two surface plasmons interfere with each other and result in symmetric and antisymmetric coupled modes, SRSP and LRSP, as we will see next.

4.3.3. Field distributions and energy flows

We consider the same phenomena observing the field distributions and energy flows near the grating surfaces. In the former we find that the total field is enhanced. In the latter we observe the symmetric (even) and anti-symmetric (odd) nature of the oscillations, which correspond to the LRSP and SRSP [Raeter 1977].

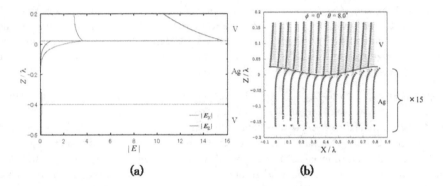

Figure 14. Field distributions (a) and energy flows (b) at $\theta = 8.0°$, which corresponds to the single dip on the $e/d = 0.4$ curve in Fig. 11.

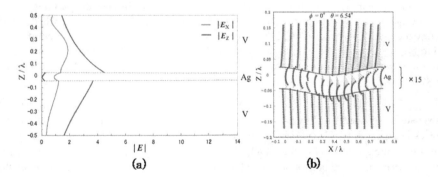

Figure 15. Field distributions (a) and energy flows (b) at $\theta = 6.54°$, which corresponds to the left dip on the $e/d = 0.08$ curve in Fig. 11.

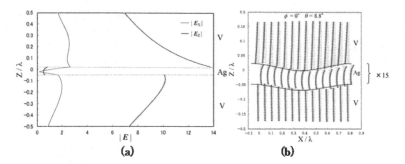

Figure 16. Field distributions (a) and energy flows (b) at $\theta = 8.8°$, which corresponds to the right dip on the $e / d = 0.08$ curve in Fig. 11.

Figures 14, 15, and 16 show the field distributions of the X- and Z-components of the total electric fields (a) and energy flows (b) in the vicinity of the silver-film grating at the incidence angles at which absorption was observed in Fig. 11. The abscissa and ordinate show the magnitude and distance in the Z direction normalized by the wavelength λ. The parallel broken lines represent the grating surfaces.

Figure 14 shows E_X and E_Z (a), S_X and S_Z (b) for the case of $e / d = 0.4$ at $\theta = 8.0°$, which corresponds to the single dip in Fig. 11(a). Figures 15 and 16 show the same thing for the $e / d = 0.08$ case. Figure 15 illustrates the results at $\theta = 6.54°$, where the left dip is observed in Fig. 11(a). On the other hand, Fig. 16 depicts the results at $\theta = 8.8°$, corresponding to the right dip in Fig. 11(a). In Figs. 14 to 16, we observe strong enhancement of E_X and E_Z (note that the magnitude of the incident radiation is 1), which is observed at the incidence angles where absorption occurs.

We observed that the total field above the grating surface decays exponentially in the Z direction and the magnitude of the total field is almost E_Z in Fig. 14(a). The state of affairs is nearly the same in the metal region except for the rapid decay. Because the grating is thick, the oscillation near the upper surface does not reach the lower surface and, hence, the field below the grating is zero. Figure 14(b) illustrates the energy flow **S**, which is magnified by 15 in the metal region. We see that the energy flow is almost in the X direction and that it goes in opposite directions in vacuum and in metal. This is commonly observed when a SISP is excited.

In Figs. 15(a) and 16(a), we again see the enhancement of E_X and E_Z on the upper and lower surface of the silver-film grating, respectively. The rate of enhancement in Fig. 15(a) is not as large as that in Fig. 16(a). E_Z is strongly enhanced at both the upper and lower surfaces of the grating and exponentially decays away from each surface. We thus observe the simultaneous excitation of surface plasmons at the surfaces. We can understand the difference of the field distributions assuming that the former and the latter refer to Figs. 15 and 16 are the results of the SRSP and the LRSP mode excitation. Figures 15(b) and 16(b) complement the understanding showing the even and odd nature of relevant oscillations.

4.4. Multilayered thin-film bi-gratings case

Next, we consider multilayered thin-film bi-gratings as shown in Fig. 1(a) indicated by $L = 4$ ($V/Ag/SiO_2/Ag/V$) that consists of a stack of silver and SiO_2 films pairs. As listed in the figure, the values of the parameters are the same as those in Fig. 11 except for $L = 4$, $n_{SiO_2} = 1.5$, $e_{SiO_2}/d = 0.3$ or 0.08 .

4.4.1. Diffraction efficiency

Figure 17 shows the (0,0)th order power reflection ρ_{00} in V_1 (Vacuum) (a) and the transmission coefficient τ_{00} in V_5 (Vacuum) (b) as functions of the incident angle θ for two different values of e_{SiO_2}/d . The curve for $e_{SiO_2}/d = 0.08$ is almost the same as for a sinusoidal silver film bi-grating (Fig. 11): the coupled surface plasmon modes, SRSP and LRSP, are excited. On the other hand, in the curve for $e_{SiO_2}/d = 0.3$, we find a new type of absorption of incident light besides the plasmon resonance absorption associated with SRSP or LRSP.

This is related to the resonant excitation of a guided wave supported by the SiO_2 film. This absorption is characterized by its occurrence over a wider range of θ . For example, in the case of $e_{SiO_2}/d = 0.3$, the extinction power is more than 50% for all angles of incidence θ ranging from $0°$ to $12°$.

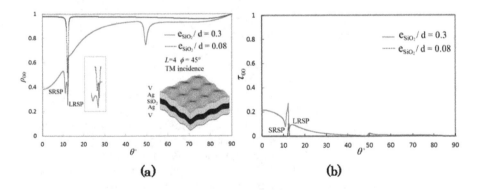

(a)　　　　　　　　　　(b)

Figure 17. The (0, 0)-th order diffraction efficiencies ρ_{00} (a) and τ_{00} (b) as functions of θ for two values of e_{SiO2}/d (L=4).

4.4.2. Field distributions

In order to examine the properties of the wide absorption found in Fig. 17, we investigated the field distributions of the total electric field E^{total} and the TM component of the (0, 0)th-order diffracted electric field $E_{l(0,0)}^{TM}$ in the vicinity of the SiO_2 film. The magnitude of E^{total} and $E_{l(0,0)}^{TM}$ (l =1,2,...5) along the Z-axis are plotted in Fig. 18 where $\theta = 0°$ and $e_{SiO_2}/d = 0.3$.

We observe in the figure that the field distributions of E^{total} inside the SiO$_2$ film indicates a standing wave pattern corresponding to the normal mode of a one dimensional cavity resonator, and that the distribution is almost close to that of $E^{TM}_{3(0,0)}$. Hence, we conclude that the wide absorption observed in the multilayered grating V/Ag/SiO$_2$/Ag/V is associated with resonance of the (0, 0)th-order diffracted wave $E^{TM}_{3(0,0)}$ in the SiO$_2$ film sandwiched by a sinusoidal silver film grating.

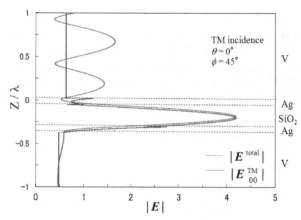

Figure 18. Standing wave pattern of the electric filed in the SiO$_2$ film.

5. Conclusions

We have investigated the resonance absorption associated with the resonant excitation of surface plasmons in bi-gratings. Calculating diffraction efficiency, expansion coefficients, field profiles, and energy flows, we examined the characteristics of the resonant excitation of surface plasmons in detail. Interesting phenomena were revealed, including the conversion of a TM (or TE) component of the incident light into a TE (or TM) component at several different incidence angles, strong field enhancement on the grating surface where surface plasmons are excited, and simultaneous resonance absorption that does not occur in the case of a singly periodic grating in general. The results presented here facilitate a clear understanding of the coupled plasmon modes, SISP, SRSP and LRSP, excited in a thin film doubly periodic metal grating.

Author details

Taikei Suyama, Akira Matsushima and Yoichi Okuno
Graduate School of Science and Technology, Kumamoto University, Kurokami, Kumamoto, Japan

Toyonori Matsuda
Kumamoto National College of Technology, Suya, Nishigoshi, Japan

Acknowledgement

This work was supported in part by Grant-in-Aid for Scientific Research from Japan Society for the Promotion of Science (Grant number 23560404). The authors thank Shi Bai and Qi Zhao for help in numerical computation and in preparation of the manuscript.

6. References

Raeter, H., (1982) Surface Plasmon and Roughness, *Surface Polaritons*, V. M. Argranovich and D. L. Mills, (Ed.), Chap. 9, 331-403, North-Holland, New York

Nevièr, M., (1980) The Homogenous Problem, *in Electromagnetic Theory of Gratings* R. Petit (Ed.), Chap. 5, 123-157, Springer-Verlag, Berlin

DeGrandpre, M.D., and L.W. Burgess, (1990) Thin film planar waveguide sensor for liquid phase absorbance measurement, *Anal. Chem.*, 62, 2012-2017

Zoran, J. and M. Jovan, (2009) Nanomembrane-Enabled MEMS Sensors: Case of Plasmonic Devices for Chemical and Biological Sensing, *Micro Electronic and Mechanical Systems*, Kenichi Takahata (Ed.), ISBN: 978-953-307-027-8, InTech

Nemetz, A., U. Fernandez and W. Knoll, (1994) Surface plasmon field-enhanced Raman spectroscopy with double gratings, *J. Appl. Phys*, 75, 1582-1585

Barnes,W. L., T. W. Preist, S. C. Kitson, J. R. Sambles, N. P. K. Cotter, and D. J. Nash, (1995) Photonic gaps in the dispersion of surface plasmons on gratings, *Phys. Rev. B*, 51, 11164 -11167

Tan, W.-C., T. W. Preist, J. R. Sambles, M. B. Sobnack, and N. P. Wanstall, (1998) Calculation of photonic band structures of periodic multilayer grating systems by use of a curvilinear coordinate transformation, *J. Opt. Soc. Am. A*, 15, 9, 2365–2372

Inagaki, T., M. Motosuga, E.T. Arakawa and J.P. Goudonnet, (1985) Coupled surface plasmons in periodically corrugated thin silver films, *Phys. Rev. B*, 32, 6238-6245

Chen, Z., I. R. Hooper, and J. R. Sambles, (2008) Strongly coupled surface plasmons on thin shallow metallic gratings, *Phys. Rev. B*, 77, 161405

Bryan-Brown, G.P., S.J. Elston, J.R. Sambles, (1991) Coupled surface plasmons on a silver coated grating, *Opt. Commu.*, 82, No. 1-2, 1-5

Davis, T. J., (2009) Surface plasmon modes in multi-layer thin-films, *Opt. Commu.*, 282, 135-140

Raeter, H., (1977) Surface plasma ocillations and their applications, *Physics of Thin Films*, G. Hass and M.H. Francombe (Ed.), Chap. 9, 145-261, Academic Press, New York

Okuno, Y. and T. Suyama, (2006) Numerical analysis of surface plasmons excited on a thin metal grating, *Journal of Zhejiang University -Science A*, 7, No. 1, 55-70

Suyama, T., Y. Okuno and T. Matsuda, (2009) Plasmon resonance-absorption in a metal grating and its application for refractive-index measurement, *Journal of Electromagnetic Waves and Applications*, 20 , No. 2, 159-168

Chen, Z. and H. J. Simon, (1988) Attenuated total reflection from a layered silver grating with coupled surafce waves, *J. Opt. Soc. Am. B*, 5, No. 7, 1396-1400

Hibbins, A. P., W. A. Murray, J. Tyler, S. Wedge, W. L. Barnes and J. R. Sambles, (2006) Resonant absorption of electromagnetic fields by surface plasmons buried in a multilayered plasmonic nanostructure, *Phys. Rev. B*, 74, 073408

Glass, N. E., A. A. Maradudin and V. Celli, (1982) Surface plasmons on a large-amplitude doubly periodic corrugated surface, *Phys. Rev. B*, 26, 5357-5365

Glass, N. E., A. A. Maradudin, and V. Celli, (1983) Theory of surface-polariton resonances and field enhancements in light scattering from doubly periodic gratings, *J. Opt Soc. Am.*, 73, 1240-1248

Inagaki, T., J. P. Goudonnet, J. W. Little and E. T. Arakawa, (1985) Photoacoustic study of plasmon-resonance absorption in a doubly periodic grating, *J. Opt Soc. Am. B*, 2, 432-439

Harris, J. B., T. W. Preist, J. R. Sambles, R. N. Thorpe and R. A. Watts, (1996) Optical response of doubly periodic gratings, *J. Opt Soc. Am. A*, 13, 2041-2049

Chen, C. C., (1973) Transmission of microwave through perforated flat plates of finite thickness, *IEEE Trans. Microwave Theory Tech.*, MTT-21, 1-6

Yasuura, K. and T. Itakura, (1965) Approximation method for wave functions (I),(II), and (III), *Kyushu Univ. Tech. Rep*, 38, 72-77, 1965; 38, 378-385, 1966; 39, 51-56, 1966

Yasuura, K.(1971). A view of numerical methods in diffraction problems, *Progress in Radio Science 1966-1969*, W. V. Tilson and M. Sauzade (Ed.), 257-270, URSI, Brussels

Okuno, Y., (1990) Mode-matching Method, *Analysis Methods for Electromagnetic Wave Problems*. E. Yamashita (Ed.), 107-138, Artech House, Boston

Hugonin, J. P., R. Petit, and M. Cadilhac, (1981) Plane-wave expansions used to describe the field diffracted by a grating, *J. Opt. Soc. Am.*, 71, No. 5, 593–598

Lawson, C. L. and R. J. Hanson, (1974) Solving Least Squares Problem, Prentice-Hall, Englewood Cliffs, N. J.

Matsuda, T. and Y. Okuno, (1993) A numerical analysis of planewave diffraction from a multilayer-overcoated grating, *IEICE*, J76-C-I, No. 6, 206–214

Matsuda, T. and Y. Okuno, (1996) Numerical evaluation of plane-wave diffraction by a doubly periodic grating, *Radio Sci.*, 31, 1791-1798

Suyama, T., Y. Zhang, Y. Okuno, Z. Q. Luo and T. Matsuda, (2010) Surface Plasmon Resonance Absorption in a Multilayered Bigrating, *PIERS Online*, 6, No. 1, 76-80

Suyama, T., Y. Okuno, A. Matsushima and M. Ohtsu, (2008) A numerical analysis of stop band characteristics by multilayered dielectric gratings with sinusoidal profile, *Progress In Electromagnetics Research B*, 2, 83-102

Hass, G. and L. Hadley, (1963) Optical properties of metals, *American Institute of Physics Handbook*, D. E. Gray, (2nd Ed.), 6-107, McGraw-Hill, New York.

Elston, S. J., G. P. Bryan-Brown and J. R. Sambles, (1991) Polarisation conversion from diffraction gratings, *Phys. Rev. B*, 44, 6393-6399

Matsuda, T., D. Zhou, and Y. Okuno, (1999) Numerical analysis of TE-TM mode conversion in a metal grating placed in conical mounting, *IEICE C-I*, J82-C-I, No. 2, 42-49 (in Japanese).

Suyama, T., Y. Okuno and T. Matsuda, (2007) Enhancement of TM-TE mode conversion caused by excitation of surface plasmons on a metal grating and its application for refractive index measurement, *Progress In Electromagnetics Research*, 72, 91-103

Ritchie, R. H., E. T. Arakawa, J. J. Cowan and R. N. Hamm, (1968) Surface-plasmon resonance effect in grating diffraction, *Phys. Rev. Lett.*, 21, 1530-1533

Localized Surface Plasmon Resonances: Noble Metal Nanoparticle Interaction with Rare-Earth Ions

V.A.G. Rivera, F.A. Ferri and E. Marega Jr.

Additional information is available at the end of the chapter

1. Introduction

Particles of sizes between 1 and 100 nm show fascinating properties with unusual characteristics that lead to the formation of unique properties in nanosystems, which are not observed in ordinary materials. These are considered hereby as nanoparticles (NPs). Additionally, metallic NPs with sizes smaller than the wavelength of light show strong dipolar excitations in the form of localized surface plasmon resonances (LSPR). LSPRs are non-propagating excitations of the conduction electrons of metallic NPs coupled to the electromagnetic field [1]. This effect has been the subject of extensive research, both fundamental and with a view to applications [2,3]. The resonance frequency of the oscillation, i.e., the surface plasmon (SP) energy, it is essentially determined by the dielectric properties of the metal and the surrounding medium, and by the particle size and shape. The collective charge oscillation causes a large resonant enhancement of the local field inside aand near the NP. This field enhancement is used in surface-enhanced Raman scattering (SERS) [4] and is currently discussed for potential applications in nonlinear optical devices [5], in optical tweezers [6], and generally for the manipulation of the local photonic density of states.

Rare-earth (RE) elements are a group of chemical elements known as Lanthanides that occur together in the periodic table. These elements are used in common consumer goods such as: computer memory, DVD's, rechargeable batteries, cell phones, car catalytic converters, magnets, fluorescent lighting and much more. Furthermore, play an essential role in modern national defense, e.g.: lanthanum in night-vision goggles, neodymium in laser range-finders, guidance systems, communications; europium in fluorescents and phosphors in lamps and monitors; erbium amplifiers in fiber-optics data transmission; samarium in permanent magnets stable at high temperatures and, others technological applications. We center our attention in the RE ions and their optical properties that are widely employed in photonic. Devices of general interest span RE ions concentrations of tens to several thousand parts per million (ppm), resulting in devices of one to tens of meters long such as channel waveguides and optical fiber, respectively. In optical devices, the RE should ideally be confined as a delta function in the center of the core for maximum gain per unit pump power. Thereby, there is a necessary tradeoff between the confinement and the RE ions concentration, such that the more confined structures require a higher rare-earth ions concentration. An important feature that distinguishes them from other optically active ions: they emit and absorb over narrow wavelength ranges, the wavelengths of the emission and absorption transitions depend the host material, the intensities of these transitions are weak, the lifetimes of metastable states are long, and the quantum efficiencies tend to be high, except in aqueous solutions. These properties lead to excellent performance of RE ions in many optical applications. Devices that provide gain (e.g. lasers and amplifiers), must have low scattering losses, and one is restricted to using single-crystal or glass hosts. Whereas in many applications crystalline materials are preferred for reasons that include higher peak cross sections or better thermal conductivities, the versatility of glasses and the broader emission and absorption spectra they provide have led to the use of RE doped glasses in many applications, eventually running into the clustering limit for the particular host glass composition [7].

Most of existing and potential future applications of NPs suffer from damping caused by metal absorption. Sudarkin and Demkovich [8] suggested increasing the propagation length of the surface plasmon polariton[1] (SPP) by creating the population inversion in the dielectric medium adjacent to the metallic surface (film). Recently, gain-assisted propagation of the SPPs at the interface between a metal and a dielectric with optical gain has been analyzed theoretically [9,10]. Thus, NPs embedded in a gain media represent a field enhancement sustained that resonant excitation can lead to a reduction in the threshold for achieving inversion in the optically active surrounding medium, and the presence of gain can counteract the inherent absorption losses in the NP [1]. While this enhancement of plasmon resonances in gain medium not is experimentally confirmed, amplification of fluorescence due to field enhancement in gain medium with metal NPs has been observed [11,12,13,14].

In the present chapter, we explore the physics of localized surface plasmons by first considering the interaction of metal NPs with an electromagnetic wave in order to arrive at the resonance condition (LSPR). After, we show studies of plasmon resonances in particles with different shapes and sizes, and the effects of interactions between particles in the ensemble. Then we will focus on the RE ions, discussing the optical properties of the trivalent ions through the principles of quantum mechanics (in terms of oscillator strengths) with special interest in the interactions between the 4f electrons themselves. Since all the other electronic shells are spherically symmetric. Finally we present diverse experimental results of the interaction of RE ions interaction with NPs, resulting in an enhancement of the

---

[1] A SPP is an electromagnetic wave propagating along the interface between two media possessing permittivities with opposite signs, such as the metal–dielectric interface.

intensity emission of the RE ions due to long-range electromagnetic interaction between LSPR and the RE ions.

In this context, it is worth mentioning that the main purpose of this chapter is to show some of concrete concept of *noble metallic nanoparticle interaction with rare-earth ions* exhibiting a field enhancement in transparent medium with noble metal NPs embedded via a *localized surface plasmon resonance*.

2. Localized surface plasmon resonance and metallic nanoparticles

First, we know that a surface plasmon resonance (SPR) can be described as the resonance collective oscillation of valence electrons in a solid stimulated by beam of light incident (electromagnetic field – see Figure 1). The resonance condition is established when the frequency of light matches the natural frequency of valence electrons oscillating against from this restoring force. SPR in nanometer-sized structures is called LSPR. For gold and silver NPs, the resonance falls into the visible region of the electromagnetic spectrum. A striking consequence of this is the bright colors exhibited by particles both in transmitted and reflected light, due to resonantly enhanced absorption and scattering. This effect has found applications for many hundreds of years, for example in the staining of glass for windows or ornamental cups[2].

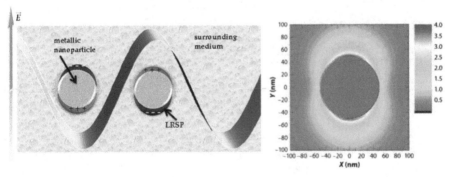

Figure 1. Left: Schematic of the electron charge displacement (valence electrons) in a metallic NP (LSPR) interacting with an incident plane wave, with electric field polarized E into a host matrix. Right: The corresponding electric field strength pattern for a 100 nm silver sphere, irradiated at a wavelength of 514 nm, this is reproduced from X. Lu et al. [15].

2.1. Noble metal nanoparticles in an electric field

For spherical NPs smaller than compared to local variations of the involved electromagnetic fields ($d<<\lambda$) in the surrounding medium, the quasi-static approximation [16], we lead to a

---

[2] Localized surface plasmons have been observed since the Romans who used gold and silver nanoparticles to create colored glass objects such as the Lycurgus Cup (4th Century A.D.). A gold sol in the British museum, created by Michael Faraday in 1857, is still exhibiting its red color due to the plasmon resonance at ,530 nm [L. M. Liz-Marzan, Mater. Today 7, 26 (2004).].

good agreement between theory and experiment. So, one can calculate the spatial field distribution (assuming as a problem of a particle in an electrostatic field), from the Laplace equation for the potential $\nabla^2\Phi = 0$, therefore the electric field $E = -\nabla\Phi$. Besides, the harmonic time dependence can then be added to the solution once the field distributions are well known. The solution this problem was obtained by Jackson [17]:

$$\Phi_{in} = -\frac{3\varepsilon_{NP}}{\varepsilon_{NP} + 2\varepsilon_m}E_0 r\cos\theta \tag{1}$$

$$\Phi_{out} = -E_0 r\cos\theta + \frac{\varepsilon_{NP} - \varepsilon_m}{\varepsilon_{NP} + 2\varepsilon_m}E_0\left(\frac{d}{2}\right)^3\frac{\cos\theta}{r^2} \tag{2}$$

E_0 is the amplitude of the electric field, ε_{NP} and ε_m are the dielectric permittivity of the NP and of the surrounding medium respectively, both are function the excitation frequency ω, r is the module of the position vector. Here Φ_{out} describes the superposition of the applied field and that of a dipole located at the NP center, i.e., an applied field induces dipole moment inside the sphere proportional to $|E_0|$. Additionally, we can arrive at

$$\alpha = 4\pi\left(\frac{d}{2}\right)^3\frac{\varepsilon_{NP} - \varepsilon_m}{\varepsilon_{NP} + 2\varepsilon_m} \tag{3}$$

Here α is the (complex) polarizability of the NP in the electrostatic approximation. Under the condition that $|\varepsilon_{NP} + 2\varepsilon_m|$ is a minimum, the polarizability shows a resonant enhancement. Thus, we can also write the Frohlich condition $\mathrm{Re}\left[\varepsilon(\omega)_{NP}\right] = -2\varepsilon_m$ that is the associated mode the dipole surface plasmon of the NP (in an oscillating field). Nevertheless, the distribution of the electric field E_{in} inside and E_{out} outside the sphere can be written as:

$$E_{in} = \frac{3\varepsilon_m}{\varepsilon_{NP} + 2\varepsilon_m}E_0 \tag{4}$$

$$E_{out} = E_0 + \frac{3n(n\cdot p) - p}{4\pi\varepsilon_0\varepsilon_m}\left(\frac{1}{r}\right)^3 \tag{5}$$

Here, $p = \varepsilon_0\varepsilon_m\alpha E_0$ is the dipole moment. Therefore, a resonance in α implies a resonant improvement of both the internal and dipolar fields, resulting in prominent applications of NPs in optical devices. Now consider a plane wave incident with $E(r,t) = E_0 e^{-i\omega t}$, this induce on NP an oscillating dipole moment $p(t) = \varepsilon_0\varepsilon_m\alpha E_0 e^{-i\omega t}$, i.e., we have a scattering of the plane wave by the NP. In this sense, the electromagnetic fields associated with an electric dipole in the near, intermediate and radiation zones are [17]:

$$H = \frac{ck^2}{4\pi}(n\times p)\frac{e^{ikr}}{r}\left(1 - \frac{1}{ikr}\right) \tag{6}$$

$$E = \frac{1}{4\pi\varepsilon_0\varepsilon_m} \left\{ k^2 (n \times p) \times n \frac{e^{ikr}}{r} + \left(3n(n \cdot p) - p \right) \left(\frac{1}{r^3} - \frac{ik}{r^2} \right) e^{ikr} \right\} \tag{7}$$

With $k = 2\pi / \lambda$ and n is the unit vector in the direction of the point P of interest. For $kr \ll 1$ (near zone) we have the electrostatic result of (5). The magnetic field present has the form $H = \frac{i\omega}{4\pi} \frac{(n \times p)}{r^2}$. Hence, in the near field the fields are predominantly electric and for static field $(kr \to 0)$, the magnetic field vanishes. For $kr \gg 1$, the dipole fields have spherical-wave form: $H = \frac{ck^2 (n \times p) e^{ikr}}{4\pi} \frac{}{r}$ and $E = \sqrt{\frac{\mu_0}{\varepsilon_0\varepsilon_m}} H \times n$.

From the viewpoint of optics, it is much interesting to note that another consequence of the enhanced polarization α in which a NP scatters and absorbs light [18]. For a sphere of volume V and dielectric function $\varepsilon_{NP} = \varepsilon_1 + i\varepsilon_2$ in the quasi-static limit, the extinction cross section $C_{ext} = C_{abs} + C_{sca}$ is:

$$C_{ext} = 9\frac{\omega}{c}\varepsilon_m^{3/2} \frac{\varepsilon_2}{\left(\varepsilon_1 + 2\varepsilon_m \right)^2 + \varepsilon_2^2} \tag{8}$$

2.2. Mie theory

For particles with larger dimensions, where the quasi-static approximation is not justified due to significant phase-changes of the driving field over the particle volume, a rigorous electrodynamics approach is required. This way, Gustav Mie solved Maxwell's equations for the case of an incoming plane interacting with a spherical particle [19]. In essence, the electromagnetic fields are expanded in multipole contributions and the expansion coefficients are found by applying the correct boundary conditions for electromagnetic fields at the interface between the metallic NP and its surrounding.

The extinction cross section of a spherical NP is given by the following expression:

$$\sigma_{ext} = \frac{\lambda^2}{2\pi} \sum_{n=0}^{\infty} (2n+1) \text{Re}\{a_n + b_n\} \tag{9}$$

Here the parameters a_n and b_n are defined as:

$$a_n = \frac{\Psi_n(\beta)\Psi'_n(m\beta) - m\Psi_n(m\beta)\Psi'_n(\beta)}{\xi_n(\beta)\Psi'_n(m\beta) - m\Psi_n(m\beta)\xi'_n(\beta)} \tag{10}$$

$$b_n = \frac{m\Psi_n(\beta)\Psi'_n(m\beta) - \Psi_n(m\beta)\Psi'_n(\beta)}{m\xi_n(\beta)\Psi'_n(m\beta) - \Psi_n(m\beta)\xi'_n(\beta)} \tag{11}$$

The size parameter β is defined as $\beta = \dfrac{\pi d m_0}{\lambda_0}$, where λ_0 is the incident wavelength with respect to vacuum, and m_0 represents the refractive index of the surrounding medium. The Ricatti-Bessel functions Ψ and ξ are defined in terms of the half-integer-order Bessel function of the first kind $(J_{n+1/2}(z))$, $\Psi_n(x) = \left(\dfrac{rx}{2}\right)^{0.5} J_{n+1/2}(x)$ and $\xi_n(x) = \left(\dfrac{rx}{2}\right)^{0.5} H_{n+1/2}(x)$, $H_{n+1/2}(x)$ is the half-integer-order Hankel function of the second kind.

We will focus our attention in silver and gold, since the localized plasmon resonance condition mentioned above is satisfied at visible light frequencies. Additional advantages of these metal NPs include simple preparation methods for a wide range of sizes and shapes and easy surface conjugation to a variety of ligands.

We now evaluate the extinction cross section using the Mie theory from equation (9), Figure 2, we take dielectric constants for silver and gold from Palik [20], and the medium dielectric constant is assumed to be 1.0 (i.e., a particle in a vacuum) and 2.0 for NPs with sizes different.

We can see from Figure 2 the dependence of resonance frequency with size of NPs (silver and gold) and the refraction index (surrounding medium). K. Lance Kelly et al. (2003), show that for spherical NPs clear differences between the quasistatic and the Mie theory results. However, the important features are retained, e.g. the frequency resonance. Although Mie theory is not a very expensive calculation, the quasistatic expressions are convenient to use when only qualitative information is needed [21].

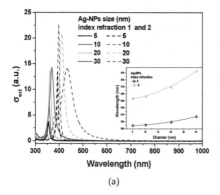 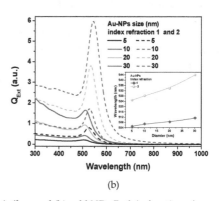

(a)   (b)

Figure 2. Extinction cross section from Mie theory for (a) silver and (b) gold NPs. Both in function of size with refraction index 1 (line curves) and 2 (dot lines). Inset figure shows a red-shift with the increment of size NPs and refraction index.

Now, Figure 3 shows clearly the dependence of resonance frequency embedded in a glass (with refraction index of: 1.99 for 400 nm, 1.88 for 630 nm, 1.75 for 900 nm and 1.70 for 1000 nm) and in a bulk with refraction index 2, in both cases with 20 nm size and a size distribution of 20 % STD, this simulation was extracted from program MiePlot v4.2.11 of Philip Laven. Also, it is possible to observe an increment of bandwidth and the intensity of peak for NPs embedded in the glass when compared with the material bulk. Therefore, LSPR results in enhanced local electromagnetic fields near the surface of the NP (Novotny & Hecht, 2006) [22], see Figure 1.

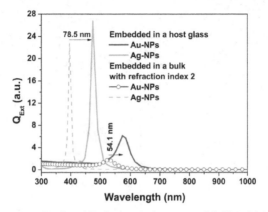

Figure 3. Extinction cross section from Mie theory via the program MiePlot v4.2.11 of Philip Laven, for silver and gold NPs with a size distribution of 20 % STD of 20 nm size embedded in a glass with refraction index in function of the wavelength and a bulk with refraction index 2 constant. In both cases, we observed a red-shift (see arrow). Inset figure shows the size distribution of NPs for 50 NPs in both cases.

Hideki Nabika and Shigehito Deki [23], show an experimental evidence of this dependence above mentioned for silver NPs of different sizes were synthesized by reducing AgNO₃ with N,N-dimethylformamide (DMF), Poly(vinylpyrrolidone) (PVP) aqueous solution (10 mL, 10 wt %) was added to DMF (80 mL), followed by an addition of AgNO₃ aqueous solution (10 mL). They obtained three sets of spherical silver NPs with a particle diameter varying from 9.7 to 27.1 nm and rod-shaped silver NPs with an aspect ratio of 1.79, its results are show Figure 4.

Thereby, the LSPR properties are sensitive to its environment in the order of its shape and size [24], the dielectric function of the glass host containing the NPs, changes the LSPR bandwidth and resonance frequency, see Figure 2-4.

We can be written as the sum of the bound and quasi-free (conduction) electron contributions [25]: $\varepsilon(\omega) = \varepsilon^{ib}(\omega) - \omega_p^2 / \omega[\omega + i\gamma]$, γ is the damping the resonance, ω the excitation frequency, ω_p the plasma frequency. The bound electron contribution ε^{ib} remains unchanged [26]. A similar expression can be used for the contribution of conduction electrons [27]: $\gamma = 1/\tau^{NP} = 1/\tau_0 + 2g_sV_F / d$. Where the first term, $1/\tau_0$ is associated to bulklike electron scattering process in the particle and the second term is a consequence of quasi-electron-free interaction with the surface and, for a sphere, V_F is the Fermi velocity, and g_s is the surface factor [25]. Nevertheless, the LSPR dependence on the matrix refractive index (n_λ) also can be calculated by [12]: $\omega_p = \sqrt{4\pi n' e^2 / \varepsilon_0 \varepsilon_d(\omega) m^*}$, where n' is the electrons density, e the electron charge, $\left(\varepsilon_d(\omega) = n_\lambda^2\right)$ the dielectric permittivity, and m^* the electron mass.

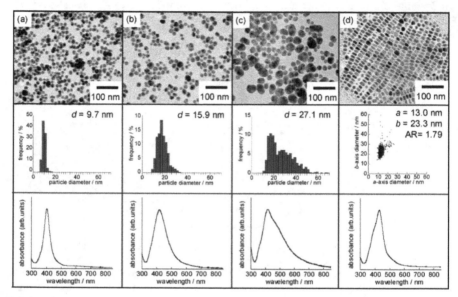

Figure 4. TEM micrographs, size distributions and optical absorption spectra of spherical silver NPs. Hideki Nabika and Shigehito Deki [23]. (a) 90 °C, AgNO₃ = 0.125 M, (b) 90 °C, AgNO₃ = 0.250 M, and (c) 110 °C, AgNO₃ = 0.250 M and (d) of the rod-shaped silver NPs.

On the other hand, in the Gans theory, the LSPR is only a function of the aspect ratio and refractive index. Thus, in certain conditions, a linear relationship between them can be resulted [28]. Nevertheless, numerical results suggest that, even when the aspect ratio is fixed and the retardation effect is weak, the position of longitudinal resonance can still depend strongly with the aspect ratio [29, 30] . Using the model of Cheng - ping Huang et al [31], we can write:

$$\lambda_p = \pi n \sqrt{10\kappa\left(2\delta^2 + r^2 \ln\left[\kappa\right]\right)} \tag{12}$$

Where $\kappa = l/2r$ is the aspect ratio of the NPs (with the inner radius r and in a cylindrical region with the length l), and δ is the skin depth. Cheng - ping Huang considered the problem as using an LC circuit model without solving the Laplace or Maxwell equations, and show that its results overcomes the deficiency of Gans theory and provides a new insight into the phenomenon. This way, we can observe resonance wavelength change with the aspect ratio of NP, Figure 5. This means a breakdown of the linear behavior presenting oscillations electrons originating from the amorphous geometry of the NPs associated with the inertia of electrons, see inset Figure 2 as well.

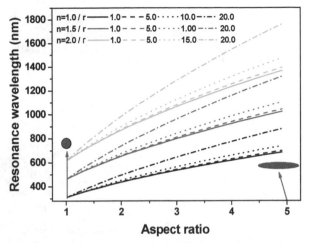

Figure 5. Resonance wavelength with aspect ratio from Cheng - ping Huang prediction for radius different and embedded in a bulk with refraction index 1.0 , 1.5 and 2.0.

Figure 6 illustrates this further, via experiments of X. Lu et al. [15] and Mock et al. [32]. Panel (a) the nanobar Ag NPs and the corresponding dark-field light scattering spectra, in broad agreement with the simulated results shown in [15]. Panel (b) shows the dipolar plasmon lineshapes of colloidal silver NPs of different shapes [32].

2.3. Coupling between metallic nanoparticles

We can say that the localized plasmon resonance frequency of a single metallic NP can be shifted through of alterations in shape, size and surrounding medium from the Frohlich condition, section 2.1. Nevertheless, in a NP ensembles we can obtain additional shifts due to electromagnetic interactions between the localized modes, see Figure 3 and 4 (c). These interactions are basically of a dipolar nature (when $d \ll \lambda$). So, the NP ensemble can be treated as an ensemble of interacting dipoles (in a first approximation). Those NPs can be embedded into a host matrix ordered or random, in one-, two- or three-dimensional arrays with interparticle spacing D. Electromagnetic coupling of those arrays shows interesting localization effects for closely spaced particles such as, enhancement process due to field localization in NP junctions. Assuming a dipolar approximation the NPs can be treated as point dipoles. In this sense, two regimes have to be distinguished (i) For $D \ll \lambda$, near-field interactions with a distance dependence of D^{-3} dominate, equation (5), and the NP array is described as an array of point dipoles coupling via their near-field, see section 2.1 and Figure 7. These arrays can serve as hot-spots for field enhancement, e.g. in a context of surface-enhanced Raman scattering (SERS). (ii) For larger NP separations, far-field dipolar coupling with a distance dependence of D^{-1} dominates, see section 2.1.

Therefore, the maximum field enhancement is determined by the shortest distance between two equipotential particles.

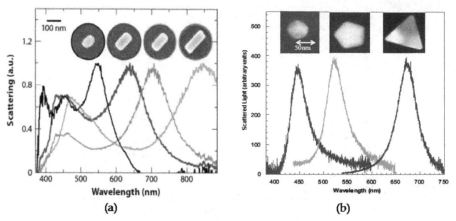

(a)   (b)

Figure 6. Panel (a) scattering spectra. In the insets are SEM images of the individual Ag nanobars. X. Lu et al. [15]. For simple, highly symmetric shapes (sphere, cube and triangular plate), the spectra are dominated by a single peak, but with the peak position sensitive to the shape, and ranging from 400 to up 800 nm. Panel (b) Scattering spectra of single silver nanoparticles of different shapes obtained in dark-field configuration. Mock et al. [32].

Figure 7 shows the near field coupling between NPs. Here, the restoring force acting on the oscillating electrons of each NP in the chain is either increased or decreased by the charge distribution of neighboring particles. Depending on the polarization direction of the exciting light, this leads to a blue-shift of the plasmon resonance for the excitation of transverse modes, Figure 7 (a), and a red-shift for longitudinal modes, Figure 7 (b).

Figure 7. Schematic of near-field coupling between metallic NPs for two different polarizations.

Today, there are several theoretical methodologies available to describe the interaction among the metallic NPs with electromagnetic radiation [33,34,35,36,37,38,39,40].

3. Rare-earths

Rare earths ions in solids are either divalent or trivalent. Their electronic configuration is $4f^N5s^25p^6$ or $4f^{N-1}5s^25p^6$, respectively. By far the most common valence state of the RE ions in solids is the trivalent one. Those ions have a long history in optical and magnetic applications. We have special interest in the devices luminescent using crystal, powders, and glasses. On the other hand, divalent RE ions have also been used in laser devices, but only in relatively exotic ones for cryogenic operation [41,42]. The most frequently used laser-active RE ions and host media together with typical emission wavelength ranges are shown in the table 1:

Ion	Common host media	Important emission wavelengths (μm)
Neodymium (Nd^{3+})	YAG, YVO$_4$, YLF, silica	1.03–1.1, 0.9–0.95, 1.32–1.35
Ytterbium (Yb^{3+})	YAG, tungstates, silica	1.0–1.1
Erbium (Er^{3+})	YAG, silica, tellurite, chacogenetos glasses	1.5–1.6, 2.7, 0.55
Thulium (Tm^{3+})	YAG, silica, fluoride glasses	1.7–2.1, 1.45–1.53, 0.48, 0.8
Holmium (Ho^{3+})	YAG, YLF, silica	2.1, 2.8–2.9
Praseodymium (Pr^{3+})	silica, fluoride glasses	1.3, 0.635, 0.6, 0.52, 0.49
Cerium (Ce^{3+})	YLF, LiCAF, LiLuF, LiSAF, and similar fluorides	0.28–0.33

Table 1. Common laser-active rare earth ions and host media and important emission wavelengths.

The 4f electrons are not the outermost ones. They are shielded from external fields by two electronic shells with larger radial extension (5s^{2}5p^6), which explains the atomic nature of their spectra absorption/emission, Figure 8. Thus the 4f electrons are only weakly perturbed by the charges of the surrounding atoms. Important characteristics that distinguish them from other optically active ions are: (i) the wavelengths of the emission and absorption transitions are relatively insensitive to host material, (ii) the intensities of these transitions are weak, (iii) the lifetimes of metastable states are long and, (iv) the quantum efficiencies tend to be high. This is why the RE ions are such a useful probe in a solid; the crystal environment constitutes only a small perturbation on the atomic energy levels, and many of the solid state, and hence spectroscopic, properties can be understood from a consideration of the free ions. In turn, the wavefunctions of the free ions constitute a good zero order approximation for a description of solid-state properties, Figure 8.

The solutions to this problem can then befactored into a product of a radial and angular function. Whereas the radial function depends on the details of the potential the spherical symmetry ensures that the angular component is identicalwith that of a hydrogen atom and can be expressed as spherical harmonics. Except for Ce^{3+} and Yb^{3+}, which have only one electron, the solutions of the central-field problem are products of one-electron states that are antisymmetric under the interchange of a pair of electrons, as required by the Pauli exclusion principle.

In describing the state of a multielectron atom, the orbital angular momenta and the spin angular momenta are added separately. The sum of the orbital angular momenta is designated by the letter L, and the total spin is characterized by S. The total angular momentum J of the atom may then be obtained by vector addition of L and S. The collection of energy states with common values of J, L, and S is called a term. Here, the Russell-

Saunders notation for the energy levels (terms) is used, $^{2S+1}L_J$, this takes into account the spin-spin coupling, orbit-orbit coupling and spin-orbit coupling. The energy levels of a free RE ion are usually interpreted by considering only interactions between the 4f electrons themselves. Since all the other electronic shells are spherically symmetric, Figure 8, their effect on all the terms of a configuration is the same in first order, and therefore do not contribute significantly to the relative positions of the 4f energy levels, we can write:

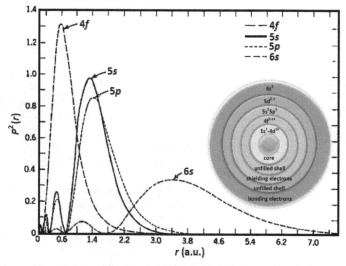

Figure 8. Square of the radial wavefunctions for the 4f, 5s, 5p and 6s energy levels from Hartree-Fock. It was a calculation for Gd$^+$ by Freeman e Watson (1962) [46]. Picture inset schematically shows that the 4f orbital is within the 6s, 5p and 5s levels.

$$H = -\frac{\hbar^2}{2m}\sum_{i=1}^{N}\Delta_i - \sum_{i=1}^{N}\frac{Z^*e^2}{r_i} + \underbrace{\sum_{i<j}^{N}\frac{e^2}{r_{ij}}}_{H_C} + \underbrace{\sum_{i=1}^{N}\zeta(r_i)s_i \cdot l_i}_{H_{SO}} + V_{EF} \tag{13}$$

Where $N=1...14$ is the number of the 4f electrons, Z^*e the screened charge of the nucleus because we have neglected the closed electronic shell, V_{EF} treats the interaction of the ion with the electromagnetic field, and $\zeta(r_i)$ the spin-orbit coupling function [43],

$$\zeta(r_i) = \frac{\hbar^2}{2m^2c^2r_i}\frac{dU(r_i)}{dr_i} \tag{14}$$

Where $U(r_i)$ is the potential in which the electron i is moving. The first two terms of the Hamiltonian (equation (13)) are spherically symmetric and therefore do not remove any of the degeneracies within the configuration of the 4f electrons, therefore we can neglect. The next two terms, which represent the mutual Coulomb interaction of the 4f electrons (H_C) and their spin-orbit interaction (H_{SO}) are responsible for the energy level structure of the 4f electrons. Details of the matrix element calculations of H_C and H_{SO} can be found in ref. [44,45].

Figure 8 shows the radial distribution functions 4f, 5s, 5p, and 6s electrons for Gd^+ as obtained from Harteww-Fock calculation by Freeman and Watson [46]. We can see that the 4f electrons are inner electrons with relatively small $<r^n>$ values. As we shall see, the crystal field interaction will be small in REs. It is also evident that the 4f wavefunctions do not extend very far beyond the 5s and 5p shells. Thus, all quantities of a solid depend on an overlap of the 4f wavefunctions with those of a neighboring ion have to be small in RE compounds.

3.1. Trivalent ions in a static crystal field

The 4f shell of the RE ions are an unfilled shell and therefore have a spherical charge distribution. If the ion is introduced into a crystal, the ion experiences an inhomogeneous electrostatic field, the so-called crystal field, which is produced by the charge distribution into the crystal. This crystal field distorts the closed shells of the RE ion. Producing an effect on the energy level 4f, i.e., removes to a certain degree the degeneracy of the free ion 4f levels, thus producing a major modification of the energy levels (but this depends on the crystal symmetry). In the luminescence spectrum are observed additional transitions, which originate from the excited crystal field levels of the ground term is evident, and the spectra look more complicated. On the other hand, these spectrums can be used to determine the crystal field energies of the ground term.

We shall now try to understand the behavior of these additional transitions. To do so we draw on our treatment of the Coulomb interaction of the 4f electrons. The crystal field interaction comes by the interaction of the 4f electrons with all the charges of the crystal with all charges of the crystal, except for trivial factors, it is therefore given by one over the radius vector between the 4f electrons and the crystal charges. This function is given, apart from radial factors, of two spherical harmonics Y_{kq}, one containing the coordinates of the 4f electrons, the other containing coordinates of the crystal charge. The latter has to be integrated over the whole crystal, whereupon it gives the strength of the crystal potential at the site of the RE ion. Finally, this shows the existence of the splitting telling us that the development of the crystal potential into spherical harmonics contains a second order term.

3.1.1. Crystal field splitting

Crystal field splitting has two aspects: (i) symmetry, namely, the number of the levels into which an ion is free the J term are split in a crystal field of a given symmetry; (ii) the actual size of the crystal field splitting.

This way, the point charge model can be used for illustrative purposes but not for a quantitative description of the crystal field interaction. A realistic description of this interaction must take into account that the crystal field is built up of the spatially extended charge clouds of the individual ions. Thus, these charge clouds can penetrate each other and thereby interact. Hence, we can considered the following elements:

1.   The ions are considered static in the crystal, i.e., we can neglected the lattice vibrations and their effect on the energy levels.
2.   We regard the 4f electrons of one RE ion, as representative of those of all the RE ions in the crystal and, thus the interaction of 4f electrons of adjacent ions is neglected.
3.   The crystal consists of the extended charge distributions. This produces an overlap of the charge distributions of the neighboring ions and the 4f electrons. Aditionally, a charge transfer between 4f electrons and the electrons of the ligands can take place. Both contribute to the crystal field interaction.
4.   The 4f electrons of one ion are considered to be independent of each other, that is correlation effects play no significant role.

Therewith, we can calculate the crystal potential $\Phi(r_i, \varphi_i, \theta_i)$ at the site of the 4f electrons and the potential energy of the 4f electrons in this potential. If the crystal has charge density $\rho(R)$ and the 4f electrons have radius r_i, we have:

$$V = -\sum_i \int \frac{e_i \rho(R)}{|R - r_i|} d\tau = -\sum_{k,i} e_i \int \rho(R) P_k \left(\cos[R, r_i] \right) \frac{r_<^k}{r_>^{k+1}} d\tau \tag{15}$$

Where $r_<$ and $r_>$ are, respectively, the smaller and larger value of r and r_i. Here $P_k(\cos[x])$ are the Legendre polynomials. Also, the equation (15) can be re-write as:

$$V = -\sum_{k,q,i} B_{k,q} C_{k,q}(\theta_i, \varphi_i) \tag{16}$$

Where the crystal field parameter $B_{k,q}$ ($k \leq 6$ for f electrons) and the summation i is carried out over all the 4f electrons of the ions. $B_{k,q}$ have the form:

$$B_{k,q} = -e \int (-1)^q \rho(R) C_{k-q}(\theta, \varphi) \frac{r_<^k}{r_>^{k+1}} d\tau \tag{17}$$

Here $C_{k,q}$ is called a tensor operator and is defined as: $C_{k,q} = \left(\frac{4\pi}{2k+1} \right)^{0.5} Y_{k,q}$

In early treatments of the crystal fiel interaction it was assumed that in RE compounds the point charge contribution would be the dominant part of the crystal field interaction. Then, in equation (16) the integral over the lattice can be replaced by a sum over all lattice points and $r_<^k$ can be replaced by r_i^k. The latter replacement can be performed as long as the charge distribution of the crystal does not enter that of the 4f electrons (as long as $r_i < R$), which implies that the potential acting on the 4f electrons obeys the Laplace equation $(\Delta\Phi(r, \varphi, \theta) = 0)$ at the position of the 4f electrons.

Figure 9 further illustrates the effect of spin–orbit and crystal field interactions on the energy levels of the Er^{3+} ion. This figure were extracted from A.J. Kenyon [47]. A further splitting of the energy levels comes about when the ion is placed in a silica host. The $4f^N$ configuration is composed for a number of states where the quantum numbers (L, S, J, and another arbitrary one) define the terms of the configuration, all of which are degenerate in the central-field approximation, as illustrated in Figure 9. Next in the hierarchy is spin–orbit, the strongest of the magnetic interactions. Spin–orbit lifts the degeneracy in total angular momentum and splits the LS terms into J levels, Figure 9.

We can see that the environment provided by silica hosts destroys the spherically symmetric environment that Er^{3+} ions enjoy in the vapor phase. Thus the degeneracy of the 4f atomic states will be lifted to some degree. This splitting is also referred to as stark splitting, and the resulting states are called Stark components (of the parent manifold). The even-k terms in the expansion split the free ion J multiplets into Stark components generally separated by 10–100 cm^{-1} (see Figure 9). The ion–lattice interaction can mix multiplets with different J values (J mixing), although it usually remains a good quantum number. The odd-k terms admix higher lying states of opposite parity [e.g., $4f^{N}15d^{1}$] into the $4f^{N}$ configuration. This admixture does not affect the positions of the energy levels, but it has a very important effect on the strengths of the optical transitions between levels.

3.2. Optical properties

In recent years most of the interest in luminescent RE ions has concentrated on the trivalent erbium (Er^{3+}), and in particular its emission band around 1550 nm. This emission is within the telecommunication windows, and fortuitously coincides with the 1550 nm intra-4f $^{4}I_{13/2} \rightarrow {}^{4}I_{15/2}$ transition of the Er^{3+} ion.

For this reason exists today a great interest in utilizing erbium-doped materials for gain elements and sources in telecommunications systems [48,49,50]. The development of the erbium-doped fiber amplifier (EDFA) in the late 1980s [51,52] exploited the $^{4}I_{13/2} \rightarrow {}^{4}I_{15/2}$ transition and allowed the transmission and amplification of signals in the 1530–1560 nm region without the necessity for expensive optical to electrical conversion [53]. Hence, it is originated a driving force for research in RE doped fibers and integrated optics waveguides has been their use for amplifying weak signals in optical communications systems at 1300 and 1600 nm. This may be achieved by simply splicing a section of RE-doped fiber into the transmission one and injecting pump light through a fiber coupler. The signal generated within the RE emission band stimulates emission of radiation at the same frequency, amplifying the optical communication signal with high gain, high efficiency and low noise, which is highly advantageous for optical communications [54]. There are five main RE candidates for use as dopants in fiber or waveguide amplifiers for optical communications systems: Er^{3+}, Tm^{3+}, Nd^{3+}, Pr^{3+} and Dy^{3+}. The Er^{3+} and Tm^{3+} ions are the choice for the 1400–1600 nm window centered at 1550 nm, based on the $^{4}I_{13/2} \rightarrow {}^{4}I_{15/2}$ transition of Er^{3+} ion and the $^{3}H_{4} \rightarrow {}^{3}F_{4}$ transition of Tm^{3+} ion. The $^{4}F_{3/2} \rightarrow {}^{4}I_{13/2}$ emission of Nd^{3+} ion, the $^{1}G_{4} \rightarrow {}^{3}H_{5}$ transition of Pr^{3+} ion and the $^{6}F_{11/2}(^{6}H_{9/2}) \rightarrow {}^{6}H_{15/2}$ transition of Dy^{3+} ion are all potentially useful for the 1300 nm telecommunication window.

Further, the 3+ ions all exhibit intense narrow-band intra-4f luminescence in a wide variety of hosts, and the shielding provided by the $5s^{2}$ and $5p^{6}$ electrons (see Figure 8), which means that rare-earth radiative transitions in solid hosts resemble those of the free ions and electron–phonon coupling is weak. Although some of the divalent species also exhibit luminescence (e.g., samarium and europium), it is the trivalent ions that are of most interest.

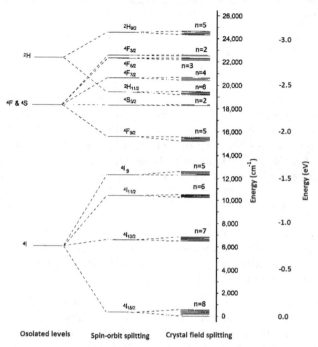

Osolated levels   Spin-orbit splitting   Crystal field splitting

Figure 9. The effect of spin-orbit and crystal field splitting on the energy levels of the Er^{3+} ion in silica matrix. The energy diagram shows the hierarchy of splitting resulting from electron-electron and electron-host interactions. Figure extracted from A.J. Kenyon [47].

As mentioned previously, the intra-4f transitions are parity forbidden and are made partially allowed by crystal field interactions mixing opposite parity wavefunctions, resulting in luminescence lifetimes long (often in the millisecond range), and linewidths narrow. The selection of an appropriate ion with intense and narrow-band emission can be obtained across much of the visible region and into the near-infrared. Figure 10 shows energy level diagrams for the isolated 3+ ions of each of the 13 lanthanides with partially filled 4f orbitals.

Almost all the RE ions their emission is due to optical transitions within the f-manifold (e.g. Tb^{3+} (4f^8), Gd^{3+} (4f^7) and Eu^{3+} (4f^7). The 4f-electrons are well shielded from the chemical environment and therefore have almost retained their atomic character. Nevertheless, for a number of RE ions, also broad emission bands are known. Prominent examples are Eu^{2+} and Ce^{3+}. Here, the emission is from to 5d-4f optical transitions. As electrons participate in the chemical bonding, the d-f emission spectra consist of broad bands. These transition types are allowed and are consequently very fast (a few μs or less).

Some line emission is not a specific property of RE ions and in addition to that, in the case of RE ions, broad emission spectra can be obtained as well, depending on the optical transitions involved. In this respect, RE ions are not unique. The unique properties of the RE ions originate from the fact that the spectral position of the emission lines is almost independent of the host lattice, in contrast to line emission generate by the emission of metal ions.

3.2.1. Radiative transitions: Intensities of optical transitions

The terms shown in equation (13) are time-dependent; thereby, they do not lead to stationary states of the system. This way, they are treated using time-dependent perturbation theory resulting in transitions between the states established by the static interactions. In luminescent devices the most important term is V_{EF} which gives rise to the emission and absorption of photons from decay radiative of the RE ions. This involves the interaction between the electron charge and the electric field and, the interaction between the electron spin and the magnetic field.

The experimental data on spectra of RE ions show that the radiation is mostly electric dipole (ED) nature, though some cases of the magnetic dipole radiation are also observed. Since the optical transitions take place between levels of a particular $4f^N$ configuration, the electric dipole radiation is forbidden in first order, because the electric dipole operator has uneven parity and the transition matrix element must have even parity (Laporte selection rules). Van Vleck [55] pointed out the dipole electric radiation can only occur because the $4f^N$ states have admixtures of $4f^{N-1}nl$ configuration (nl will be mostly 5d), thus the $4f^{N-1}nl$ has to be chosen such that it has opposite parity from $4f^N$. The admixture is produced via interactios that have odd parity and this depend of the host matrix of the RE ions. However, we can have four dominant sources of optical radiation in RE ions spectra:

i.   Forced electric dipole induced by odd terms of the crystal field.
ii. Forced electric dipole radiation induiced by lattices radiation.
iii. Allowed magnetic dipole transition.
iv. Allowed electric quadrupole radiation.

In free atoms, magnetic dipole is about six orders of magnitude weaker than the electric dipole radiation. This latter occurs only as a consequence of a perturbation, both kinds of radiation shows up in the RE spectra with about intensity. Then, quadrupole radiation are less probable still in comparison with the magnetic dipole.

There are different types of transitions between levels (called electric dipole, electric quadrupole, and magnetic dipole) and therefore the transitions are divided into allowed ones (with a high probability) and so-called forbidden transitions (for transitions where the electric dipole transition is quantum mechanically forbidden).

For the first order allowed magnetic dipole radiation the free atom selection rules are still quite valid: $\Delta S = \Delta L = 0$ and $\Delta J = 0,\ \pm 1\ \left(0 \neq 0\right)$. In the electric dipole case we have: $\Delta l = \pm 1$, $\Delta S = 0$, $\left|\Delta L\right|$, $\left|\Delta J\right| \leq 2l$. However, since ED transitions is induced by the crystal field, the free

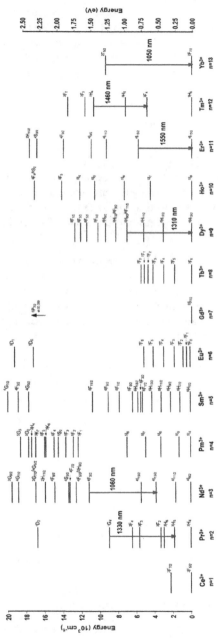

Figure 10. Energy levels of the triply charged lanthanide ions. Besides, the most technologically important radiative transitions are labeled. Figure adapted from A.J. Kenyon [47].

atom selection rules break down almost completely, i.e., the selection rules with regard to the L, S and J quantum numbers are now essential governed by the crystal field interaction yielding $|\Delta J| \leq 6$ (approximately). For the electric quadrupole case the rules selection is: $\Delta S = 0$, $|\Delta L|, |\Delta J| \leq 2$.

We now outline the Judd-Ofelt theory for the determination of the intensities in RE crystal spectra [56,57]. It is in essence a quantification of the ideas formulated by van Vleck [55] under optical radiation sources. Intensities are often expressed in terms of oscillator strengths, where the total oscillator strength for a transition from an energy level a to an energy level b is given:

$$f(a,b) = \frac{8\pi^2 m \upsilon}{3h(2J+1)} \left[\frac{(n^2+2)^2}{9n} S_{ed} + n S_{md} \right] \tag{18}$$

Here h is the Planck constant, υ the frequency of the transition $a \to b$, n the refraction index of host. S_{ed} and S_{md} are the electric dipole and magnetic dipole intensities respectively and are defined as:

$$S_{ed}(\alpha J, \alpha' J') = \sum_{\lambda = 2,4,6} \Omega_\lambda \left\langle f^N \alpha [SL] J \left\| U^{(\lambda)} \right\| f^N \alpha' [S'L'] J' \right\rangle^2$$

$$S_{md}(\alpha J, \alpha' J') = \beta^2 \sum_{\lambda = 2,4,6} \Omega_\lambda \left\langle f^N \alpha [SL] J \left\| L + 2S \right\| f^N \alpha' [S'L'] J' \right\rangle^2 \tag{19}$$

$U^{(\lambda)}$ is a tensor operator of rank λ, and the sum runs over the three terms values 2, 4 and 6 of λ. With:

$$\Omega_\lambda = (2\lambda + 1) \sum_{k,q} \frac{\left| B_{k,q}^* \right|^2 \left| Y_{k,\lambda} \right|^2}{2k+1} \tag{20}$$

The Ω_λ parameters have so far been assumed to arise solely from crystal field; however, they also contain contributions from admixtures by the lattice vibrations. The asterisk means that for $B_{k,q}^*$ the radial integral is to be taken between states 4f and nl instead of 4f and 4f.

Also, it is interesting to determine the coefficient for spontaneous light from state $a(\alpha J)$ to state $b(\alpha' J')$. Those are defined as:

$$A(\alpha J, \alpha' J') = A_{ed} + A_{md} = \frac{64\pi^4 \upsilon^3}{3hc^2(2J+1)} \left[\frac{n(n^2+2)^2}{9} S_{ed} + n^3 S_{md} \right] \tag{21}$$

Here c velocity of light in vacuum. The Judd-Ofelt formalism has been applied to the analysis of a number of systems. In most of these analyses the crystal field splitting of the terms is neglected; therefore, the total absorption intensities between the ground term and the excited terms are analyzed with only three empirical parameters $\Omega_\lambda (\lambda = 2,4 \text{ and } 6)$.

If b is an excited state that decays only by the emission of photons, its observed that relaxation rate is the sum of the probabilities for transitions to all possible final states. The total rate is the reciprocal of the excited-state lifetime τ_a

$$\tau_a = \frac{1}{A(\alpha J, \alpha' J')} \tag{22}$$

The branching ratio $\beta_{a,b}$, for the transition $b \rightarrow a$ is the fraction of all spontaneous decay processes that occur through that channel and is defined as follows:

$$\beta_{a,b} = \frac{A(a,b)}{\sum_c A(a,c)} = A(a,b)\tau_a \tag{23}$$

The branching ratio, which has an important influence on the performance of a device based on a particular transition, appears often in the discussion of specific ions. It has a significant effect on the threshold of a laser and the efficiency of an amplifier.

4. Metallic nanoparticle embedded in a gain media: LSPR interaction radiative transitions

The RE-doped laser crystals and glasses are among the most popular solid-state gain media. In order of maintaining the efficiency of these materials in which the RE dopant is uniformily dispersed, as in multi-component glasses. In this sense, glasses have good optical, mechanical, and thermal properties to withstand the severe operating conditions of optical amplifier. Desirable properties include hardness, chemical inertness, absence of internal strain and refractive index variations, resistance to radiation-induced color centers, and ease of fabrication. Furthermore, the matrix host composition affects the solubility of the RE dopant, this affect the lifetime, absorption, emission, and excited state absorption cross sections of the dopant transitions [47]. This competing absorption phenomenon can seriously diminish the efficiency of an optical device. For example, a decrease in the excited-state absorption for Er^{3+}-doped fibers going from a germane-silicate host to an alumina-silicate host has been verified [58], demonstrating the importance of host selection for a given RE ion or laser transition. Host glasses compatible with this relatively high concentration of RE without clustering require the open, chain-like structure of phosphate glasses or the addition of modifier ions (Ca, Na, K, Li, or other) to open the silicate structure and increase solubility [48]. The limitation owing to clustering in a predominantly silica host without modifier ions has been well documented [19]. The maximum erbium concentration in silica for optimum amplifier performance has been suggested to be lesser to 100 ppm [59]. However a 14.4-dB gain, 900-ppm erbium-doped silica fiber amplifier has been reported, indicating that higher concentrations can produce useful devices [48]. Nevertheless, the interactions between host and RE ions, it is necessary to consider background losses from impurity absorption and scattering mechanisms that decrease the efficiency of the optical device. For example, depending on the phonon energies of the host matrix, some of the level lifetimes can be strongly quenched by multi-phonon transitions. Such effects are minimized in low-phonon-energy host media such as fluoride fibers. The effect of internal loss is most dramatic in distributed amplifiers, where pump light must travel long distances in the process of distributing gain.

Additionally, into the glass matrix can happen diverse kinds of interactions, in particular dipole–dipole interactions between the RE ions similar or of different species, allow energy transfer between those. This is exploited e.g. in Er^{3+}–Yb^{3+}-codoped fibers, where the pump radiation is dominantly absorbed by Yb^{3+} ions and mostly transferred to Er^{3+} ions [48].

In this way, RE-doped glasses shown are excellent materials that provide gain media. This makes them ideal candidates for embedding metallic NPs, resulting in new materials for the fabrication of devices optical with large enhancements even when the gain is saturated due light localization effects. It has been shown theoretically and experimentally that homogeneous aggregates of structures supporting LSPR can lead to extremely large enhancement of local field amplitudes exceeding those of single structures [60,61]. Besides, inorganic glasses are the host matrix for metallic NPs formation. The wide temperature range of glass viscosity growth provides the possibility to control over the NPs size within the wide range by means of modifying the temperature and duration of thermal processing. In fact, only such kind of matrix makes it possible to control and investigate all the stages of NPs formation, including the starting stage [14,62].

The LSPR in metallic NPs is predicted to exhibit a singularity when the surrounding dielectric medium has a critical value of optical gain [63]. This singularity is obtained from the equation (4), for $\mathrm{Re}\left[\varepsilon_{NP}(\omega)\right] \approx -2\varepsilon_m$, the NP absorption presents a maximum. This is so-called the Frohlich condition is associated with dipole mode [1].

Therefore, the strong local electric field induced by the NP (LSPR) can enhance the total electric field, and can also improve the quantum yield of the luminenescence of the RE ions from the NP (this strong enhancement of electromagnetic fields is essential in nonlinear optics effects) [64,65]. Thus, the enhancement obtained with these NPs is due to the formation of EDs which generates a polarization given by $\vec{P}_{NP} = qeN'\vec{x}$, where $\vec{x}$ is the distance between each NP , $q = \pi d^3/6$ is the NP specific volume, N' is the conduction electrons density, and e is the electronic charge [12]. Therefore, we obtained a modification in the local electric field by these dipoles (local field correction). Using the results obtained by O.L. Malta et al [66], the effective electric field can be written as:

$$\vec{E}_{eff} = (\varepsilon_0 + 2)\left[1 + q\omega_p^2 / \left\{3\varepsilon_0\left[(1-q)\left(\omega_p^2/3\varepsilon_0\right) - \omega^2 + i\gamma\omega\right]\right\}\right]\vec{E}_0/3 \qquad (24)$$

Where ε_0 is the dielectric constant in the presence of an external electromagnetic field of amplitude E_0. In the presence of an electromagnetic wave we have LSPR which forms EDs separated by different distance $\vec{r}$, some of which will contribute to the luminescence enhancement.

Assuming that the RE ions may occupy different sites in the host, a direct coupling between the excited states of the RE ions and the NPs modifies the Stark levels energies [12,13,67]. As the NPs just give a contribution to the local field when the light is present, the oscillator strength of a spectral line, corresponding to transition from the ion ground level i to the component f of the excited level can be re-write as:

$$f(a,b) = \frac{8\pi^2 m v}{3h(2J+1)} \left[\frac{\left(n^2+2\right)^2}{9n} S_{ed} + \sum_\lambda \left| \left\langle i \left| D^{(1)}_{\lambda\ NP} \right| f \right\rangle \right|^2 + n S_{md} \right] \tag{25}$$

The second term into parenthesis is added to the equation from the Judd-Ofelt theory [56,57], and represents the ED transition due to LSPR of NP. To obtain non vanishing matrix elements of the components $D^{(1)}_{\lambda\ NP}$ it is necessary to admix into $\langle i |$ and $| f \rangle$ other states of opposite parity.

Thus, the initial and final states are: $\langle i | = \langle \phi_i | + \sum_\beta \langle \phi_i | V | \phi_\beta \rangle / \left(E_i - E_\beta \right) \langle \phi_\beta |$, and $| f \rangle = | \phi_f \rangle + \sum_\beta \langle \phi_\beta | V | \phi_f \rangle / \left(E_f - E_\beta \right) | \phi_\beta \rangle$, then:

$$\left\langle i \left| \vec{P} + \vec{P}_{NP} \right| f \right\rangle = \sum_\beta \left\{ \frac{\langle \phi_i | V | \phi_\beta \rangle \langle \phi_\beta | \vec{P} | \phi_f \rangle}{E_i - E_\beta} + \frac{\langle \phi_i | \vec{P} | \phi_\beta \rangle \langle \phi_\beta | V | \phi_f \rangle}{E_f - E_\beta} + \frac{\langle \phi_i | V | \phi_\beta \rangle \langle \phi_\beta | \vec{P}_{NP} | \phi_f \rangle}{E_i - E_\beta} \cdots \right.$$
$$\left. \cdots + \frac{\langle \phi_i | \vec{P}_{NP} | \phi_\beta \rangle \langle \phi_\beta | V | \phi_f \rangle}{E_f - E_\beta} \right\} \tag{26}$$

Where V is the crystalline field, equation (16), responsible for the Stark levels and treated as a perturbation, ϕ_i and ϕ_f have the same parity, ϕ_β has opposite parity in relation of ϕ_i and ϕ_f, $\vec{P}$ is the ED from the light interaction. For example, Er^{3+} ions presented an electronic configuration [Xe]4f¹¹, with $J = 15/2$ been the spin-orbit ground state because the f shell is more than half-filled. The order of these Stark levels depends on the strength of the crystal field and the influence of $\vec{P}_{NP}$.

This can be observed experimentally in the luminescence spectra of different RE ions shown in Figure 11 interaction with silver or gold NPs in different host matrix, as a blue- or red-shift (see equation 26) and a pronounced increase in luminescence (see equation 25).

In this scenario, the influence of $\vec{P}_{NP}$ appeared in the luminescence spectra as a blue or a red -shift of the peaks, e.g. a transfer energy from the NPs dipoles to RE ions, thus modifying the Stark energy level's bandwidth, see Figure 11. This energy transfer is obtained from the second term added in equation (25) and can be understood through the following definition: $D^{(1)}_{q'\ NP} = \sum_m r_m C^{(1)}_{\lambda\ NP} \left(\theta_m, \phi_m \right)$, where $C^{(1)}_{\lambda\ NP} \left(\theta_m, \phi_m \right)$ is the Racah tensor

and where r_m is the distance between the NPs and the RE ions ($r_m = \left| \vec{x} \right|$).

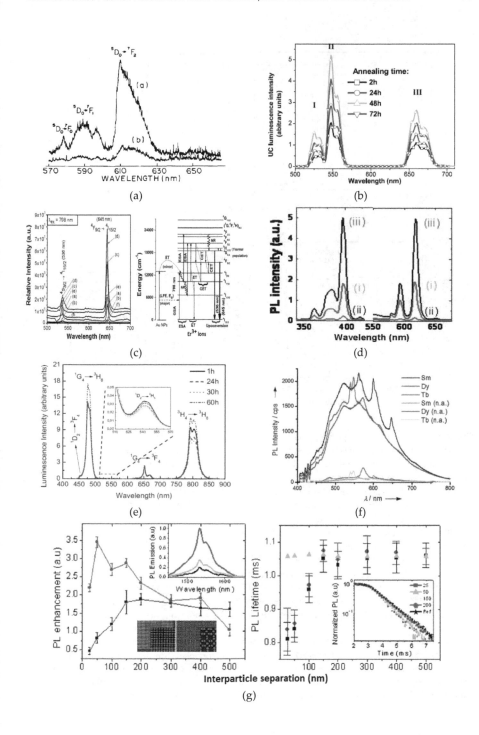

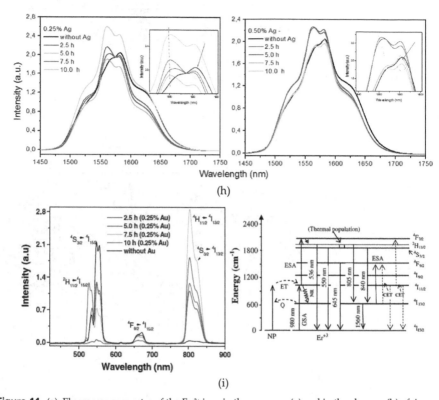

(h)

(i)

Figure 11. (a). Fluorescence spectra of the Eu³⁺ ions in the presence (a) and in the absence (b) of Ag NPs.The silver concentration is 7.5 in weight percent. The matrix glass composition can be found in [66]. (b). Frequency UC spectra for excitation at 980 nm, (for composition see Ref. [68]. After the cooling the samples were annealed for different durations (24, 48, and 72 h) in order to nucleate silver NPs [68]. (c). Upconversion spectra extracted from [69] under excitation wavelength at λ_{ex}=798 nm (for composition and for amplification ratio see Ref. [69]. The bases of the emission curves (c), (d), and (e) have been uplifted for better visibility. (d). PL spectra extracted from [70] (i) Eu(III)EDTA, 3H₂O complex, (ii) Eu complex with Au nanoparticles and (iii) Eu complex with Au–ZnO nanoparticles. (e). Upconversion spectra of Tm³⁺/Yb³⁺ codoped PbO–GeO₂ samples containing silver NPs obtained by pumping the samples with a diode laser operating at 980 nm. For more details see [71]. (f). PL spectra (λ_{exc}≈337 nm) of sy-activated with 0.037 mol% Ag containing glasses co-doped with 0.19 mol% Sm (black curve), Dy (red), and Tb (blue), respectively, and of non-activated and non-annealed (n.a.) 0.037 mol% Ag containing samples co-doped with 0.19 mol% Sm (grey), Dy (magenta), and Tb (light blue) [72]. (g). Picture on the left, Integrated PL enhancement in periodic (circle) and Fibonacci (square) nanoparticle arrays of various interparticle separations. The top inset shows the representative PL spectrum of periodic (bottom), Fibonacci (top) nanoparticle arrays with Δ_{min}= 50 nm and unpatterned area (middle),and the inset bottom shown the SEM micrograph of (left) periodic, (right) Fibonacci array Au nanocylinders. Picture on the right, PL lifetime of periodic (circle), Fibonacci (square), unpatterned (triangle). Inset Er decay of unpatterned (star) and Fibonacci arrays with different Δ_{min} as specified in the legend. For more details see Ref. [73].

(h). Picture on the left; PL of the TE025-Y samples pumped with diode laser at 980 nm, showing PL enhancement. The inserted figure shows a zoom of the peaks. The vertical dashed line is a reference for showing the blue-shift of the peaks, and the arrow indicates the enhancement due to transfer energy from EDs to Er^{3+}. Picture on the right; same for the samples TE050-Y. The enhancement of luminescence was found to be reproducible for all ours samples. These pictures was extracted and modified from Ref. [12]. These spectrums are of Er^{3+}-doped tellurite glass, more details see [12].

(i). Luminescence spectra of Er^{3+}:Au-doped tellurite glass for several annealing times, pumped at 980 nm. Picture on the left; Upconversion, for the range $400<\lambda<900$ nm. Picture on the right; Energy-levels diagram of Er^{3+} for the luminescence spectra. ET stands for energy transfer, CET for cooperative energy transfer, Q for quenching, GSA for ground state absorption, NR for non-radiative decay, and ESA for excited state absorption. For more details see Ref. [13].

Nevertheless, an interesting question arises: how these NPs are excited?

As explained above, those NPs can be excited by a predefined incident radiation through a direct coupling between the excited states of the both RE ions and NPs, resulting in: (i) a local field increase (Frohlich condition), at ω_p, (ii) a nonradiatively decay (heat generation by Joule effect) or (iii) a radiative energy release which depends on the albedo of the NPs. Thereby, the exact response of LSPR will depend on the details of the physical system (e.g. arrangement, shape, host matrix), and usually not strictly symmetric about the resonant frequency. A schematic representation of the interaction process within the RE:NP system is depicted in Figure 12 (a).

Also we can elucidate the enhancement (quenching) from the process of energy transfer as following: Er^{3+} emission promotes energy transfers into a plasmonic mode, which can decay nonradiatively by heat generation (Joule effect) or radiatively by releasing energy that depends on the albedo of the NPs. We consider two types of emission from a system of identical dipoles [74]:

$$I_1 \propto I_p \eta_p \eta_0 \tag{27}$$

$$I_2 \propto I_1 \eta_0 \eta_{LSPR} Q_{scatt} \tag{28}$$

Here, I_1 is the intensity emission for a single emitter, I_2 is the intensity emission of the NP, I_p the pump intensity, η_p is the pump efficiency, η_0 is the internal quantum efficiency of the energy transfer, η_{LSPR} is the efficiency of the energy transfer (nonradiative) to LSPR modes and the scattering efficiency of the plasmon at the emission wavelength Q_{scatt} is:

$$Q_{scatt} = Q_{scatt}^{(Ray)} = \frac{8}{3} \frac{\omega_p^4 y^4}{\left(\omega^2 - \omega_p^2 \right)^2 + \frac{4}{9} y^6 \omega_p^4} \tag{29}$$

Where $Q_{scatt}^{(Ray)}$ is the Rayleigh scattering [67,68] and, $y = \omega D/c$, with c the speed of light. Therefore the total intensity I_T, can be written as [62]:

$$I_T = I_1\left(1 - \eta_{LSPR}\right) + \eta_0 I_2 \tag{30}$$

We consider I_1 as the contribution from the emitter dipole (RE ions) uncoupled to the metal and I_2 the plasmon enhanced emission from coupled dipoles, see Figure 12 (b).

Finally, we can obtain the following conditions: $\eta_0^2 Q_{scatt}^{(Ray)} \geq 1$, where $Q_{scatt}^{(Ray)}(\omega, D)$, i.e. the luminescence enhancement depends on the incident radiation frequency and the NP size. But, as were mentioned, the incident radiation for the NP activation comes from the emitter dipole $\left(I_1 \eta_{LSPR}\right)$ and not from the pump radiation (I_P), i.e., from transition radiative from RE. The coupling efficiency η_{LSPR} is defined by: $\eta_{LSPR} = \Gamma_{LSPR} / \left(\Gamma_{rad} + \Gamma_{nonrad} + \Gamma_{LSPR}\right)$, where: Γ_{LSPR} is the energy transfer rate to the LSPR mode [73,74].

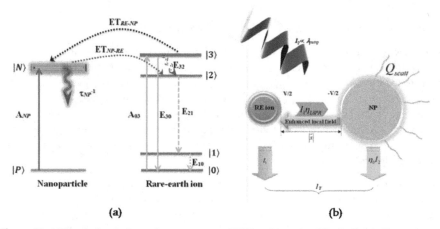

(a)   (b)

Figure 12. (a) Energy level scheme for a resonant and RE ion absorption. The dot line indicates the energy transfer between RE→NP or NP→ RE, and the vertical dot line shows the transition radiative under consideration. The curved arrows indicate non-radiative transitions. (b) Schematic representation of the system RE:NP. A monochromatic plane wave with pump intensity (I_P) which is proportional to the pump wavelength λ_{pump}, induces the following processes: (i) absorption of RE ion, I_P, (ii) activation of NPs, due the coupling RE:NP (transitions levels), $I_1 \eta_{LSPR}$ (iii) NP transmitter, via electric dipole. Such coupling depends on the coupling efficiency η_{LSPR}. Where I_1 the intensity emission of emitter, I_2 the intensity emission of the NP, η_{LSPR} the efficiency of the energy transfer (nonradiative) to LSPR modes. Equipotential surfaces (electric dipole coupling) with electric potentials -V/2 and V/2 for the NPs and the Er^{3+} ions respectively.

The strong local electric field induced by NPs (ED), increases the quantum yield η of the RE luminescence, defined by the ratio of the local field $\vec{E}_{loc}$ and the incident field $\vec{E}_i$, $\eta = \left(r_m + d\right) / d = \left|\vec{E}_{loc}\right| / \left|\vec{E}_i\right|$ [14], here $\vec{E}_{eff} = \vec{E}_{loc} + \vec{E}_i$. Thus, the maximum field enhancement is determined by the shortest distance between two equipotential particles. It corroborates well with the demonstration in the references [12,13,67,75].

Another possible mechanism for the energy transfer is due to that part of silver or gold (small silver or gold aggregates) probably remained under the form of ions, atoms, charged or neutral dimmers and multimers. Consequently, the contribution of the latter species to the RE luminescence enhancement and band shape features cannot be excluded [14], Figure 11. Moreover, the insertion of silver or gold in the samples leads to strong modification of the glassy network and consequently on the RE local environment, this is verified in reference [62].

5. Conclusion

This chapter present results where it was demonstrated the simultaneous exploitation of the enhanced local field due to NPs and energy transfer processes in order to enhance the luminescence spectra of a glassy composite material, and others. Besides, from these nanoparticles it is possible to modulate the down/up-conversion emission of the REs with applicability in areas such as optical telecommunication, including biomedical imaging and energy conversion. On the other hand, this significant enhanced fluorescence has high potential for application in photonics, optical displays, lasers and optical memory devices, amongst others. However, the success of new applications of nanoparticles depends on improvement in the understanding of the properties of LSPR and the environment around them. Further efforts and systematic studies must be realized in order to offer new developments to extend the analytical applications field of metallic nanoparticle interaction with rare-earth ions.

We hope that this chapter stimulated our readers for the development of theoretical and experimental work on plasmonics and rare-earths.

Author details

V.A.G. Rivera, F.A. Ferri and E. Marega Jr.
Instituto de Física de São Carlos, INOF/CEPOF, USP, São Carlos – São Paulo, Brazil

Acknowledgement

This work was financially supported by the Brazilian agencies FAPESP, CNPq and CEPOF/INOF. V.A.G. Rivera thanks to FAPESP for financial support (project 2009/08978-4 and 2011/21293-0) that allowed my pos-doctoral and my gratefully to Dr. Luiz Antonio Nunes of the Instituto de Fisica de São Carlos – University São Paulo - Brazil and the Dr. Yannick Ledemi and the Dr. Younnes Messaddeq of the Centre d'Optique, Photonique et laser – University Laval- Canada for the discussions on this issue.

6. References

[1] Stefan Alexander Maier. Plasmonics Funadamentals and Applications. Springer Science+Business Media LLC; 2007.

[2] Mark I. Stockman. Opt. Exp. 2011;19(22): 22029-22106.

[3] Y. Wang, E. W. Plummer and K. Kempa. Advances in Physics 2011;60(5): 799-898.

[4] K. Kneipp, H. Kneipp, I. Itzkan, R. R. Dasari, and M. S. Feld. J. Phys.: Condens. Matter 2002;14:R597–R624.

[5] Y. Takeda and N. Kishimoto. Nuc. Inst. Met. Phys. Res. B 2003;206: 620–623.

[6] Peter J. Reece. Nat. Phot. 2008;2:333-334.

[7] Desurvire E., J. L. Zyskind and C. R. Giles. IEEE/OSA J. Ligh. Technol. 1990;LT8: 1730–1741.

[8] A. N. Sudarkin and P. A. Demkovich. Sov. Phys. Tech. Phys. 1989;34: 764-766.

[9] I. Avrutsky. Phys. Rev. B 2004;70: 155416-155421.

[10] M. P. Nezhad, K. Tetz, and Y. Fainman. Opt. Express 2004;12: 4072-4079.

[11] Chau, K. J., Dice, G. D., and Elezzabi, A. Y. Phys. Rev. Lett. 2005;94: 173904-173907.

[12] V.A.G. Rivera, S. P. A. Osorio, Y. Ledemi, D. Manzani, Y. Messaddeq, L. A. O. Nunes, and E. Marega. Opt. Exp. 2010;18: 25321-25328.

[13] S.P.A. Osorio, V.A.G. Rivera, L.A.O. Nunes, E. Marega, D. Manzani, Y. Messaddeq. Plasmonics 2012;7: 53-58.

[14] V.A.G. Rivera, Y. Ledemi, S.P.A. Osorio, D. Manzani, Y. Messaddeq, L.A.O. Nunes and E. Marega. J. Non-Crys. Sol. 2012;358: 399-405.

[15] X. Lu, M. Rycenga, S.E. Skrabalak, B.Wiley, and Y. Xia. Annu. Rev. Phys. Chem. 2009;60: 167-192.

[16] U. Kreibig, M. Vollmer. Optical Properties of Metal Clusters. Springer-Verlag, Berlin; 1995.

[17] Jackson John D. Classical Electrodynamics. John Wiley & Sons, Inc., New York, NY, 3rd edition; 1999.

[18] Bohren, Craig F. and Huffman Donald R. Absorption and scattering of light by small particles. John Wiley & Sons, Inc., New York, NY, 1 edition; 1983.

[19] Mie Gustav. Ann. Phys. 1908;25: 377-345.

[20] E. D. Palik. Handbook of Optical Constants of Solids Academic. Elsevier, Orlando, FL. ISBN: 978-0-12-544415-6; 1985.

[21] K. Lance Kelly, Eduardo Coronado, Lin Lin Zhao, and George C. Schatz. J. Phys. Chem. B 2003;107: 668-677.

[22] Novotny L. and Hecht B. Principles of nano-optics, Cambridge University Press, United Kingdom, ISBN-13 968-0-521-83224-3; 2006.

[23] Hideki Nabika and Shigehito Deki. J. Phys. Chem. B 2003;107(35): 9161-9164.

[24] D. D. Evanoff, R. L. White and G. Chumanov. J. Phys. Chem. B 2004;108(37): 1522-1524.

[25] H. Baida, P. Billaud, S. Marhaba, D. Christofilos, E. Cottancin, A. Crut, J. Lermé, P. Maioli, M. Pellarin, M. Broyer, N. Del Fatti, and F. Vallé. Nano Lett. 2009;9(10): 3463-3469.

[26] C. Voisin, N. Del Fatti, D. Christofilos and F.J. Valleé. Phys. Chem. B 2001;105: 2264-2280.

[27] F. Hache, D. Ricard and C.J. Flytzanis. J. Opt. Soc. Am. B 1986;3(12): 1647-1655.

[28] S. Link, M. B. Mohamed and M. A. El-Sayed. J. Phys. Chem. B 1999;103: 3073-3077.

[29] H. Kuwata, H. Tamaru, K. Esumi and K. Miyano. Appl. Phys. Lett. 2003;83: 4625-4627.

[30] S.W. Prescott and P. Mulvaney. J. Appl. Phys. 2006;99: 123504-123510.

[31] Cheng-ping Huang, Xiao-gang Yin, Huang Huang and Yong-yuan Zhu. Opt. Exp. 2009;17(8): 6407-6413.

[32] Mock J. J., Barbic M., Smith D. R., Schultz D. A., and Schultz S. J. Chem. Phys. 2002;116(15): 6755–6759.

[33] Purcell E. M. and Pennypacker C. R. Astrophys. 1973;186: 705-714.

[34] Ruppin R. Phys. Rev. B 1982;26: 3440-3444.

[35] Pinchuk A., Kalsin A., Kowalczuk B., Schatz G. and Grzybowski B. J. Phys. Chem. C 2007;111: 11816-11822.

[36] Gerardy J. M. and Ausloos M. Phys. Rev. B 1982;25: 4204-4229.

[37] Claro F. Phys. Rev. B 1984;30: 4989-4999.

[38] Rojas R. and Claro F. Phys. Rev. B 1986;34: 3730-3736.

[39] Olivares I., Rojas R., Claro F. Phys. Rev. B 1987;35: 2453-2455.

[40] M. Chergui, A. Melikian , H. Minassian. J. Phys. Chem. C 2009;113: 6463–6471.

[41] P. P. Sorokin and M. J. Stevenson. Phys. Rev. Lett. 1960;5: 557-559.

[42] D.C. Brown. IEEE J. Sel. Top. Quantum Electron. 2005;11: 587-599.

[43] S. Hufner. Optical Spectra of Transparent Rare Earth Compounds. Academic press New York – San Francisco – London; 1978.

[44] B.R. Judd. Operator Techniques in Atomic Spectroscopy. McGraw-Hill, New York; 1963.

[45] B.G. Wybourne. Spectroscopic properties of Rare Earths, Wiley, New York; 1965.

[46] A.J. Freeman and R.E. Watson. Phys. Rev. 1962;127: 2058-2075.

[47] A.J. Kenyon. Prog. Quan. Elec. 2002;26: 225–284

[48] Michel J.F. Digonnet and Marcel Dekker. Rare-earth-doped fiber lasers and amplifiers. 2d ed., edited by, Inc. New York – Basel; 2001.

[49] V.A.G. Rivera, E.F. Chillce, E.G. Rodrigues, C.L. Cesar, L.C. Barbosa. J. Non-Crys. Sol. 2006;353: 125-130.

[50] V.A.G. Rivera, E.F. Chillce, E.G. Rodrigues, C.L. Cesar, L.C. Barbosa. Proc. SPIE 2006;6116: 190-193.

[51] P.J. Mears, L. Reekie, I.M. Jauncey and D.N. Payne. Elec. Lett. 1987;23: 1026-1028.

[52] E. Desurvire, R.J. Simpson and P.C. Becker. Opt. Lett. 1987;12: 888-890.

[53] E. Desurvire. Phys. Today 1994;47: 20-27.

[54] M. Yamane and Y. Asahara. Glasses for Photonics. Cambridge - University Press, Cambridge, United Kingdom; 2002.

[55] J.H. van Vleck. J. Phys. Chem. 1937;41: 67-80.

[56] B.R. Judd. Phys. Rev. 1962;127: 750-761.

[57] G. S. Ofelt. J. Chem. Phys. 1962;37: 511-520.

[58] Arai K., H. Namikawa, K. Kumata, T. Honda, Y. Ishii, T. Handa. J. Appl. Phys. 1986;59: 3430–3436.

[59] Shimizu M., M. Yamada, M. Horigucho and E. Sugita. IEEE Pho. Tech. Lett. 1990;2: 43–45.

[60] V.A. Markel, V. M. Shalaev, E. B. Stechel, W. Kim, and R. L. Armstrong, Phys. Rev. B 1996;53: 2425.

[61] V. M. Shalaev, E. Y. Poliakov and V. A. Markel. Phys. Rev. B 1996;53: 2425-2449.

[62] V.A.G. Rivera, S.P.A. Osorio, D. Manzani, Y. Messaddeq, L.A.O. Nunes, E. Marega Jr. Opt. Mat. 2011;33: 888-892.

[63] N. M. Lawandy. Appl. Phys. Lett. 2004;85: 5040-5042.

[64] S. Kim, J. Jin, Y. Kim, I. Park and Y. Kim. Nature 2008;453: 757-760.

[65] S. Kuhn, U. Hakanson, L. Rogobete and V. Sandoghdar. Phys. Rev. Lett. 2006;97: 017402.

[66] O.L. Malta, P.O. Santa-Cruz, G.F. de Sá and F. Auzel. J. of Lumin. 1985;33: 261-272.

[67] T. Som and B. Karmakar. J. Appl. Phys. 2009;105: 013102.

[68] L.R.P. Kassab, F.A.Bonfim, J.R. Martinelli, N.U. Wetter, J.J.Neto and Cid B. Araujo. Appl. Phys. B 2009;94: 239-242.

[69] T. Som and B. karmakar. J. Opt. Soc. Am. B 2009;26(12): B21-B27.

[70] Krishna Kanta Haldar and Amitava Patra. App. Phys. Lett. 2009;95: 063103.

[71] Thiago A. A. Assumpção, Davinson M. da Silva, Luciana R. P. Kassab and Cid B. de Araújo. J. App. Phys. 2009;106: 063522.

[72] Maik Eichelbaum and Klaus Rademann. Adv. Funct. Mater. 2009;19: 2045–2052.

[73] A. Gopinath, S. V. Boriskina, S. Yerci, R. Li and L. Dal Negro. Appl. Phys. Lett. 2010;96: 071113.

[74] J. R. Lakowicz. Anal. Biochem. 2005;337: 171-194.

[75] F. Le, D. W. Brandl, Y. A. Urzhumov, H. Wang, J. Kundu, N. J. Halas, J. Aizpurua, and P. Nordlander. ACS Nano 2008;2: 707-718.

Chapter 4

A Treatise on Magnetic Surface Polaritons: Theoretical Background, Numerical Verification and Experimental Realization

Yu-Hang Yang and Ta-Jen Yen

Additional information is available at the end of the chapter

1. Introduction

The polariton is a kind of coupling between electromagnetic waves (photons) and elementary excitations such as phonons, plasmons and magnons, It includes two modes of surface and bulk polariton that can be excited by means of semiconductors, metals, ferromagnets, antiferromagnets, and so on. The coupling causes an intensity enhancement of the electromagnetic field, which is very useful for nano-technology such as biosensing, waveguide applications, nano-antenna device. The majority of the literature is concentrating on the magnetic surface polariton supported by magnetic materials. The outline of paper is as follows. In sections 2 we introduce the fundamental of the polaritons first, and then the general dispersion equation of the magnetic surface polaritons (MSPs) mode by considering the full form of Maxwell's equations in section 3. Next, in sections 4 and 5, we present numerical and experimental results of realizing the MSPs, by naturally existing materials — in section 4, by ferromagnetic and antiferromagnetic materials and in section 5, by the effective media of ferromagnetic and anti-ferromagnetic superlattices, respectively. By employing metamaterials, artificially constructed materials whose properties mainly stem from structures rather than their constitutive elements, one can also achieve MSPs mode with a greater engineering freedom; thus, in section 6, we start from interpreting what metamaterials and their operation rationales are, and then how metamaterials support the MSPs mode. Finally, in section 7, there comes a conclusion. Note that in this study we moderately modify the quoted definitions in the diagrams to correspond to our definitions.

2. Fundamental of polaritons

The well-known surface plasmon polaritons (SPPs) denote that a collective oscillation of electrons couples with the transverse magnetic wave, and they propagate along the interface between two media (ex: a metal and a dielectric) with an exponential decay into two media. Due to the polarization dependence of SPPs, it is a reasonable perception to realize a magnetic analog by a transverse electric wave. In addition, we have known the fact that the wave equation can be expressed by either electric or magnetic component of the electromagnetic wave, so that it implies that all electromagnetic phenomena chould be symmetric. In recent years, more and more researchers have paid intensive attention to the electric SPPs and their applications, in particular for bio-sensing [15] and nanophotonic applications [27].

In contrast, the magnetic surface polaritons (MSPs) mode did not attract much attention yet, and that is because magnetic responses are typically weak and their resonant frequencies are usually below the far infrared region. In our opinion, it is useful and interesting to understand the mechanism of the MSPs even it still does not yet become a protagonist. Throughout this study, we will show there is a potential material to promote the inherent magnetism in natural materials.

At first, let us briefly define the well-known magnetic polariton. In general, the magnetically ordered media support three kinds of elementary modes, spin waves (or call magnons), magnetostatic modes, and magnetic polaritons. Three modes are classified by the dominant restoring force as shown in Fig. 1 [21]. When the wavevector (k) in vacuum is greater than 10^8 m^{-1}, the exchange interaction is important and one calls this mode as magnon. The equation of motion is usually needed to solve this mode. For magnetostatic mode (magnetostatic limit, $k \ll \frac{\omega}{c}$), $10^7 \leqslant k \leqslant 10^8$ m^{-1}, the exchange and dipolar interactions may be both important; $3 \times 10^3 \leqslant k \leqslant 10^7$ m^{-1}, the dipolar terms are mainly dominant. For this region, we need a Hamiltonian treatment for such a magnetostatic mode. For magnetic polaritons, $k \leqslant 3 \times 10^3$ m^{-1}, it is recognized as electromagnetic region where the full form of Maxwell's equations including retardation corrections have to be considered. As an example, the dispersion relation of a ferromagnetic insulator [21] is plotted in Fig. 1, in where ω_0 and ω_m will be defined later. In this literature, we just concentrate on MSPs, and study three situations: a pure ferromagnets (or antiferromagnets), ferromagnetic (or antiferromagnetic) superlattices, and magnetic metamaterials.

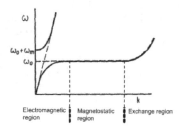

Figure 1. The dispersion diagram. In general, $k > 10^8$ m^{-1} is an exchange region; $3 \times 10^3 \leqslant k \leqslant 10^8$ m^{-1} is a magnetostatic region; $k \leqslant 3 \times 10^3$ m^{-1} is an electromagnetic region.

3. General dispersion equation for magnetic surface polaritons

It is worthy to derive a general dispersion relation in advance before we begin studying the real cases. It should help one to understand all definitions that are used in our whole literature. First, the geometry considered in this study consists of two semi-infinite linear media, including the magnetic and nonmagnetic media shown in Fig. 2. Second, we assume

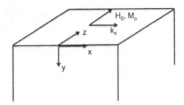

Figure 2. The geometry considered in this study. The half space above the $y > 0$ plane is a magnetic medium; below the $y < 0$ plane is a nonmagnetic medium. The external static magnetic field (H_0) is parallel to the easy axis, $\hat{z}$.

that the magnetic permeability tensor of the magnetic medium can be written by

$$\overset{\leftrightarrow}{\mu}_{eff}(\omega) = \begin{pmatrix} \mu_{xx} & i\mu_{xy} & 0 \\ -i\mu_{xy} & \mu_{yy} & 0 \\ 0 & 0 & \mu_{zz} \end{pmatrix}, \tag{1}$$

and the electric permittivity can be given by

$$\overset{\leftrightarrow}{\varepsilon}_{eff}(\omega) = \begin{pmatrix} \varepsilon_{xx} & 0 & 0 \\ 0 & \varepsilon_{yy} & 0 \\ 0 & 0 & \varepsilon_{zz} \end{pmatrix}. \tag{2}$$

Note that we are interesting in the dispersion relation itself so that the damping and spatial dispersion can be ignored here. It will be worthy to consider the damping only when the reflectivity calculation is executed; spatial dispersion is meaningful only for magnons. The constitutive relations for the linear medium read

$$\vec{D}(\omega) = \overset{\leftrightarrow}{\varepsilon}_{eff}(\omega)\vec{E}(\omega), \quad \vec{B}(\omega) = \overset{\leftrightarrow}{\mu}_{eff}(\omega)\vec{H}(\omega). \tag{3}$$

Next, we follow our previous work [37] and ready to derive rigorously a general dispersion equation. We are looking for MSPs mode. Therefore, without loss of generality, the transverse electric fields of the MSPs mode above and below the plane $y = 0$ can be respectively written by

$$\begin{aligned} \vec{E}^+(\omega) &= \hat{z}\, E_z^+ e^{ik_x x} e^{-\alpha^+ y}, \\ \vec{E}^-(\omega) &= \hat{z}\, E_z^- e^{ik_x x} e^{\alpha^- y}. \end{aligned} \tag{4}$$

where k_x is the direction of the MSPs mode in vacuum. α^+ and α^- stand for attenuation coefficients of the MSPs mode for $y > 0$ and $y < 0$, respectively. Faraday's law can solve the corresponding magnetic induction as follows,

$$\begin{aligned} \vec{B}^+(\omega) &= \frac{1}{i\omega}\left[(-\alpha^+)\hat{x} - (ik_x)\hat{y}\right]\vec{E}^+(\omega), \\ \vec{B}^-(\omega) &= \frac{1}{i\omega}\left[(\alpha^-)\hat{x} - (ik_x)\hat{y}\right]\vec{E}^-(\omega). \end{aligned} \tag{5}$$

Next, the magnetic fields can be solved by Eq. 3:

$$\begin{aligned} \vec{H}^+(\omega) &= \frac{-1}{\mu_{xy}^2 - \mu_{xx}\mu_{yy}}\left[(-\mu_{yy}\alpha^+ + \mu_{xy}k_x)\hat{x} + (-i\mu_{xy}\alpha^+ - i\mu_{xx}k_x)\hat{y}\right], \\ \vec{H}^-(\omega) &= (-\alpha^+)\hat{x} + (-ik_x)\hat{y}. \end{aligned} \tag{6}$$

Applying Ampere's law with Eqs. 1-3, we can solve the following wave equations:

$$\left[\mu_{xx} \frac{\partial^2}{\partial x^2} + \mu_{yy} \frac{\partial^2}{\partial y^2} + \left(\mu_{xx}\mu_{yy} - \mu_{xy}^2 \right) \frac{\omega^2}{c^2} \varepsilon_{zz} \right] \vec{E}^+ (\omega) = 0,$$

$$\left(\frac{\partial^2}{\partial x^2} + \frac{\partial^2}{\partial y^2} + \frac{\omega^2}{c^2} \varepsilon_{zz} \right) \vec{E}^- (\omega) = 0.$$

(7)

Using Eqs. 4 and 7, we can solve the attenuation coefficients as follows,

$$\alpha^+ = \sqrt{ \frac{\mu_{xx}}{\mu_{yy}} k_x^2 - \left(\frac{\mu_{xx}\mu_{yy} - \mu_{xy}^2}{\mu_{yy}} \right) \frac{\omega^2}{c^2} \varepsilon_{zz}, }$$

$$\alpha^- = \sqrt{ k_x^2 - \frac{\omega^2}{c^2} }.$$

(8)

Note that it is also straightforward to solve the bulk polaritons, propagating perpendicular to the $\hat{z}$ axis. All we do is just to replace $\alpha^\pm$ by ik_y. Then, one is given by

$$\frac{c^2 \left(\mu_{xx} k_x^2 + \mu_{yy} k_y^2 \right)}{\omega^2} = \left(\mu_{xx}\mu_{yy} - \mu_{xy}^2 \right) \varepsilon_{zz},$$

$$k_x^2 + k_y^2 = \frac{\omega^2}{c^2}.$$

(9)

The continuity of the tangential components of the electric field (E_x) and the magnetic filed (H_x) at the interface ($y = 0$) yields

$$E_x^+ = E_x^-, \quad \frac{\mu_{yy} B_x^+ + i\mu_{xy} B_y^+}{\mu_{xx}\mu_{yy} - \mu_{xy}^2} = \alpha^-.$$

(10)

Using the Eqs. 5, 6 and 10, the general dispersion equation of the MSP's mode reads

$$\frac{-\alpha^+ \mu_{yy} + \mu_{xy} k_x}{\mu_{xx}\mu_{yy} - \mu_{xy}^2} = \alpha^-.$$

(11)

Note that Eq. 11 itself implies a non-reciprocal dispersion relation due to $\pm |k_x|$. Now, we have enough equations to begin our discussion for the MSPs mode.

4. Realization of MSPs mode by magnetic materials

Over the past 50 years, Damon and Eshbach [8] first reported the surface magnetostatic modes without retardation effect for the case of a ferromagnetic slab; effective media for superlattices have also done [17]. Later Hartstein et al. reported more detailed discussion of the MSPs for semi-infinite medium of a pure magnetic materials [1]. Before discussing the semi-infinite gyromagnetic uniaxial medium (gyromagnetic ratio, $\gamma \neq 0$), we have to assume several limitations for this study: (a) we neglect the local effects of the surface on the spin wave (i.e. pinning effects) and (b) the exchange interaction is ignored as well.

4.1. Semi-infinite ferromagnets

In this part, we consider the case, a pure ferromagnet and vacuum. For simplicity, we treat the electric permittivity of a ferrromagnet $[\overleftrightarrow{\varepsilon}(\omega)]$ as scalar value, 1. Also, we neglect damping of $\overleftrightarrow{\mu}(\omega)$ so that μ_{xx}, μ_{zz}, and μ_{xy} in Eq. 1 are real. The external static magnetic filed (H_0) is along the easy axis $+\hat{z}$ shown in Fig. 2. The half space $y < 0$ is vacuum, and $y > 0$ is a pure ferromagnet. In the Voigt geometry, the magnetic permeability satisfies the equation, $\mu_v(\omega) = \mu_{xx} - \mu_{xy}^2/\mu_{xx}$ plotted in Fig. 3 [1]. We also define $\omega_v = \gamma\sqrt{B_0 H_0}$ at resonance and $\omega_1 = \gamma B_0$ at $\mu_v(\omega) = 0$. For a uniaxial ferromagnet ($\mu_{xx} = \mu_{yy}$), the Eqs. 8 and 11 are rewritten by

$$\alpha^+ = \sqrt{k_x^2 - \mu_v \frac{\omega^2}{c^2}}, \quad \alpha^- = \sqrt{k_x^2 - \frac{\omega^2}{c^2}}. \tag{12}$$

and

$$\frac{-\alpha^+ \mu_{xx} + \mu_{xy} k_x}{\mu_{xx}^2 - \mu_{xy}^2} = \alpha^-. \tag{13}$$

For the MSPs mode, the α^+ and α^- in Eq. 12 have to be positive that yield to an exponential

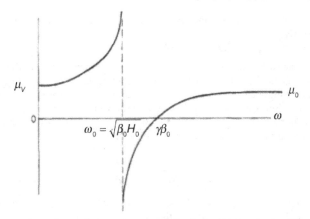

Figure 3. Frequency dependence of magnetic permeability in the Voigt geometry. [1]

decay into both magnetic and nonmagnetic materials (vacuum). The dispersion diagram [1] is plotted in Fig. 4 in where the shaded region makes the existence of the MSPs mode possible. For a uniaxial ferromagnet, the components of the magnetic permeability tensor in Eq. 1 can be written by

$$\mu_{xx} = \mu_{yy} = \mu_0 \left(1 + \frac{\omega_m \omega_0}{\omega_0^2 - \omega^2}\right),$$

$$\mu_{zz} = \mu_0, \tag{14}$$

$$\mu_{xy} = \mu_0 \frac{\omega_m \omega}{\omega_0^2 - \omega^2}.$$

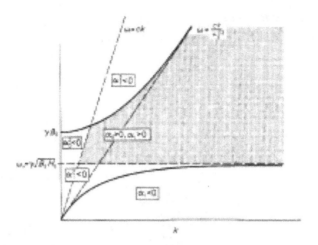

Figure 4. The dispersion relation of a pure ferromagnet. [1]

Here, two angular frequencies (ω_m and ω_0) are written in terms of the magnetization (M_0) and external field (H_0):

$$\omega_m = 4\pi\gamma M_0, \quad \omega_0 = \gamma H_0, \tag{15}$$

where ω_0 is angular frequency of the ferromagnetic resonance, and γ is the gyromagnetic ratio. Note that μ_0 in Eq. 14 is caused by other magnetic dipole excitation (such as optical magnons) at higher frequency (ex: $\mu_0 = 1.25$ in YIG). Before discussing a detailed calculation of dispersion equation, we would like to return to Eq. 6 that can simply predict a non-reciprocal dispersion relation. One write down the polarization of the magnetic field of the MSPs at a fixed point by investigating the components in Eq. 6:

$$\frac{H_y^+}{H_x^+} = i\frac{\mu_{xy}\alpha^+ + \mu_{xx}k_x}{\mu_{yy}\alpha^+ - \mu_{xy}k_x}. \tag{16}$$

When we take the magnetostatic limits ($k_x \to \pm\infty$), $\alpha^\pm$ will be very close to $|k_x|$ [see Eq. 12]. Then, the Eq. 16 reduces to $H_y^+ = iH_x^+$ for the $+k_x$ direction and $H_y^+ = -iH_x^+$ for the $-k_x$. Therefore, the polarization of magnetic field of the MSPs mode has a different rotation for different $\pm k_x$ directions, leading to a non-reciprocal result. Next, we are ready to investigate it in detail by solving the Eq. 13 at the magnetostatic limit $k_x \to \pm\infty$. One is given by

$$\mu_{xx} \pm \mu_{xy} \equiv \mu_{\rho\pm}(\omega) = -1. \tag{17}$$

Using Eq. 14, we write

$$\mu_{\rho\pm} = \mu_0\left(1 + \frac{\omega_m}{\omega_0 \mp \omega}\right) = -1. \tag{18}$$

Eq. 18 describes a condition to which the MSPs mode have a magnetostatic analog. Plotting a dispersion relation of the surface polariton [Eq. 13] is a good method to observe the condition shown in Fig. 5. Note that an applied static field is already considered in Fig. 5. It is very clear to observe the non-reciprocal nature of the MSPs mode (solid curves). Next, we consider

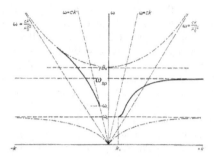

Figure 5. Non-reciprocal relation. The solid curves are the surface polaritons, and dot-dash curves are bulk polaritons. [1].

four situations: (a) For the magnetostatic limit $k_x \to +\infty$, we take $\mu_{\rho+}$ in Eq. 18, and then the frequency of the asymptotic surface polaritons reads

$$\omega_{sp} = \gamma \left(\frac{H_0 + \mu_0 B_0}{1 + \mu_0} \right), \tag{19}$$

where $B_0 \equiv H_0 + 4\pi\gamma M_0$. This is just the frequency of the unretardated surface magnon mode (also termed as DE mode [8]). (b) For the magnetostatic limit $k_x \to -\infty$, we find that $\mu_{\rho-}$ is always positive, yielding to no magnetostatic angalog. In general, the MSPs mode is called as real mode if magnetostatic limit is valid; otherwise, it is a virtual mode. Therefore, a non-reciprocal nature of the MSPs mode in a pure ferromagnet is acknowledged according to a magnetostatic mode. The investigation is in a good agreement with abovementioned polarization of magnetic field. (c) Next, let us consider a special case, $\mu_0 = 1$, and for the moment we concentrate on the $+k_x$ solution. The frequency of the MSPs mode starts at

$$\omega_v = \gamma\sqrt{B_0 H_0}, \tag{20}$$

and from Eq. 13 one obtains a corresponding wavevector

$$k_v = \frac{\omega_v}{c}\sqrt{\frac{B_0}{B_0 - H_0}} > \frac{\omega_v}{c}, \tag{21}$$

where the retardation corrections are already involved. The ω_v is also the frequency for the bulk magnon excitation under a condition, $\mu_v(\omega) \to \pm\infty$ [see Fig. 3]. If there is no external field, the ω_v reduces to zero so that the dispersion relation of the MSPs mode begins from the zero wavevector. The situation just likes electric SPPs supported by a dielectric and a metal [24]. (d) Now, let us consider the $-k_x$ solution including the retardation corrections. The retarded MSPs mode begins at the light line where $\mu_{xx}(\omega_-) = 1$ [see Fig. 5]. We solve Eq. 14 to yield the beginning frequency:

$$\omega_- = \gamma\sqrt{\frac{H_0^2 + \mu_0 H_0 B_0}{\mu_0 - 1}}. \tag{22}$$

Such a virtual mode terminates on the upper branch of the bulk polaritons. In addition, we see that from Eq. 22 the MSPs mode never starts at $\mu_0 \to 1$ (i.e. $\omega_- \to \infty$). Compared

Fig. 3 with Fig. 5, we note that the MSPs mode in the $-k_x$ direction still can exist even if $\mu_v > 1$. It is a very different from isotropic medium ($\mu_{xx} = \mu_{yy} = \mu_{zz}$) because its MSPs mode is only found within a gap where $\mu_v < 1$. From Eq. 21, the light line cannot excite any MSPs mode due to lack of the momentum conservation. In order to excite the MSPs mode in a pure ferromagnet, the optical coupler is need to transfer the momentum of the light to MSPs mode. Here, we quote the reference [[14]] to show the numerical calculations of the reflectivity spectra of the MSPs mode. Note that the damping factor is no longer neglected when one calculates a reflectivity spectrum. In Fig. 6, the calculated reflectivity spectra result from Otto ATR [1] configuration (optical coupler) for a pure ferromagnet (YIG). The figures 6(a) and 6(b) correspond to $+k_x$ and $-k_x$ directions, respectively. "h" is the air gap between a prism and a YIG. In Fig. 6(a), it shows that a larger damping ($1/\tau$) of the MSPs has larger loss, leading to much difficulty for determining the resonance frequency. In Fig. 6(b), the resonance frequency of the MSPs mode is blue-shift among increasing incidence angles. The asymmetric reflectivity deep results from an appreciable photon content in both figures 6. In general, the resonance frequency of the MSPs mode for a pure ferromagnet lies in a microwave region so that one needs a big sample for measuring the MSPs mode. There is, as we know, not a direct experimental verification.

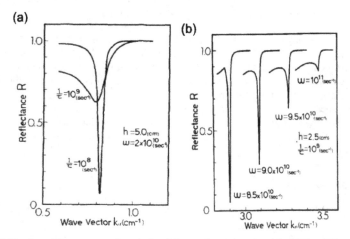

Figure 6. (a) The Otto ATR spectra at the $+k_x$ for different time relaxation (τ). (b) The ATR at the $-k_x$ for different incidence angle within the prism. [14]

4.2. Semi-infinite antiferromagnets

In this case, we focus on a gyromagnetic uniaxial antiferromagnetic medium in the Voigt configuration. Again, we neglect the exchange interactions and damping of $\overleftrightarrow{\varepsilon}(\omega)$ and $\overleftrightarrow{\mu}(\omega)$, and the external static magnetic field is applied in the $\hat{z}$ direction. The magnetization (M_0) splits the ferromagnetic resonant frequency (ω_0) into two separable frequencies given as $\omega_{\pm} = \omega \pm \omega_0$. The effective magnetic permeability can be derived by considering Bloch's

---

[1] attenuated total reflectance

equations [20], which reads

$$\mu_{xx} = \mu_{yy} = \mu_0 \left(1 + \frac{\omega_m \omega_A}{\omega_{an}^2 - \omega_+^2} + \frac{\omega_m \omega_A}{\omega_{an}^2 - \omega_-^2} \right),$$

$$\mu_{zz} = \mu_0, \tag{23}$$

$$\mu_{xy} = \mu_0 \left(\frac{\omega_m \omega_A}{\omega_{an}^2 - \omega_+^2} - \frac{\omega_m \omega_A}{\omega_{an}^2 - \omega_-^2} \right),$$

where ω_{an} is the antiferromagnetic resonance frequency and is determined by the anisotropy field (H_A) and exchange field (H_E),

$$\omega_{an} = \sqrt{\omega_A (2\omega_E + \omega_A)} = \gamma \sqrt{H_A (2H_E + H_A)}, \tag{24}$$

where $\omega_A = \gamma H_A$ and $\omega_E = \gamma H_E$. Next, one just follows the Eq. 8, yielding to Eq. 25,

$$\alpha^+ = \sqrt{k_x^2 - \left(\frac{\mu_{xx}^2 - \mu_{xy}^2}{\mu_{xx}} \right) \frac{\omega^2}{c^2} \varepsilon_{zz}},$$

$$\alpha^- = \sqrt{k_x^2 - \frac{\omega^2}{c^2}}. \tag{25}$$

Also, the dispersion relation from Eq. 11 can read

$$\frac{-\alpha^+ \mu_{xx} + \mu_{xy} k_x}{\mu_{xx}^2 - \mu_{xy}^2} = \alpha^-. \tag{26}$$

In the absence of the applied field ($H_0 = 0$), one has $\mu_{xy} = 0$ [see Eqs. 15 and 23]. Then, Eqs. 25 and 26 reduce more compact forms:

$$\alpha^+ = \sqrt{k_x^2 - \varepsilon_{zz} \mu_{xx} \frac{\omega^2}{c^2}},$$

$$\alpha^- = \sqrt{k_x^2 - \frac{\omega^2}{c^2}}. \tag{27}$$

$$\frac{-\alpha^+}{\mu_{xx}} = \alpha^-. \tag{28}$$

Using Eqs. 27 and 28 together, a simpler dispersion relation of surface polariton can be given by

$$k_x^2 \left(1 - \mu_{xx}^2 \right) = \frac{\omega^2}{c^2} \mu_{xx} \left(\varepsilon_{zz} - \mu_{xx} \right). \tag{29}$$

The simpler dispersion relation [5] is plotted in Fig. 7. In Fig. 7, we can observe the reciprocal property of the MSPs mode. Also, it is straightforward to calculate the frequency of the

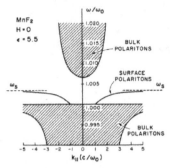

Figure 7. The shaped areas are bulk polaritons, and two solid curves are surface polaritons. Without an external static magnetic field, the dispersion relation is reciprocal. ($H_0 = 0$, $\varepsilon_{zz} = 5.5$) [5]

MSPs mode at the magnetostatic limit. One just solves $\mu_{xx} = -1$ [see Eq. 28] and yields the asymptotic frequency of MSPs shown in Fig. 7 as follows,

$$\omega_s = \sqrt{\omega_{an}^2 + \omega_m \omega_A}. \tag{30}$$

On other hand, with an applied static field, Eq. 26 has to be solved numerically shown in Fig. 8 [5]. Interestingly, one can clearly see the non-reciprocal dispersion relation of the MSPs mode that is just like a pure ferromagnet case with an applied static field. Note that the third MSPs mode starts at near $\omega/\omega_0 = 1.01$ and has no magnetostatic analog (i.e. virtual mode). However, two MSPs modes at lower frequency region have magnetostatic limits (i.e. ω_{sm+} and ω_{sm-}).

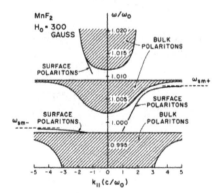

Figure 8. Dispersion relation for MSPs mode on MnF$_2$ at an applied field of 300 G. There are two real modes and virtual mode that are all non-reciprocal modes. [5]

Now, we would like to talk about the experimental verification for the surface polaritons. Let's simply describe the numerical results. There are typically two kinds of couplers: ATR and diffraction gratings. Here, we quote reference [32] in where they considered the diffraction gratings. In Fig. 9, the sample is MnF$_2$, a uniaxial antiferromagnet with the parameters $H_E = 7.87\,KG$, $H_A = 550\,HG$, $M_0 = 0.6\,KG$, and $\varepsilon_{zz} = 5.5$. Here, the Ω is our previous definition of antiferromagnetic resonance frequency, ω_{an}. There two solid lines are light lines ($k_x = \omega_0/c$), and two dotted vertical lines correspond to the wavevectors induced by the gratings ($q = k_x \pm s$). It is clear to observe that light lines never interact any of the branch of MSPs mode (dash curves). However, wavevector induced by gratings ($q = k_x + s$) indeed interacts a MSPs mode at the direction ($+k_x$), leading to the excitation of the MSPs mode. Note that when Eq. 1 includes damping terms, the leaky MSPs modes appear as shown by solid curves in Fig. 9. The difficulty to observe the surface polaritons for a pure antiferromagnet should be easier than a pure ferromagnet. The reason is that the resonance frequency of the MSPs mode lies at far infrared regions and the sample need not be so large. M.R.F. Jensen et al reported the first direct experimental evidence [18] for a pure antiferromagnet (FeF$_2$, $\omega_{an} \sim 1.57$ THz).

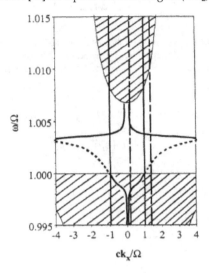

Figure 9. The shaped areas are bulk modes and the dashed lines are surface modes at no applied static field. The dotted vertical lines are the grating induced lines ($q = k_x \pm s$), and the light line is solid line ($q = k_x \pm s$). "s" is a reciprocal spatial period of the gratings. [[32]]

5. Realization of MSPs by effective media of magnetic superlattices

There are many researches on superlattice's surface polariton such as lateral superlattices [23, 36]. However, the lateral superlattice cases will be excluded in this study. In this section, the period of the superlattices will only to be along $\hat{y}$ axis. The superlattices considered here is composed of two alternating uniaxial magnetic materials and uniaxial magnetic (or nonmagnetic) materials ($\mu_{xx}^{(j)} = \mu_{yy}^{(j)}$, $j = $ medium 1 or 2). Proposed that the condition, $kd << 1$ (period, $d = a + b$, see Fig. 10) is valid, and then the effective medium theory can adequately approximate the bulk and surface polaritons.

The magnetic permeability of each layer in a magnetic superlattices still can be respectively described by Eq. 1, and then we quote references [2, 25] that rigorously derived the effective magnetic permeability and electric permittivity of the magnetic superlattices. Therefore, one

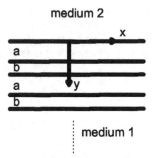

Figure 10. The region $y > 0$ is the magnetic superlattices composed of two alternating materials. The symbol "a" stands for the thickness of a ferromagnetic (or antiferromagnetic) material; "b" stands for the thickness of a nonmagnetic materials. $d = a + b$ is a period of two alternating magnetic materials.

is given by

$$\bar{\mu}_{xx} = \frac{(a+b)^2 \mu_{xx}^{(1)} \mu_{xx}^{(2)} + ab \left[\left(\mu_{xx}^{(1)} - \mu_{xx}^{(2)} \right)^2 - \left(\mu_{xy}^{(1)} - \mu_{xy}^{(2)} \right)^2 \right]}{(a+b) \left(a\mu_{xx}^{(2)} + b\mu_{xx}^{(1)} \right)},$$

$$\bar{\mu}_{yy} = \frac{(a+b) \mu_{xx}^{(1)} \mu_{xx}^{(2)}}{a\mu_{xx}^{(2)} + b\mu_{xx}^{(1)}}, \qquad (31)$$

$$\bar{\mu}_{xy} = \frac{a\mu_{xy}^{(1)} \mu_{xx}^{(2)} + b\mu_{xy}^{(2)} \mu_{xx}^{(1)}}{a\mu_{xx}^{(2)} + b\mu_{xx}^{(1)}},$$

$$\bar{\varepsilon}_{zz} = \frac{a\varepsilon_{zz}^{(1)} + b\varepsilon_{zz}^{(2)}}{a+b}.$$

Most works on magnetic superlattices have almost considered the special case, that is, medium 2 is nonmagnetic ($\mu_{xx}^{(2)} = 1$ and $\mu_{xy}^{(2)} = 0$). Here, we also concentrate on such special case. Accordingly, we substitute Eq. 14 into $\mu_{xx}^{(1)}$ and $\mu_{xy}^{(1)}$ in Eq. 31 for ferromagnetic/nonmagnetic superlattices; for antiferromagnetic/nonmagnetic superlattices, Eq. 23 may be chosen. For both cases, an external static magnetic filed (H_0) is applied along the easy axis $+\hat{z}$.

5.1. Ferromagnetic superlattices

In the ferromagnetic superlattices case, we see $\bar{\mu}_{xx} \neq \bar{\mu}_{yy}$ and $\bar{\mu}_{xy} = a\mu_{xy}^{(1)} / \left(a + b\mu_{xx}^{(1)} \right)$ so that the attenuation coefficients and dispersion equation of the MSPs mode can be given by [see Eqs. 8 and 11]

$$\alpha^+ = \sqrt{\frac{\bar{\mu}_{xx}}{\bar{\mu}_{yy}} k_x^2 - \left(\frac{\bar{\mu}_{xx}\bar{\mu}_{yy} - \bar{\mu}_{xy}^2}{\bar{\mu}_{yy}} \right) \frac{\omega^2}{c^2} \bar{\varepsilon}_{zz}},$$

$$\alpha^- = \sqrt{k_x^2 - \frac{\omega^2}{c^2}}, \qquad (32)$$

and

$$\frac{-\alpha^+ \bar{\mu}_{yy} + \bar{\mu}_{xy} k_x}{\bar{\mu}_{xx} \bar{\mu}_{yy} - \bar{\mu}_{xy}^2} = \alpha^-. \tag{33}$$

Again, Eq. 33 directly depends on the direction of k_x, leading to non-reciprocal dispersion relation. We would like to quote the results of the reference [4], and then the dispersion relation (Eq. 33) is rewritten as follows,

$$\left(\alpha^-\right)^2 \bar{\mu}_{xx} \bar{\mu}_v - 2\left(\alpha^-\right) \bar{\mu}_{xx} \bar{\mu}_v - k_x^2 + \bar{\mu}_{xx} \frac{\omega^2}{c^2} = 0. \tag{34}$$

where we substitute α^+ in Eq. 32. The numerical calculation of Eq. 34 is shown in Fig. 11 at the condition, $a = b$. Note that we re-define the expression in the Eq. 15 for the reason to be consistent with reference [4]:

$$\omega_0 = g\mu_0 \gamma_0 H_0, \quad \omega_m = g\mu_0 \gamma_0 M_0, \tag{35}$$

where g is a Lande factor, μ_0 is a magnetic permeability in the vacuum, and $\gamma_0 = e/2m$ with $e > 0$ and m being the electron charge and electron mass. The numerical results show no magnetostatic analog (real mode) for this configuration (i.e. $a = b$) at $k_x \to \pm\infty$, and only a virtual mode exists for $-k_x$. For $a > b$, the magnetostatic analog can occurs at $-k_x$ (the evidence is not shown here). In addition, if the retardation corrections are included, then the virtual MSPs mode still exists for $a < b$. Note that Fig. 11 shows that the surface polaritons exist only in a restricted range.

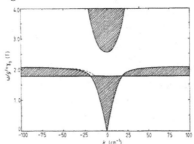

Figure 11. Ferromagnetic (Fe)/nonmagnetic superlattices. $a = b = 5 \times 10^{-4}$ cm. $\mu_0 H_0 = 1$ T. $\mu_0 M_0 = 2.15$ T. $g^{Fe} = 2.15$. The shaped region is the bulk mode; the dotted line is the surface mode. [4]

5.2. Antiferromagnetic superlattices

Let us consider the superlattices that are composed of medium 1 (uniaxial antiferromagnetic) alternating with medium 2 (non-magnetic). In the absence of an applied static magnetic field ($H_0 = 0$), the component of the magnetic permeability of two media can read from Eq. 23,

$$
\begin{aligned}
\mu_{xx}^{(1)} &= \mu_{yy}^{(1)} = 1 + \frac{2\omega_m \omega_A}{\omega_{an}^2 - \omega^2} = \frac{\omega_1^2 - \omega^2}{\omega_{an}^2 - \omega^2} \equiv \mu_1, \\
\mu_{xx}^{(2)} &= \mu_{yy}^{(2)} = 1 \equiv \mu_2, \\
\mu_{zz}^{(1)} &= \mu_{zz}^{(2)} = \mu_0 = 1, \\
\mu_{xy}^{(1)} &= \mu_{xy}^{(2)} = 0.
\end{aligned}
\tag{36}
$$

where $\omega_1 = \sqrt{\omega_{an}^2 + 2\omega_m \omega_A}$ at $\mu_1 = 0$. Here, we do not attempt to derive the non-reciprocal dispersion equation at an applied static field, but just show the numerical results later. Using Eqs. 31 and 36, the component of the effective magnetic permeability can be given by

$$\bar{\mu}_{xx} = \frac{\bar{\omega}_1^2 - \omega^2}{\omega_{an}^2 - \omega^2},$$
$$\bar{\mu}_{yy} = \frac{\omega_1^2 - \omega^2}{\bar{\omega}_{an}^2 - \omega^2}, \tag{37}$$

where $\bar{\omega}_1^2 = (b\omega_{an}^2 + a\omega_1^2)/(a+b)$ and $\bar{\omega}_{an}^2 = (a\omega_{an}^2 + b\omega_1^2)/(a+b)$. Reference [3] showed the results of Eq. 37 in Fig. 12 for $a > b$.

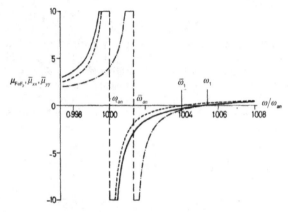

Figure 12. A FeF$_2$/ZnF$_2$ superlattices with $\frac{a}{a+b} = 0.75$. μ_{FeF_2} (solid line), $\bar{\mu}_{xx}$ (dash line), and $\bar{\mu}_{yy}$ (dash-dot line). ω_{an}, $\bar{\omega}_{an}$, $\bar{\omega}_1$ and ω_1 are marked. [3]

Now the attenuation coefficients can be given by Eq. 8

$$\alpha^+ = \sqrt{\frac{\bar{\mu}_{xx}}{\bar{\mu}_{yy}} k_x^2 - \bar{\varepsilon}_{zz} \cdot \bar{\mu}_{xx} \frac{\omega^2}{c^2}},$$
$$\alpha^- = \sqrt{k_x^2 - \frac{\omega^2}{c^2}}. \tag{38}$$

And, the general dispersion equation of the MSPs mode reads from Eq. 11

$$\frac{-\alpha^+}{\bar{\mu}_{xx}} = \alpha^-, \tag{39}$$

Note that Eq. 39 implies a reciprocal dispersion relation of the surface polaritons. Substitution of Eq. 38 into Eq. 39 can solve the explicit form

$$k_x = \frac{\omega}{c} \sqrt{\frac{\bar{\mu}_{yy}(\bar{\mu}_{xx} - \bar{\varepsilon}_{zz})}{\bar{\mu}_{xx}\bar{\mu}_{yy} - 1}}. \tag{40}$$

Eqs. 39 and 40 describes the condition for the existence of the MSPs mode. First, we investigate the magnetostatic limit ($k_x \rightarrow \pm\infty$), yielding to

$$\bar{\mu}_{xx} = -1, \; \bar{\mu}_{xx}\bar{\mu}_{yy} = 1. \tag{41}$$

We define $\bar{\mu}_{yy} < 0$ as real mode and $\bar{\mu}_{yy} > 0$ as virtual mode. In a result, only the real mode has a magnetostatic analog, and the virtual mode does not. Here, we simply persent two kinds of situations: (1) If $a > b$, the real mode lies in a frequency region, $\bar{\omega}_{an} < \omega < \omega_{sp}$ (i.e. $\bar{\mu}_{yy} < 0$) and virtual mode in another frequency region, $\omega_{an} < \omega < \bar{\omega}_{an}$ (i.e. $\bar{\mu}_{yy} > 0$) [see Fig. 12]. At $k_x \rightarrow \pm\infty$, Eq. 41 yields the asymptotic frequency of the MSPs mode

$$\omega_{sp} = \sqrt{\frac{\omega_{an}^2 + \omega_1^2}{2}}, \tag{42}$$

where $\omega_1 = \sqrt{\omega_{an}^2 + 2\omega_m\omega_A}$ at $\mu_1 = 0$ [see Eq. 36]. Note that the resonance frequency of the MSPs mode is independent of a and b and, consequently, is just the same as a pure antiferromagnet (see Eq. 30). (2) For $a < b$, there is only the virtual mode to exist at the frequencies, $\omega_{an} < \omega < \bar{\omega}_1$.

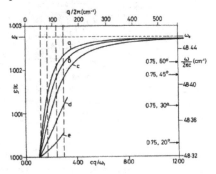

Figure 13. The dispersion relation versus f_1 at no applied static magnetic field ($H_0 = 0$). The curves from a to e represent $f_1 = 1$, 0.75, 0.5, 0.25, and 0.1, respectively. [3]

For the FeF$_2$/ZnF$_2$ system, reference [3] has summarized the results of the dispersion curves of the MSPs modes versus $f_1 \equiv a/(a+b)$ various at $+k_x$ shown in Fig. 13. Figure 13 marks four black arrows for $f_1 = 0.75$ that respectively corresponds to four incidence angle: 20°, 30°, 45°, and and 60°. These angles are in a position to transfer momentum of light into the MSPs mode, leading to a reflectivity deep in the Otto ATR spectra (like Fig. 6). Note that one can clearly observe that only real modes (magnetostatic analog) can exist at the conditions, $a > b$.

Under an applied static magnetic field, on other hand, the numerical calculations should be required for this discussion shown in Fig. 14 [17]. Figure 14(a) shows a pure antiferromagnet (i.e. $f_1 = 1$); Fig. 14(b) shows $f_1 = 0.75$; Fig. 14(c) is $f_1 = 0.25$. For $f_1 = 1$ and $f_1 = 0.75$, one can observe that there are two magnetostatic modes and one is virtual mode at $\pm k_x$. As for $f_1 = 0.25$, there are all virtual modes at $\pm k_x$.

Finally, If $a = b$ ($f_1 = 0.5$), the dispersion relation of the antiferromagnetic superlattices (MnF$_2$) [4] is shown in Fig. 15. It shows the zero and non-zero applied static magnetic field, respectively. Fig. 15(a) shows the reciprocal dispersion relation; Fig. 15(b) shows

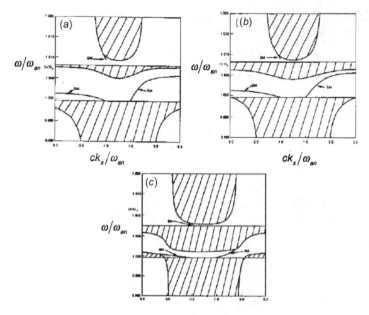

Figure 14. The dispersion relation at an applied static magnetic field. The solid curves represent the surface modes, and shaded regions are bulk modes. (a) a pure antiferromagnet (MnF$_2$) (b) antiferromagnetic superlattices, MnF$_2$/ZnF$_2$ ($f_1 = 0.75$). (c) antiferromagnetic superlattices, MnF$_2$/ZnF$_2$ ($f_1 = 0.25$). $\omega_{an} = 260$ GHz. $H_0 = 200$ G, $H_A = 7.85$ kG, $H_E = 550$ kG, $M_0 = 0.6$ kG, $\varepsilon_{zz}^{(1)} = 5.5$, and $\varepsilon_{zz}^{(2)} = 8$. [3]

non-reciprocal dispersion relation. The MSPs modes only exist in a restricted region that bulk modes are not excited. Here, for consistence with reference [4], we recall the Eq. 35 and change the expression in Eq. 24 as following:

$$\omega_{an} = g\mu_0\gamma_0\sqrt{H_A\left(H_A + 2H_E\right)}. \tag{43}$$

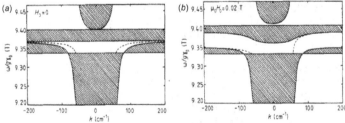

Figure 15. The antiferromagnetic (MnF$_2$)/nonmagnetic superlatticess. $a = b = 5 \times 10^{-4}$ cm, $\mu_0 H_E = 55$ T, $\mu_0 H_A = 0.787$ T, $\mu_0 M_0 = 0.754$ T, and $g = 2$. (a) $\mu_0 H_0 = 0$. (b) $\mu_0 H_0 = 0.02$ T. The shaped region is the bulk polaritons; the dotted curve is the surface polaritons mode. [4]

6. Realization of MSPs by metamaterials

6.1. Introduction of metamaterials

In 1968, a Russian physicist V. G. Veselago theoretically calculated that a medium might bend light to the "wrong" pathway whenever the medium could "give" a negative refractive index [35]. In his theory, the negative refractive index medium (NRIM) is composed of both negative electric permittivity and negative magnetic permeability at the same time but unfortunately, Veselago never experimentaly demonstrated such an NRIM mainly because he could not find out a material with negative magnetic permeability. Three decades later, British physicist J.B. Pendry suggested that the two-dimensional metallic wires [12] and metallic split-ring resonators (SRRs) [11] yield to negative permittivity and negative permeability, respectively. These innovative electromagnetic responses were artificially created by a periodic array of sub-wavelength unit cells that afterwards, were termed as metamaterials and led to a variety of rare and even unprecedented electromagnetic properties such magnetic magnetism at terahertz region [33], inverse optical rules [e.g., inverse Snell's law [22], inverse Doppler shift [30] and inverse Cerenkov radiation [28]] , superlensing effect [31], invisibility cloaking [7], and others.

In 2001, Shelby et al. integrated the metallic wires and metallic SRRs into a prism structure [see Fig. 16], and for the first time demonstrated the negative refractive index [22]. This result demonstrated that metamaterials were recognized to be an effective medium due to the sub-wavelength characteristics. Therefore, the magnetic metamaterials (like SRRs) can be expected to process the MSPs mode as same as that for the aforementioned magnetic superlattices. In fact, some groups have already theoretically predicated the surface mode of the ferromagnetic metals [1] and the metamaterials [26].

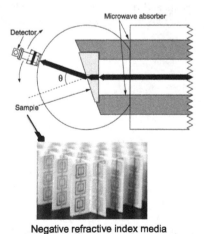

Negative refractive index media

Figure 16. Experimental verification of the negative refractive index medium. The sample consisted of a periodic array of the metallic wires and the SRRs is so called a NRIM whose geometrical shape is a prism now. One can observe that the NRIM bends light into a "wrong" pathway, and Snell's law verifies the negative refractive index. [22]

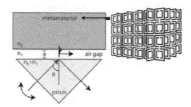

Figure 17. The Otto ATR configuration. The magnetic metamaterials is composed of a two-dimensional grid of the SRRs. "a" is gap between prism and metamaterials. [16]

6.2. Uniaxial magnetic metamaterials

Now, we quote an interesting work done by J. N. Gollub et at. who constructed a two-dimensional array of the SRRs into a physical three-dimensional bulk [16] shown in Fig. 17. Such the physical three-dimensional bulk metamaterials is considered as a semi-infinite medium due to the subwavelength SRR structure. According to their geometry, we regard the effective electric permittivity and magnetic permeability of the bulk as anisotropic and uniaxial, respectively. Therefore, it is straightforward to write

$$
\overleftrightarrow{\varepsilon}_{eff}(\omega) = \begin{pmatrix} \bar{\varepsilon}_{xx} & 0 & 0 \\ 0 & \bar{\varepsilon}_{yy} & 0 \\ 0 & 0 & \bar{\varepsilon}_{zz} \end{pmatrix},
$$
$$
\overleftrightarrow{\mu}_{eff}(\omega) = \begin{pmatrix} \bar{\mu}_{xx} & 0 & 0 \\ 0 & \bar{\mu}_{xx} & 0 \\ 0 & 0 & \bar{\mu}_{zz} \end{pmatrix}.
$$
(44)

Also, the attenuation coefficients and dispersion equation of the surface mode of this bulk medium can be written by setting $\bar{\mu}_{xy} = 0$ and $\bar{\mu}_{xx} = \bar{\mu}_{yy}$ into Eqs. 8 and 13:

$$
\alpha^{+} = \sqrt{k_x^2 - \bar{\varepsilon}_{zz}\bar{\mu}_{xx}\frac{\omega^2}{c^2}},
$$
$$
\alpha^{-} = \sqrt{k_x^2 - \frac{\omega^2}{c^2}},
$$
(45)

and

$$
\frac{-\alpha^{+}}{\bar{\mu}_{xx}} = \alpha^{-}.
$$
(46)

They are as same as those for a pure antiferromagnet without an applied static field [Eqs. 27 and 28]. The magnetic permeability of the bulk medium is calculated by means of the standard retrieval process [6, 34] shown in Fig. 18 [16]. There exists a $\bar{\mu}_{xx} < 0$ between 12 and 14 GHz in where the MSPs mode is expected. From Eq. 46, one knows that the momentum never be conserved without an optical coupler (diffraction gratings or prism). The experimental result of the Otto ATR spectra [16] is shown in Fig. 19 , where the dash curve is numerical calculation and the solid curve is experimental measurement. The experimental result is in a good agreement with numerical calculation, and demonstrated successfully that the MSPs mode is excited occurring at reflectivity deep (12.2 GHz).

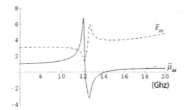

Figure 18. The effective permittivity and permeability of the bulk metamaterials. $\bar{\mu}_{xx} < 0$ for 12 and 14 GHz. [16]

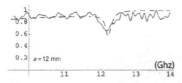

Figure 19. The experimental results of the ATR spectra, and $a = 12\ mm$. The dash curve is numerical calculation, and the solid curve is experimental measurement. [16]

6.3. Biaxial magnetic metamaterials

Here, we shortly re-describe what we just talked about the conditions for an excitation of the MSPs at magnetostatic limit. First, for a pure uniaxial ferromagnet, the MSPs can be excited at the condition, $\mu_{xx} \pm \mu_{xy} = -1$ [Eq. 17] with an applied static magnetic field. Note that the surface mode might exist whether or not the μ_v in Fig. 3 is positive or negative. Second, for a pure antiferromagnet without an applied static field, the condition is $\mu_{xx} = -1$ at magnetostatic limit, which is also for uniaxial magnetic metamaterials. Third, for a ferromagnetic superlattices, the condition is not considered in this chapter, but we present a special numerical result shown in Fig. 11. Fourth, for an antiferromagnetic superlattices, the necessary condition is $\bar{\mu}_{xx} < 0$. It collocates with $\bar{\mu}_{yy} < 0$ being a real mode and $\bar{\mu}_{xy} > 0$ being a virtual mode, respectively.

Now, we would like to introduce our previous work [37] that considered a situation as same as the antiferromagnetic superlattices. In a word, the magnetic metamaterials have an effect biaxial tensors are given by

$$
\overset{\leftrightarrow}{\varepsilon}_{eff}(\omega) = \begin{pmatrix} \bar{\varepsilon}_{xx} & 0 & 0 \\ 0 & \bar{\varepsilon}_{yy} & 0 \\ 0 & 0 & \bar{\varepsilon}_{zz} \end{pmatrix},
$$
$$
\overset{\leftrightarrow}{\mu}_{eff}(\omega) = \begin{pmatrix} \bar{\mu}_{xx} & 0 & 0 \\ 0 & \bar{\mu}_{yy} & 0 \\ 0 & 0 & \bar{\mu}_{zz} \end{pmatrix}.
$$

(47)

The proposal structure of our work [shown in Fig. 20] is a periodic array of the sandwich unit cells. Fig. 20(b) shows there are three layers in a sandwich unit cell that includes two SRRs and one spacer (Roger RT5800). The dimension of the unit cell is $6 \times 6\ mm^2$ and thickness is $2\ mm$, which is smaller than our operating wavelength (23 mm or 13.22 GHz). Therefore, it is safe to consider a sandwich unit cell as an effective medium. Its effective electric permittivity and magnetic permeability are retrieved shown in Fig. 21. At 13.22 GHz, we have

$$\bar{\mu}_{xx} = -4.20, \ \bar{\mu}_{yy} = -0.252, \ \bar{\varepsilon}_{zz} = 3.46. \tag{48}$$

Therefore, our design only possesses the real mode. Using COMSOL Multiphysics finite element based electromagnetic solver [2], we simulate the field distribution of the MSPs mode shown in Fig. 22 where one can clearly see the exponential decay of the MSPs propagating at the interface. By substituting Eq. 48 into Eq. 40, the wavelength of the MSPs mode can be evaluated as 3.947 mm that is in a good agreement with the simulation result (not shown). Note that such a biaxial magnetic metamaterials depends on designed geometry to determine the resonance frequency of the MSPs mode, instead of inherent magnetism.

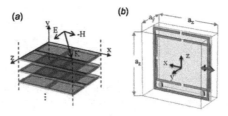

Figure 20. The sandwich unit cell of the biaxial magnetic metamaterials. $a_x = a_y = 6 \ mm, a_z = 2 \ mm.$
[37]

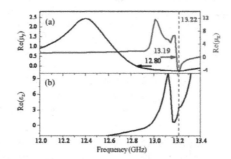

Figure 21. The retrieval result of the biaxial magnetic metamaterials. (a) Red curve is $\bar{\mu}_{xx}$; black curve is.
(b) Black curve is $\bar{\varepsilon}_{zz}$. At 13.22 GHz, we have $\bar{\mu}_{xx} = -4.2, and \bar{\mu}_{yy} = -0.252, \bar{\varepsilon}_{zz} = 3.46.$ [37]

6.4. One-dimensional magnetic photonic crystal

In last two sections, the samples are not composed of magnetic materials, and so there need not be an applied static magnetic field. Now, we basically introduce that a magnetic metamaterials is composed of YIG rods [29] shown in Fig. 23(a). The structure is like a one-dimensional photonic crystal (PC), and arranged periodically in a square lattice embedded within the air. The lattice constant is $a = 8 \ mm$ and the radius of the YIG rod is $r = 2 \ mm$ where axis of the rods is at $\hat{z}$ axis. Again, an applied magnetic static field is along the easy axis $\hat{z}$. The operating frequency is 5 GHz (i.e. wavelength = 60 mm) that is enough

---

[2] http://www.comsol.com/

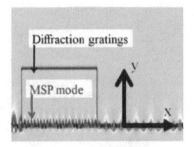

Figure 22. The field distribution of the MSPs mode for a biaxial magnetic metamaterials at 13.22 GHz.
[37]

lager to be regarded this structure as an effective medium. For the reflectivity simulation, the
magnetic permeability of the YIG needs a damping factor that can be modified from Eq. 14:

$$\omega_0 \rightarrow \omega_0 + i\beta\omega \tag{49}$$

The retrieved effective permittivity and permeability, and band diagram of the bulk
polaritons are plotted in Figs. 23(c) and 23(b), respectively. Note that the condition of the
magnetostatic mode is as same as that for a pure ferromagnet [i.e. see Eq. 17]. In a word, the
condition is $\mu_{eff} = -1$ [see in Fig. 23(c)]. In this work, Liu et al. planed to apply a static
magnetic field upon the PC, which yields to the modeling of the reflectivity as shown in Fig.
23(a). One can observe the reflectivity has "own" direction at the resonance frequency of the

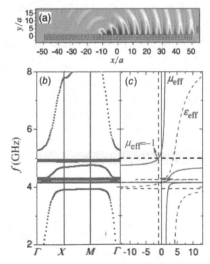

Figure 23. (a) The line source is transverse magnetic wave with the electric field polarized along the rod,
and is placed near the surface of the magnetic metamaterials ($\frac{y}{a} = 1$; $\frac{x}{a} = -10$) (b) The photonic band
diagram of the magnetic metamaterials with $H_0 = 900\ Oe$. The yellow region is photonic band gap. (c)
The retrivevd effective constitutive parameters. The blue dashed curve is ε_{eff} and red solid curve is μ_{eff}.
Note that $\mu_{eff} = 1$ at 5 GHz. [29]

MSPs. The main mechanism is non-reciprocal dispersion relation of the MSPs because the left-propagating direction of the MSPs is forbidden.

7. Conclusions

In this study, we review a few important papers that help one to understand the conditions about the existence of the magnetic surface polaritons (MSPs) modes. Note that ferromagnetic resonance frequency is typically in the microwave frequency region, and antiferromagnetic resonance frequency in the far infrared. However, Metamaterials resonance frequency can be artificially determined.

The MSPs mode can further be considered as two kinds of modes, real mode (magnetostatic analog) and virtual mode. First, for real mode, we basically summarize those aforementioned conditions. (a) For a pure uniaxial ferromagnet with $\overleftrightarrow{\varepsilon} = 1$, at an external static magnetic field (H_0), the condition is $\mu_0 \left[1 + \omega_m/(\omega_0 - \omega)\right] = -1$ where ω_{msp} is locating within the region, $\mu_V < 1$. In addition, the dispersion relation of the MSPs mode is non-reciprocal. On the other hand, without H_0, the condition is $\mu_0 \left[1 - \omega_m/\omega\right] = -1$ and magnetostatic analog starts at $k_x = 0$ with non-reciprocal dispersion relation. (b) For a pure uniaxial anti-ferromagnet, without H_0, the condition is $\mu_{xx} = -1$ where ω_{msp} is living within the region, $\mu_V < 1$. The dispersion relation of the MSPs is reciprocal. On other hand, with H_0, numerical calculation shows there is a possibility to have a magnetostatic analog with non-reciprocal dispersion relation. (c) For a ferromagnetic superlattices, with H_0, there is a magnetostatic analogs for $a/b > 0.5$. The dispersion relation of the MSPs is non-reciprocal. (d) For an anti-ferromagnetic superlattices, without H_0, the condition is $\bar{\mu}_{xx} = -1$ and $\bar{\mu}_{xx}\bar{\mu}_{yy} = 1$. The dispersion relation of the MSPs mode is reciprocal for $a > b$. On other hand, with H_0, numerical calculation shows that the magnetostatic analog can exist for $a > b$.

Second, for the virtual modes, the numerical calculation is necessary, and it is not easy to give a exact condition. However, for magnetic superlattices, if the ratio of "a" to "b" is less than 0.5, then the virtual modes should exist. Finally, let us comment the conditions for magnetic metamaterials. The magnetic metamaterials are ability to design their constitutive parameters by means of a specific the geometry of the sub-wavelength unit cell. Theoretically, one can well control the real modes or virtual modes, yielding to the versatile devices such as optical bistability [9, 10, 13], second harmonic generation [19]. Although over the past researches have less attention to the MSPs modes due to the experimental difficulty compared to electric SPPs, we expect the MSPs mode will get more attention by means of the magnetic metamaterials in the future.

Acknowledgments

The authors would like to gratefully acknowledge the financial support from the National Science Council (98-2112-M-007-002MY3, 99-2923-M-007-003-MY2, 99-ET-E-007-002-ET, 100-ET-E-007-002-ET, 100-2120-M-002-008, 100-2120-M010-001).

Author details

Yu-Hang Yang and Ta-Jen Yen
Department of Materials Science and Engineering, National Tsing Hua University, Hsinchu, Taiwan

8. References

[1] A. Hartstei, E. Burstein, A. M. R. B. & Wallis, R. F. [1973]. Surface polaritons on semi-infinite gyromagnetic media, *Journal of Physics C-Solid State Physics* 6(7): 1266–1276.

[2] Almeida, N. S. & Mills, D. L. [1988]. Effective-medium theory of long-wavelength spin-waves in magnetic superlattices, *Physical Review B* 38(10): 6698–6710.

[3] Almeida, N. S. & Tilley, D. R. [1990]. Surface-polaritons on antiferromagnetic superlattices, *Solid State Communications* 73(1): 23–27.

[4] Barnas, J. [1990]. Spin-waves in superlattices .3. magnetic polaritons in the voigt configuration, *Journal of Physics-Condensed Matter* 2(34): 7173–7180.

[5] Camley, R. E. & Mills, D. L. [1982]. Surface polaritons on uniaxial antiferromagnets, *Physical Review B* 26(3).

[6] D. R. Smith, S. Schultz, P. M. & Soukoulis, C. M. [2002]. Determination of effective permittivity and permeability of metamaterials from reflection and transmission coefficients., *Phys. Rev. B* 65: 195104.

[7] D. Schurig, J. J. Mock, B. J. J. S. A. C. J. B. P. A. F. S. & Smith, D. R. [2006]. Metamaterial electromagnetic cloak at microwave frequencies, *Science* 314(5801): 977–980.

[8] Damon, R. W. & Eshbach, J. R. [1961]. Magnetostatic modes of a ferromagnet slab, *Journal of Physics and Chemistry of Solids* 19(3-4): 308–320.

[9] G. D. Xu, T. Pan, T. C. Z. & Sun, J. [2008]. Optical bistability with surface polaritons in layered structures containing left-handed metallic magnetic composites, *Applied Physics B-Lasers and Optics* 93(2-3): 551–557.

[10] I. L. Lyubchanskii, N. N. Dadoenkova, A. E. Z. Y. P. L. & Rasing, T. [2008]. Optical bistability in one-dimensional magnetic photonic crystal with two defect layers, *Journal of Applied Physics* 103(7).

[11] J. B. Pendry, A. J. Holden, D. J. R. & Stewart, W. J. [1999]. Magnetism from conductors and enhanced nonlinear phenomena, *Ieee Transactions on Microwave Theory and Techniques* 47(11): 2075–2084.

[12] J. B. Pendry, A. J. Holden, W. J. S. & Youngs, I. [1996]. Extremely low frequency plasmons in metallic mesostructures, *Physical Review Letters* 76(25): 4773–4776.

[13] J. Bai, S. F. Fu, S. Z. & Wang, X. Z. [2011]. Reflective optical bi-stability of antiferromagnetic films, *European Physical Journal B* 83(3): 343–348.

[14] J. Matsuura, M. F. & Tada, O. [1983]. Atr mode of surface magnon polaritons on yig, *Solid State Communications* 45(2).

[15] J. N. Anker, W. P. Hall, O. L. N. C. S. J. Z. & Duyne, R. P. V. [2008]. Biosensing with plasmonic nanosensors, *Nature Materials* 7(6): 442–453.

[16] J. N. Gollub, D. R. Smith, D. C. V. T. P. & Mock, J. J. [2005]. Experimental characterization of magnetic surface plasmons on metamaterials with negative permeability, *Physical Review B* 71(19).

[17] M. C. Oliveros, N. S. Almeida, D. R. T. J. T. & Camley, R. E. [1992]. Magnetostatic modes and polaritons in antiferromagnetic nonmagnetic superlattices, *Journal of Physics-Condensed Matter* 4(44): 8497–8510.

[18] M. R. F. Jensen, T. J. Parker, K. A. & Tilley, D. R. [1995]. Experimental-observation of magnetic surface-polaritons in fef2 by attenuated total-reflection, *Physical Review Letters* 75(20): 3756–3759.

[19] M. W. Klein, C. Enkrich, M. W. & Linden, S. [2006]. Second-harmonic generation from magnetic metamaterials, *Science* 313(5786): 502–504.

[20] Mills, D. L. & Burstein, E. [1974]. Polaritons - electromagnetic modes of media, *Reports on Progress in Physics* 37(7): 817–926.

[21] Pincus, P. [1962]. Propagation effects on ferromagnetic resonance in dielectric slabs, *Journal of Applied Physics* 33(2): 553.

[22] R. A. Shelby, D. R. S. & Schultz, S. [2001]. Experimental verification of a negative index of refraction, *Science* 292: 77.

[23] R. E. Camley, M. G. C. & Tilley, D. R. [1992]. Surface-polaritons in antiferromagnetic superlattices with ordering perpendicular to the surface, *Solid State Communications* 81(7): 571–574.

[24] Raether, H. [1988]. *Surface plasmons on smooth and rough surfaces and on gratings,* Springer-Verlag.

[25] Raj, N. & Tilley, D. R. [1987]. Polariton and effective-medium theory of magnetic superlattices, *Physical Review B* 36(13): 7003–7007.

[26] Ruppin, R. [2000]. Surface polaritons of a left-handed medium, *Physics Letters A* 277(1): 61–64.

[27] S. Lal, S. L. & Halas, N. J. [2007]. Nano-optics from sensing to waveguiding, *Nature Photonics* 1(11): 641–648.

[28] S. Xi, H. S. Chen, T. J. L. X. R. J. T. H. B. I. W. J. A. K. & Chen, M. [n.d.]. Experimental verification of reversed cherenkov radiation in left-handed metamaterial, *Physical Review Letters* 103(19).

[29] S. Y. Liu, W. L. Lu, Z. F. L. & Chui, S. T. [2011]. Molding reflection from metamaterials based on magnetic surface plasmons, *Physical Review B* 84(4).

[30] Seddon, N. & Bearpark, T. [2003]. Observation of the inverse doppler effect, *Science* 302(5650): 1537–1540.

[31] Smolyaninov, II, Y. J. H. & Davis, C. C. [2007]. Magnifying superlens in the visible frequency range, *Science* 315(5819): 1699–1701.

[32] Stamps, R. L. & Camley, R. E. [1989]. Greens-functions for antiferromagnetic polaritons .2. scattering from rough surfaces, *Physical Review B* 40(1): 609–621.

[33] T. J. Yen, W. J. Padilla, N. F. D. C. V. D. R. S. J. B. P. D. N. B. & Zhang, X. [2004]. Terahertz magnetic response from artificial materials, *Science* 303(5663): 1494–1496.

[34] T. M. G. Xudong Chen, B.-I. W., Pacheco, J. J. & Kong, J. A. [2004]. Robust method to retrieve the constitutive effective parameters of metamaterials, *PHYSICAL REVIEW E* 70: 016608.

[35] Veselago, V. G. [1968]. Electrodynamics of substances with simultaneously negative values of sigma and mu, *Soviet Physics Uspekhi-Ussr* 10(4): 509.

[36] Wang, X. Z. & Tilley, D. R. [1995]. Retarded modes of a lateral antiferromagnetic nonmagnetic superlattice, *Physical Review B* 52(18): 13353–13357.

[37] Y.H. Yang, I. W., U. H. L. & Yen, T. [2012]. Magnetic surface polariton in a planar biaxial metamaterial with dual negative magnetic permeabilities, *Plasmonics* 7(1): 87–92.

Chapter 5

Photoacoustic Based Surface Plasmon Resonance Spectroscopy: An Investigation

K. Sathiyamoorthy, C. Vijayan and V.M. Murukeshan

Additional information is available at the end of the chapter

1. Introduction

Surface Plasmon resonance effect (SPR) has very important applications in many fields. [1-3] The most significant of them perhaps is in the field of real-time bio-sensing. It is important for bio-sensors to be sensitive enough to detect even the smallest amounts of bio-molecules. There are several instances when conventional spectroscopy becomes inadequate, even for clearer and transparent materials. Such a situation arises when one attempts to measure a very weak absorption, which in turn involves the measurement of significantly small changes in the intensity of the strong and unattenuated transmitted signal.

Surface plasmons (SP) are the quanta of elementary excitations involving the oscillations of the electron cloud against the background ionic arrangement in a metal surface. SP can be excited at the metal-dielectric interface using light by matching the momenta between the incident photon and the surface plasmon. Momentum matching can be achieved by various techniques.[4-7] The simplest way to achieve momentum matching between excitation photon and plasmon is utilization of high index prism over the metal medium. SP excitation using prism coupling technique is based on either Kretschmann or Otto configuration.[8, 9] The present study is based on Kretschmann configuration in which the metal film has direct contact with the prism base. The magnitude of the wavevector of SP depends on the dielectric constants of both the metal and the surrounding dielectric medium. It is extremely sensitive to properties of the dielectric medium which is in contact with the metal. As SP circumvents diffraction, it can be confined in nanosize domain and hence have interaction with material structures of dimension from sub-micrometer to a few nanometers. The most common technique widely examined for investigating the effects of the excitation of SP is the measurement of reflection coefficient of incident optical beam. But such an approach is mostly not suitable for the sample with weak absorption characteristics. Higher detection sensitivities than those possible with such an approach are always desirable for improving sensing performance. The measurement of phase variation during SP excitation is expected to provide higher sensitivity to SPR sensors. However, the complexity of the technique associated with phase detection impedes the targeted development. Moreover, most of the optical detection systems (such as linear photo arrays, CCD etc.) currently employed for measuring the reflection remain off the interrogation region due to their sophistication thereby increasing the size and the cost of the system.

The significance of the photo acoustics techniques is in this context that it may pave the way for the design of bio-sensors with greater sensitivity as they are capable of detecting temperature rises of 10^{-6} to $10^{-5}$of a degree in the sample, corresponding to a thermal power generation of about $10^{-6}calories/cm^3$. Furthermore, since the sample itself constitutes the electromagnetic radiation detector enabling studies over a wide range of optical and electromagnetic wavelengths without the need to change detector systems. Besides, the compactness of the PA detection system enables easy integration with the existing Kretschmann configuration based SPR system for applications such as on-chip detection.

Photoacoustic (PA) spectroscopy is a very powerful analytical tool for examining the optical absorption properties of solids as it measures directly the energy absorbed by the material on exposure to light. Conventional optical absorption/transmission spectroscopy requires a homogenous and partially transparent sample and further, it cannot be used with highly scattering samples, where the scattering affects significantly the accuracy of the measurement of optical absorption coefficient. The only other method of obtaining some spectroscopic information from opaque samples is diffuse optical reflection spectroscopy where special treatment is required for sample surface preparation. Such problems are absent in the case of photoacoustic spectroscopy and spectral information can be obtained from a variety of samples including opaque materials, powders etc.

The photoacoustic effect was first reported by Alexander Graham Bell in 1880 though the wide use of this technique for spectroscopic purposes started only almost after a century. From then, there has been a steady progress in the development of PA spectrometers in view of their potential application particularly in biological and environmental studies.[10, 11] A major improvement in the sensitivity of photoacoustic method has been brought about by the availability of laser sources, highly sensitive microphones and other efficient acoustic detection systems.[10-12] Markus Nagele et al developed a novel type of highly sensitive multipass resonant PA cell for trace-gas detection making use of new compact low-power laser sources such as quantum-cascade lasers.[12] Frank Muller et al presented a transportable, highly sensitive photoacoustic spectrometer based on a continuous-wave dual-cavity optical parametric oscillator.[13] Innovations in the design of inexpensive and high performance photoacoustic spectrometers continue to attract recent attention, rendering this form of spectroscopy increasingly versatile.[10, 11, 14-16]

The principle and theory of standard PA technique are well established by Rosencwaig who described the fundamental principles and investigated several possible applications of the PA technique.[17] Patel and Tam reviewed the physics of the technique and extended the technique to many other kinds of applications.[18-21] The most frequently used methods are microphonic detection of the periodic pressure variations developed due to nonradiative deexcitation of the absorbed light by the sample enclosed in an air tight cell. In the case of solid samples, the sample is placed inside a specially designed and acoustically insulated gas filled chamber. The sample is then illuminated with monochromatic light, whose intensity is modulated by means of a mechanical chopper. Light absorption by a material usually leads to two types of deexcitation processes namely radiative and nonradiative. The nonradiative deexcitation processes result in heating the sample. Since the incident radiation is intensity modulated by the chopper, the sample gets heated up repetitively. The resulting periodic heat flow from the solid absorber to the surrounding gas medium creates pressure fluctuations in the cell, which are detected by the microphone. Therefore the depth of specimen responsible for the PA signal is restricted within a thermal diffusion length defined by.[22]

$$\mu = \sqrt{\frac{k}{\pi \rho f c}} \qquad (1)$$

Where k is the thermal conductivity, ρ the density, c the specific heat and f is the chopping frequency. The effect of scattered light on measurement is completely avoided by placing the microphone at remote place. This high sensitive microphone is connected to the main chamber containing experimental sample through a small tunnel. Thus PAS is immune towards the stray light from the highly scattering sample. The photo-exciting system consists of a high power optical source, a monochromator and light focusing set up. The detection system consists of a high sensitive microphone, preamplifier and a dual phase lock-in amplifier. The uses of lock-in amplifier effectively increases S/N ratio and enables the simultaneous measurement of phase and amplitude of the PA signal with respect to light modulation frequency. Other than spectral measurements, the photoacoustic technique can also be used in several other applications such as measurement of thermal diffusivity, detection of phase transition, depth profiling, subsurface imaging etc [7, 8, 9, 10].[23-25] The main aspect of cell design is the positioning of the microphone with respect to incident light to avoid scattered light and to improve performance [11, 12, 13].[26-28] In the present study the PA cell designed to study SP excitation and provision is made to study samples of different properties in different gas environments.

Both PA and reflectance techniques are established in view of simultaneous measurement of SP excitation. The conventional way of detection of reflectance is by either a linear array of photodiodes or a goniometer. A single photodetector can also be used to measure reflectance by using a second rotation stage moving in conjunction with rotation of the prism. The system employing these types of configurations is found to be expensive and hence two-prism concepts has been proposed in the recent past.[29, 30] The aim of the earlier reported two-prism configuration is to avoid above mentioned expensive detection techniques.[29] The second prism is integrated to counter compensate the displacement that caused due to the beam walk over the metal coupled face of the first prism during SPR angular measurement.[29] The main aim of usage of second prism in the earlier study is to make the reflected beam stationary and incident normally with respect to a single photodiode over wide angular rotation.[29] But in the case of SPR measurement, it is imminent that the surface plasmon excitation spot i.e., the interrogation spot should remain stationary during SPR angular measurement as the beam walk over the metal face of the prism affects the SPR measurement sensitivity significantly. A two-prism technique with a modified optical configuration is designed to improve SPR measurement. The proposed optical configuration involves two prisms which are positioned in such a way that the SP excitation spot and reflected light remain stationary during angular displacement and rotation. Finally these two concepts such as two prisms configuration and PA measurement are integrated to find application as SPR sensor based on photoacoustic detection

This chapter in this context describes objectives, scope and methodology for this photoacoustic based surface plasmon resonance spectroscopy. The main objective is to carry out photoacoustic investigation of surface plasmons, so as to explore the possibility of employing photoacoustic techniques in real time sensing based on plasmon resonance spectroscopy. Such sensors exploit the high sensitivity of the surface plasmon frequency to the refractive index of layers of adsorbed molecules on the surface. Generally the resonance conditions are influenced by the material adsorbed onto the thin metal film, which will be explored in detail. An in house developed photoacoustic technique is employed as the detection system for spectroscopic investigations. Two types of set-ups are proposed to investigate SP assisted sensing applications. The first set up is based on a closed photoacoustic cell in which dye coated glass and Au substrates used as sensor element. The second sensor is based on photoacoustic based measurement of resonance angle of prism-based SP excitation. The setup is also equipped with photodetecting system to measure the reflectance of the incident light. The applications of two proposed methodologies for possible sensor application are explained by taking porphyrin as sensor element as it finds wide applications in the biomedical field. There has been a lot of recent interest in supramolecular aggregation effects of porphyrins leading to the formation of self organized nanorods, nanowheels and nanotubes in view of their scope in sensor applications as well as the interesting modifications in their physical properties. The use of metalloporphyrin arrays as sensors is a new concept in artificial olfaction. Its application as 'Electronic nose' applications based on visualizing the color changes associated with the interaction of vapors have been reported. Besides, it properties on Au substrate is also investigated to examine the effect of SP on its optical properties and on the improvement of the sensitivity of the measurement.

2. Review of theory

2.1. Theoretical investigation on prism based SP excitation

The principle of prism based SPR sensor is the measurement of attenuated total reflection (ATR) which is an experimental technique often used with infrared (IR) spectroscopy.

Attenuated total reflection spectroscopy utilizes total internal reflection (TIR) which occurs when a beam of radiation from a denser medium enters a rarer medium above the critical angle θc. The ATR method in IR spectroscopy involves using a crystal as the high refractive index material to excite the test sample. Infra-red light is passed through the ATR crystal in such a way that it should reflect at least once off the internal surface in contact with the sample. This reflection forms an evanescent wave which extends into the sample, typically by a few micrometres. The attenuated energy from each evanescent wave is passed back to the infra-red beam, which then exits the opposite end of the crystal. A detector can be used to measure the exiting beam to detect any changes to the beam.

In the same context SPR sensor also uses a high index prism to excite surface plasmon in the metal film. Here surface plasmon not an evanescent wave as in FT-IR spectroscopy interacts with the test sample but the measurement method is based on same ATR principle. In the SPR sensor, ATR is due to excitation of SP at the SPR angle. It is observed above the critical angle when the momentum of the excitation light matches with that of the surface plasmon i.e., when $k_{sp} = k_o n_p \sin \theta_{sp}$, where k_{sp} and k_o are the wavevectors of the incident photon and excited plasmon respectively, n_p is the refractive index of the prism and θ is the angle of incident beam with respect to normal to the prism surface]. As the excitation of surface plasmon depends on the nature of the surrounding dielectric medium, any change in property of the surrounding dielectric will be reflected as shift in SPR angle.

Fresnel's equations are used to explain the excitation of SP using high index prism. The Fresnel's equations describe how light behaves when it moves through media of differing refractive indices. It describes the reflection and transmission of electromagnetic waves at a dielectric interface. As the SPR experiment setup utilizes an ATR method and the ATR method is applied by using a prism with Kretschmann configuration, where the metal layer is in direct contact with the ATR crystal, Fresnel equations for a three layer model (prism-metal-air) is applicable for the present configuration.

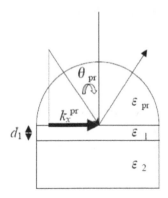

Figure 1. Schematic of three-layer prism setup[31]

Fresnel's equation of three layer model, i.e. prism-metal-air, is defined as

$$R = \left| r^p_{pr12} \right|^2 \tag{2}$$

where R is the reflectance of the incident light, and r^p_{pr12} is the reflected portion of incident light, the superscript P denotes a P-polarized model, and subscript pr12 denotes a prism-medium 1-medium 2 model.

The reflected portion of incident light is given by[31]

$$r_{pr12}^p = \frac{r_{pr1}^p + r_{12}^p e^{2ikz1d1}}{1 + r_{pr1}^p r_{12}^p e^{2ikz1d1}} \tag{3}$$

where r_{pr1}^p denotes the reflected potion of light from prism to medium 1, r_{12}^p denotes the reflected portion of light from medium 1 to medium 2, $kz1d1$ denotes the phase factor.

The reflected portion of light from prism to medium 1 is given by[31]

$$r_{pr1}^p = \frac{\cos\theta_{pr}/n_{pr} - (\varepsilon_1 - n_{pr}^2 \sin^2\theta_{pr})^{1/2}/\varepsilon_1}{\cos\theta_{pr}/n_{pr} + (\varepsilon_1 - n_{pr}^2 \sin^2\theta_{pr})^{1/2}/\varepsilon_1} \tag{4}$$

where θ_{pr} denotes the incident angle of the light entering the prism, n_{pr} denotes the refractive index of the prism, and ε_1 denotes the dielectric constant of medium 1.

The reflected portion of light from medium 1 to medium 2 is given by[31]

$$r_{12}^p = \frac{(\varepsilon_1 - n_{pr}^2 \sin^2\theta_{pr})^{1/2}/\varepsilon_1 - (\varepsilon_2 - n_{pr}^2 \sin^2\theta_{pr})^{1/2}/\varepsilon_2}{(\varepsilon_1 - n_{pr}^2 \sin^2\theta_{pr})^{1/2}/\varepsilon_1 + (\varepsilon_2 - n_{pr}^2 \sin^2\theta_{pr})^{1/2}/\varepsilon_2} \tag{5}$$

where ε_1 denotes the dielectric constant of medium 1, n_{pr} denotes the refractive index of the prism, θ_{pr} denotes the incident angle of the light entering the prism and $\tilde{\varepsilon}_2$ denotes the dielectric constant of medium 2.

Phase factor is given by(Yamamoto, 2008)

$$kz1d1 = \frac{\omega}{c}(\varepsilon_1 - n_{pr}^2 \sin^2\theta_{pr})^{1/2} \tag{6}$$

where $\omega = \frac{2\pi c}{\lambda}$ is the frequency, c denotes the speed of light, λ denotes the wavelength of incident light, ε_1 denotes the dielectric constant of medium 1, n_{pr} denotes the refractive index of the prism and θ_{pr} denotes the incident angle of the light entering the prism.

These equations relate reflectance (intensity of exiting beam) to the refractive indices of the prism, metal and air, thickness of metal layer, wavelength of incident light and the dielectric constants of the metal and air.

2.2. Theory of photoacoustic spectroscopy

The PA effect is based on the generation of acoustic signal when a sample enclosed in an airtight chamber is heated periodically on irradiation with an intensity modulated beam. The corresponding pressure fluctuation produced by the sample in the ambient gas is easily detected by a sensitive microphone. Assuming a optically opaque sample and negligible heat flux into the air in contact with the irradiated surface of the sample, the pressure fluctuation inside the chamber is given by[22, 32]

$$Q = \frac{vP_0I_0(\alpha_s\alpha_g)^{1/2}}{2\pi l_g T_0 k_s f \sinh(l_s\sigma_s)} e^{j(\omega t - \frac{\pi}{2})} \tag{7}$$

where γ is the ratio of specific heat capacity of air, P_0 is the ambient pressure, T_0 is the ambient temperature, I_0 is the radiation intensity, f is the modulation frequency, $\alpha_i, k_i l_i$ are thermal diffusivity, thermal conductivity and length of the medium $i = g$ refers to gas and $i = s$ refers to solid sample. The complex quantity $\sigma_i = (1 + j)a_i$, where $a_i = \left(\frac{\pi f}{\alpha}\right)^{\frac{1}{2}}$ is the thermal diffusion coefficient of the medium i. The expression in equation (7) can be solved based on whether the sample is thermally thick or thermally thin. A sample is considered to be thermally thick when the heat generated due to the absorption of light in the sample reaches the surface and enters the surrounding air column of the photo acoustic cell. For thermally thick sample, the product $l_s a_s$ will become greater than $1(i.e., (l_s a_s \gg 1))$. In that case equation (7) of the PA signal reduces to,

$$Q = \frac{vP_0I_0(\alpha_s\alpha_g)^{1/2}}{\pi l_g T_0 k_s} \frac{\exp[-l_s(\frac{\pi f}{\alpha_g})^{\frac{1}{2}}]}{f} e^{j(\omega t - \frac{\pi}{2} - l_s a_s)} \tag{8}$$

From the above expression it is clear that the photo acoustic signal varies exponentially with the modulation frequency according to $[(\frac{1}{f}) \exp(-b\sqrt{f})]$, where $b = l_s \left(\frac{\pi}{\alpha_s}\right)^{\frac{1}{2}}$ and the phase decreases linearly with $b\sqrt{f}$. Thus, thermal diffusivity α_s can be obtained either from the amplitude or phase data with respect to the modulation frequency for the optically opaque and thermally thick sample.

For the thermally thin sample, the product $l_s a_s$ will become much lesser than $1(i.e., (l_s a_s \ll 1)$. The equation 7 in this case simplifies to

$$Q = \frac{vP_0I_0(\alpha_g)^{1/2}\alpha_s}{(2\pi)^{3/2}l_g l_s T_0 k_s} \frac{e^{j(\omega t - 3\pi/4)}}{f^{3/2}} \tag{9}$$

The above expression implies that the PA signal amplitude from a thermally thin sample under the heat transmission configuration varies as $f^{-1.5}$ and the phase is insensitive to the variation in the modulation frequency. Hence in the case of thermally thin and optically opaque samples, the PA signal is independent of the optical absorption coefficient of the sample, but depends on the wavelength of the incident radiation.

3. Experimental set-up and characterization

3.1. Protocol for sample preparation

Fe(III) tetraphenylporphyrin is prepared according to standard procedures reported in literature.[33] About 50 mg of 5,10,15,20 tetraphenyl porphine Fe(III) chloride procured from

Aldrich, is dissolved in chloroform and shaken with $1M$ hydrochloric acid for four hours. The chloroform layer was separated and dried with unhydrous sodium sulphate. The iron porphyrin was purified by recrystallization from chloroform-ethanol mixture and the optical absorption spectrum was checked with standard spectrum. The HCl treatment is to remove any μ-oxo compounds which are present in iron porphyrin samples. The samples are prepared in different physical forms, namely, liquid, powder and film. The porphyrin solutions are prepared at two different concentrations 10^{-4} M and 10^{-6} M respectively in chloroform. The glass substrate of dimension 1 cm x $1cm$ is ultrasonically cleaned using acetone and then double distilled water a couple of times and vacuum dried. The solution with higher concentration is used to coat a film on the glass plate and Au coated BK7 prism. A few drops of the porphyrin solution are deposited over the glassplate using a 2 ml syringe to form a porphyrin film. Similarly phorphyrin film is formed on Au film coated on the prism base. The deposited film is allowed to dry under vacuum. As the solvent evaporates, the porphyrin film shrinks resulting in the formation islands of porphyrin microclusters on the glass substrate at scattered positions. This approach has been used recently to prepare porphyrin samples where aggregation effects have been observed.[34, 35] Size characterization of the aggregates is done with Atomic Force Microscopy. Contact AFM measurements were done using a Digital Instruments Multimode head attached to a Nanoscope-IV controller. Standard Si_3N_4 cantilevers were used for the normal topography and friction mode imaging. Photoacoustic (PA) studies on powder samples are done by smearing the porphyrin on the grease coated glass substrate.

3.2. Two-prism configuration to excite and measure SP excitation

A schematic diagram representing the experimental set-up is shown in figure 2. Laser beam is focused on to the prism base using the lens and reflected light is measured using an ultrasensitive photodiode. The polarizer is set to polarize the incident beam as P polarized in the direction parallel to the plane of incidence, in this case in the horizontal plane. SP is excited using Kretschmann configuration in which the metal film is in direct contact with the prism base. Lock-in-amplifier and chopper are integrated with the proposed set-up to improve signal to noise ratio (SNR) of the measurement. A chopper is used to modulate the intensity of the incident light and the lock-in amplifier is employed to measure the data with reference to the chopping frequency. Au film is used as the metal film, which is coated on the base facet of BK7 prism.

The prism is fixed on the rotation table and rotated in steps of $0.1°$ to change the momentum of incident light with respect to dielectric-metal interface. SP will be excited once the momentum between the incident light and surface plasmon is matched. Maximum energy transfer between photon to surface plasmon takes place at SPR angle and is very sensitive to dielectric property of the material adjacent to Au film and is given by $K_{sp} = \dfrac{\omega}{c}\sqrt{\dfrac{\varepsilon_m \varepsilon_d}{\varepsilon_m + \varepsilon_d}}$,

where ω is the frequency, c is the velocity of the light, and ε_m and ε_d are dielectric constants of metal and dielectric medium respectively.

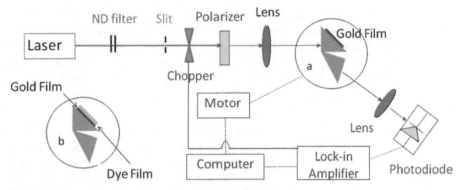

Figure 2. Schematic diagram of experimental set-up for SPR sensor

The point at which SP is excited is called the interrogation point/spot. It is represented by a symbol S in figure 3. This spot should not walk over the prism base during SPR measurement and is avoided by positioning the prism with respect to rotation axis (P) of the prism table.[36]

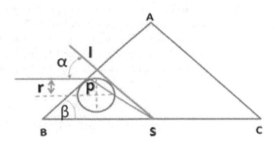

Figure 3. The configuration for incident ray with respect to axis of rotation of the prism table (P) to keep the coupling spot remains practically stationary during SPR measurement

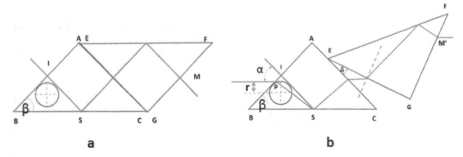

Figure 4. The two-prism configuration with (a) angle = 0° and (b) angle = Δ° between two prisms and ray traces within the prisms

Figure 3 represents a single prism marked by ABC. The solid line (IS) represents the normal incidence of the excitation beam. The centre of the dotted cross represents the axis of the rotation. A circle is drawn with axis of rotation as center and $r = l_p / n_p$ radius, where l_p is the length of the light path inside the prism from the entrance face to the coupling spot and n_p is the refractive index of the prism. It touches the prism surface at I and the beam which strikes at the point of contact (I) always ensures stationary SP excitation spot (S) during SPR angular rotation.

Though the coupling spot remains stationary at the prism base, the reflected light from it always rotates with respect to rotation axis. In such configuration, a linear photodiode array or goniometer is used to collect reflected light from the prism base and is quite expensive. Hence the significance of the concept of employing the two prism configuration.[29, 30] In the two prism technique, the first prism (ABC) is used to excite surface plasmon and the second prism (GEF) is employed to counter the orientation of reflected light from the first prism to opposite direction. This configuration always ensures that the reflected light from both prisms always remains fixed in orientation with respect to the incident beam direction and hence a single photodiode is employed to record the reflected light. Figure 4 represents the two-prism configuration with ABC and GEF as the two prisms. In figure 4 (a) each prism has been rotated by $90°$ ($\Delta = 0$) and the combination act as a parallel plate. The beam which is incident normally (IS) on two prism combination will also emerge out normally at **M**. Though the interrogation spot remains stationery during rotation, the light getting reflected from the second prism (GEF) will makes a walk-over on the prism base. Hence there will be a small shift (δ) in the exit beam (**M'**) with respect to fixed point M and the amount of shift depends on angular separation Δ between the two prisms. Figure 4 (b) represents the shift of the exit beam M' due to angular separation Δ and is given by

$$\delta = \frac{l}{2} \cos\left(\sin^{-1}\left(\frac{1}{n_p} \sin\left(\frac{1}{n_p} (\sin\alpha\cos\beta + \left(n_p^2 - \sin^2\alpha\right)^{1/2} \sin\beta) + \Delta \right) \right) \right) \qquad (10)$$

where α is the angle of rotation of the prism with normal incident angle, β is the angle of the prism and Δ is angle between the prisms, n_p is the refractive index and l is the length of the prism facet,.

Figure 5 represents the shift (δ) as a function of incident angle α. The plot A represents the displacement curve for the prism configuration $\Delta = 0$. It shows minimum at $90°$ rotation and exhibits displacement of about 0.035 mm with respect to **M** during the angular rotation between $40°$ to $120°$ which is found to be lower than the earlier reported two prism configuration.[29] The shifts in displacement for different angular separations (Δ) are also investigated. The shift in minimum towards higher degrees due to increase in angular separation (Δ) is observed. Displacement studies for different angular separations (Δ) are represented by the graphs A, B, C and D. The shift is about $2°$ per $3°$ angular separation is observed.

In the present SPR sensor study, the angle (Δ) between the two prisms is selected as $3°$ to get the reflected beam well deviated from the direction of the incident beam for interference free detection. Besides this, it is ensured that the reflected light remains well within the active area of the photodiode during the SPR angular detection. Both amplitude and phase of the reflected light are measured simultaneously (figure 6). The figure 6(a) represents the measurement of reflectance as the function of the angular rotation. The dip in the graph is due to excitation of SP and has an SPR angle at 43.6 °. The narrowness of the resonance dip indicates the good confinement of SP at the metal-dielectric interface. The experiment is repeated for improved signal to noise ratio and the obtained signal is theoretically analyzed.

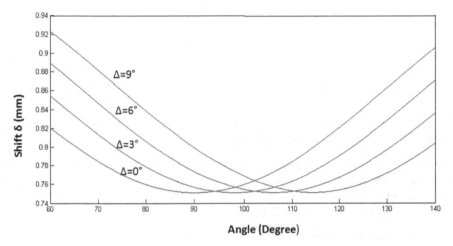

Figure 5. Theoretical simulation of the expression (10), Shift (δ) in terms of Angle (α) for various angular separations (Δ)

The quantitative description of the reflectivity R for the P wave is given by Fresnel's equations for systems comprising three or more layers as,

$$R = \left| r^p_{012} \right|^2 = \left| \frac{r^p_{01} + r^p_{12} \exp(2ik_{z1}d)}{1 + r^p_{01} r^p_{12} \exp(2ik_{z1}d)} \right|^2 \tag{11}$$

where d is the thickness of the metal film, and the reflection coefficient of the P wave is given by

$$r^p_{ik} = \left(\frac{k_{zi}}{\varepsilon_i} - \frac{k_{zk}}{\varepsilon_k} \right) \Big/ \left(\frac{k_{zi}}{\varepsilon_i} + \frac{k_{zk}}{\varepsilon_k} \right) \tag{12}$$

r^p_{ik} is the Fresnel coefficient of p- polarized light between the ith layer and the kth layer in the glass prism(layer 1)/metal (layer 2)/sample (layer 3) configuration (refer figure 1b).

Moreover $r^p_{012} = \mathrm{Re}\left(r^p_{012} \right) + \mathrm{Im}\left(r^p_{012} \right)$. Hence, the phase of the reflected light can be expressed as

$$\phi = \tan^{-1}\left(r^p{}_{012}\right) = \tan^{-1}\left(\operatorname{Im}\left(r^p{}_{012}\right) / \operatorname{Re}\left(r^p{}_{012}\right)\right) \tag{13}$$

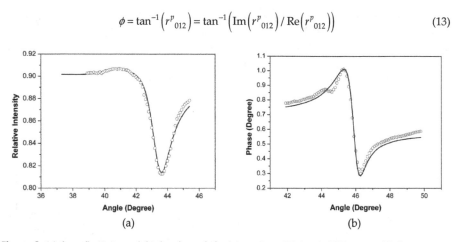

(a)   (b)

Figure 6. (a) the reflectivity and (b) the phase shift of the reflected light of a SPR sensor. Circles represent the experimental value and solid lines represent the theoretical simulation

Using the expression (11), theoretical fit on the experimental data (circles) is done to get the refractive index and thickness of the Au film. The solid line in figure 6(a) represents the theoretical fit. The refractive index and thickness of Au film are found to be0.1726 + i3.3418 and 85nm respectively at wavelength 532 nm.

Theoretical simulation is also performed on the obtained phase data (figure 6(b)) by using expression (13) to validate the results obtained through amplitude data (figure 6(a)). The solid line in figure 6(b) represents the theoretical fit done on the measured phase data pointed   represented by circles. It may be noted from figure 6(b) that the theoretical fit matches exactly with the experimentally obtained phase data.

3.3. Variable volume closed photoacoustic cell

Figure 7 shows a sketch of the designed variable volume closed PA cell and corresponding experimental set-up. The cell is made from an aluminium rod of diameter 80 *mm*. The main chamber is cylindrical in shape with a volume that can be varied between 160to 640 *mm³*. The microphone is connected to the main chamber through a narrow tunnel of diameter 1.2*mm*. A half inch 50*mV/Pa* sensitive microphone from G.R.A.S is used as PA detecting element. The head of the microphone is acoustically isolated from the surroundings by a proper protective cover made of aluminium, screwed to the side of the PA cell. This cover also prevents dust particles from reaching the microphone diaphragm. PA cell is designed in such a way that its cell volume could be changed to study samples of different properties in different gas environments. The sample holder is made of a brass rod of 1*cm* thick and 1 *cm* diameter with a depression on the flat surface. Sample holders with depression of various depths are also made to accommodate samples of different absorption capacities. Particularly samples with low absorption coefficient are taken in relatively large amounts in the case of PA studies. Since the PA signal varies inversely as the gas volume and the thermal conductivity of the gas ($\sqrt{k_g} \dfrac{P_0}{T_0}$ where k_g is the thermal conductivity of the sample, P_0 the pressure and T_0 temperature of the coupling gas medium), the distance between the sample and the cell window must be greater than the thermal diffusion of the gas since the boundary layer of the gas acts as an acoustic piston generating the pressure wave. The distance of the sample surface from the optical widow is varied by screwing the threaded cylindrical rod, on which sample holder rests, connected at the base of the PA cell. (see figure.1). This alters the cell volume, which is an important design parameter. The source used to excite optically the sample is xenon lamp of 500 W from M. Watnabe & Co Ltd. An image of the source is condensed within the size of the entrance slit of the monochromator using a proper combination of optical components to maximize the amount of light entering the monochromator to increase intensity of the monochromatised light. Since the intensity of the PA signal is proportional to the intensity of the incident light, the monochromatised light from the exit face of the monochromator is tight focused on the sample surface. The intensity of the light falling on the sample is modulated by a mechanical chopper of frequency 10 Hz (SR540 chopper) and the reference output of the chopper controller is connected to the lock-in amplifier (SR830 dual-phase digital lock-in amplifier). The lock-in amplifier analyses the PA signal with respect to chopping frequency. The designed PA spectroscopic setup is characterized by measuring frequency response of the spectrometer using carbon black and the result agrees well with the Rosencwaig's theory [figure8]. Then the photoacoustic spectrometer is standardized by recording the spectrum of Holmium oxide.[37]

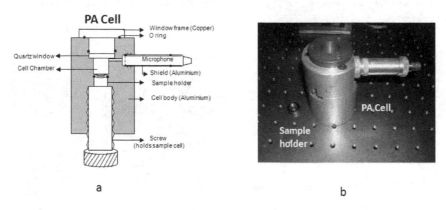

Figure 7. (a) Schematic diagram of Photoacoustic cell and (b) Experimental set-up.

The entire PA system is computer interfaced. Wavelength tuning of the monochromator is achieved by a stepper motor rotated by a driving circuit which contains switching circuit powered by 2N3055 power transistor and speed of the scanning is controlled through LPT1 port of the computer. The lock-in amplifier is controlled through the RS232 port.

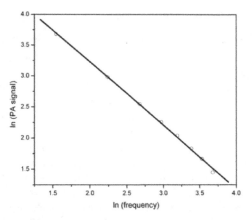

Figure 8. Frequency response of the PA cell.

3.4. Photoacoustic based surface plasmon resonance spectroscopy

The photoacoustic set-up to measure resonance angle of SP excitation in prism coupling technique is shown in the following figure.[38] The light from the source [DPSSL, SUWTECH LDC-1500] was coupled to an optical fibre and guided to an assembly consisting of a polarizer and an optical chopper. Another optical fibre is used to collect the polarised beam to guide through the focusing lens onto the prism base. The prism was mounted on the PA chamber containing microphone. The PA chamber was fabricated in such a way that a small air cell separates the gold-layer and the transducer chamber so as to enable the creation of sound waves.

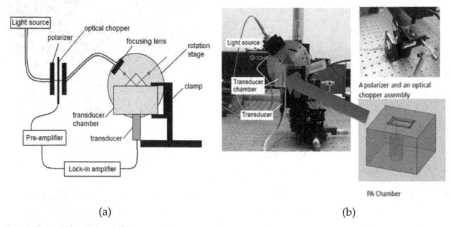

(a)   (b)

Figure 9. (a) Schematic of SPR experiment setup measuring photoacoustic signal (side view) and (b) the corresponding experimental set-up

The technical drawings for the transducer chamber can be found in figure 9 (b). It encloses chamber of depth 1 mm. PA chamber is connected to the microphone by a tunnel of diameter 1.2 mm and a length of about 14 mm. The contact areas between the prism and transducer chamber as well as between the transducer chamber and the transducer were sealed using vacuum grease to prevent any leakage of the acoustic piston effect out of the air pocket. The transducer chamber is positioned in such a way that the beam focuses on a stationary spot on the gold nano-layer on the prism throughout the scanning range of incident angles. The beam spot on the gold nano-layer is ensured to be stationary for the same reasons mentioned earlier in SPR experiment conducted for reflectivity measurement. For this investigation, a pre-amplifier with filter is used in addition to the lock-in amplifier to further improve the signal to noise ratio. The transducer is connected to the source input, while the optical chopper is connected to the reference input channels of the lock-in amplifier. The rotation stage is again jogged at intervals of 0.1°, and the photoacoustic signal at each jog step is recorded. Due to the sensitivity of the transducer, Fast Fourier Transform (FFT) filter smoothing was done on the photoacoustic signals to eliminate high frequency noise.

Carbon black is a strong absorber of the light where $(1/\alpha)< 1$ and $\mu> 1$ (where α is the optical absorption coefficient) and hence a strong PA signal is expected. The strength of the signal is inversely proportional to the chopping frequency according to Rosencwaig's theory.[22, 39] Thus a log-log plot of the PA signal versus chopping frequency is expected to be a straight line of slope-1.

The PA cell is tested by recording the spectrum of carbon black as the standard sample. A Helium-Neon laser is used as the source of excitation. The PA signal is recorded as the function of modulation frequency obtained using a mechanical chopper. Figure 10 represents the measured frequency response of the PA cell. The data fit well to a straight line of slope -1.

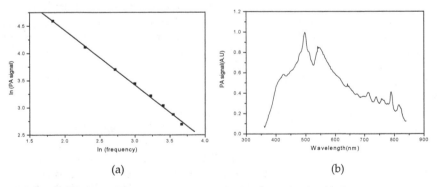

(a)   (b)

Figure 10. (a) Frequency response of PA cell and (b) PA Spectrum of carbon black.

Fig 10 (b) shows the PA spectrum of carbon black recorded with the Xenon lamp as the source. In the case of thermally thin and optically opaque samples, the PA signal is independent of the optical absorption coefficient of the sample, which in turn depends on the wavelength of the incident radiation. Thus the carbon black sample, which satisfies the above condition, should exhibit the spectrum of the source used. The observed spectrum shows the intensity variation of the xenon lamp as a function of wavelength.

4. Results and discussion

The application of the proposed configuration is illustrated by taking porphyrin as sample (dielectric medium). There has been a lot of recent interest in supramolecular aggregation effects of porphyrins in view of their scope in sensor applications. The use of metalloporphyrin arrays as sensors is a new concept in artificial olfaction.

The metalloporphyrin selected for the present study is FeTPPCl. Figure 11 shows the AFM pictures of the film containing FeTPP aggregates on the glass plate. Microclusters of the FeTPP in the size range of 0.8 to 1.6 *microns* are clearly seen in the picture. The AFM pictures indicate an inhomogeneous distribution of porphyrin molecules obtained on the glass surface.

FeTPP aggregates on the glass plate are studied using optical absorption (OA) and photoacoustic spectroscopy. Figure 12 shows the PA spectrum of FeTPP in three physical forms, namely, solution at low concentration (curve A), powder (curve B) and film on glass plate (curve C). The PA spectrum of the solution at low concentration is obtained by depositing a couple of drops of porphyrin solution on the tissue paper. In the case of powder samples, only a few micrograms are taken to avoid signal saturation. The PA spectrum of aggregates on the glass plate is compared with those of porphyrin solution at low concentration deposited on tissue paper and porphyrin powder in order to distinguish between the effects due to aggregation and concentration.

The Soret band is seen in figure 12 at around 450 nm and the Q bands in the region of 500-600 *nm*. These features are similar to those reported for the solution spectrum.[40] The peak corresponding to the Soret band is found to be red shifted to 450 nm in the case of films from the corresponding value of 444 nm for the spectra of the powder as well as the dilute solution. This shift and the accompanying broadening occur only in the film and not in the cases of the dilute solution. These effects are not observed even in the case of the powder sample where the concentration is high. Thus the observed spectral features appear to result from aggregation effects rather than concentration effects. The aggregates could be J-aggregates which are characterized by shifting of bands to larger wavelengths with respect to long wavelength absorption band of the monomers at about 444nm. Figure 12(i) shows the OA spectrum of the films on the glass substrate. No specific feature is seen in the Q band region in the optical spectrum. However, the PA spectrum, shown in figure 12ii(C) clearly exhibits a pronounced two-peak structure between 500-600 nm. Such features are usually attributed to hyper spectra due to metal to ligand charge transfer.[33, 41] These arise from the considerable mixing of the metal d_π orbitals with the LUMO of the Porphyrin, which renders Fe porphyrins of considerable interest for photochemical studies. The broadening, almost nonzero extinction coefficient throughout the visible region and spectral shifts of the bands are indicative of formation of aggregates during the preparation of films on glass substrate. [35]

The Q bands in OA spectrum are not well pronounced as compared to PA spectrum of the film, apparently due to the inhomogeneous distribution of film on the glass substrate and low absorption of the sample. The broad spectral features of FeTPP in liquid form at low concentration, film form and powder form appear to be similar. However, the process of preparing a solution-dried film on glass plates leads to some amount of aggregation as judged by the shift and the broadening in the bands. The hyper spectral features in Q band region of the FeTPP are clearly resolved in PA spectra, particularly in the case of aggregates. This could be due to the fact that the efficiency of nonradiative thermal processes is much larger in aggregates (than in monomeric species) to which PAS is very sensitive. The increased metal-ligand energy transfer in the case of aggregate films could be the reason for the observation of well pronounced hyper spectra.

PA studies at low concentration solution and powder are done to bring out the effect of concentration on shifting and broadening of the peaks in PA spectra. The PA spectrum of porphyrin solution at low concentration is studied by depositing a few drops of porphyrin on the tissue paper, which is a convenient way of recording the spectra of liquid samples using the microphone-based PA spectrometer. The experiment is also repeated by depositing low concentrated solution on a glass plate to avoid aggregation. However, such samples did not produce any appreciable PA signal due to the low concentration. Aggregation processes are also absent in powder samples and occur when the porphyrin films are deposited on the glass substrate. The PA spectra reveal the aggregation effects in a more detailed manner compared to OA spectra. PA spectra of powder samples can be obtained using only at a few micrograms of the sample smeared over the grease coated glass plate as signal saturation effects are observed at higher density.[42]

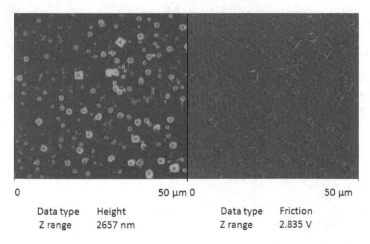

0	50 μm 0	50 μm
Data type	Height	Data type	Friction
Z range	2657 nm	Z range	2.835 V

Figure 11. AFM picture of FeTPP aggregates on glass substrate

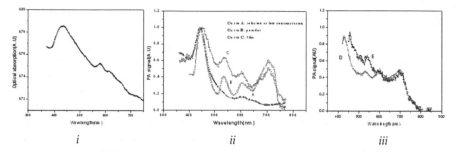

i   ii   iii

Figure 12. (i) Optical absorption spectrum of FeTPP on glass substrate, (ii) Photoacoustic spectrum of FeTPP.[(a) curve A -solution at low concentration, curve B -powder and curve C- film on glass substrate.] and (iii) Photoacoustic spectrum of FeTPP.[(d) Curve D -dye film on glass plate and curve E –dye film on Au film.]

AFM studies on porphyrin deposited on Au film show no distinct features as observed in the case of porphyrin film formed on glass substrate. PA studies on porphyrin film on Au substrate {Figure 12(iii) show improved PA signal as compared to porphyrin film on glass substrate [Figure 12(iii) D]. The proposed concept based on PA cell with porphyrin-glass and porphyrin-Au substrates is expected to find applications as gas sensor.

Investigation of SPR excitation is performed using the two-prism technique. Figure 13shows surface plasmon excitation in Au film of thickness 60 nm using prism in Kretschmann configuration. The graph is a plot of reflectivity against angular rotation of the prism. It exhibits a dip at 43.95 degrees due to excitation of surface plasmon. The angle at which SP is excited is called surface plasmon resonance angle. From the experimental data, it can be observed that a peak in reflectivity occurs before a dip. At the peak, the ratio of incident light intensity to reflected light intensity is at a maximum. This would mean that there is minimum loss of incident light and the peak is a result of TIR. The peak is at the angle where the highest amount of energy of the incident light is passed, through the evanescent wave along the metal and crystal interface, back to the reflecting light.

Therefore, the subsequent minimum reflectivity would give the SPR angle, since the angle is after the critical angle for TIR, and is where the greatest amount of energy is lost due to maximum excitation of surface plasmons. After the SPR angle, reflectivity increases as incident angle increases. This indicates that the amount of energy lost through surface plasmon excitation is decreasing as the consequence of deviation from momentum match between incident photon and excited SP, with the light intensity returning to initial levels. The profile of reflectivity against incident angle forms a basis for comparison with photoacoustic signals.

The experimental set-up used to study SPR is described in the experimental section titled photoacoustic based surface plasmon resonance spectroscopy. Figure 13 shows the graph of photoacoustic signal obtained against incident angle for gold nano-layer of 60 nm thickness. The graph contains two photoacoustic profiles for two modulation frequencies such as 7 Hz and 11 Hz trespectively. From the data, it can be seen that at both frequencies, a peak occurs at the SPR angle. This result is consistent with earlier reported data by different groups.[43-46] For 7 Hz modulation frequency, the photoacoustic signal is larger at every incident angle as compared to 11 Hz. These indicate that lower frequencies resulted in larger photoacoustic signals.

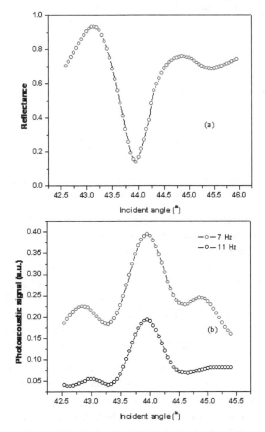

Figure 13. Comparison of (a) reflectivity profile and (b) photoacoustic profile

Figure 13 shows the comparison of the reflectivity profile with photoacoustic signal profile against incident angle. As can be seen from the graph, peaks in reflectivity correspond with dips in photoacoustic signal, and vice versa. It is thus established that reflectivity and photoacoustic signal have an inverse relationship. According to Inagaki et al[43], the absorptivity for p-polarized photons is affected at each angle of incidence into two parts, one of which contributes to the photoacoustic signal by decaying nonradiatively into heat, and the other part which escapes from the sample by decaying into photons or into emission of photoelectrons without contributing to the photoacoustic signal.

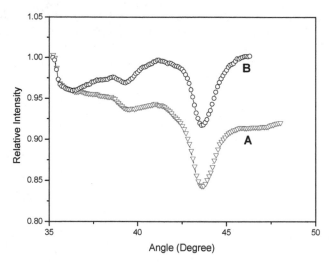

Figure 14. Reflectivity response of a Kretschmann surface plasmon arrangement as a function of angle for 532 nm wavelength. (A) Bare Au film and (B) Au-Dielectric (porphyrin medium).

At the critical angle where reflectivity is the highest, the absorption of p-polarized photons is the lowest since there is little difference between the incident intensity of light and reflected intensity of light. As such, the least amount of p-polarized photons decays into heat. Fluctuation of thermal expansion is then at its least, resulting in low amplitudes of the photoacoustic piston effect, i.e. minimal photoacoustic signal. Conversely, At the SPR angle, reflectivity is at its lowest due to maximum absorption of p-polarized photons. The resulting amount of p-polarized photons decaying into to heat is at a maximum. This fluctuation of thermal expansion would then be at its maximum, resulting in high amplitudes of the photoacoustic piston effect, i.e. maximum photoacoustic signal.

SPR measurement is repeated on Au-porphyrin dielectric medium to study the response of SPR angle due to adsorption of dye molecules for possible sensor applications. The plot B in Figure 14 represents the corresponding SPR measurement. The result shows a lowering in the reflection minimum of SP excitation as compared bare Au film (plot A in figure 14). The excited surface plasmon in the metallic film suffers significantly from strong damping caused by internal absorption and radiation losses. The damping of surface plasmon is characterized by an exponential decay of its intensity away from the metal surface. The radiative loss is due to plasmon re-emission of light. The metal dissipation loss in this case is given by[47]

$$\gamma_m = \frac{2\pi}{\lambda_e} \frac{\varepsilon_m''}{(\varepsilon_m')^2} \left(\frac{\varepsilon_m' \varepsilon_d'}{\varepsilon_m' + \varepsilon_d'} \right)^{3/2} \tag{14}$$

where and denote the real and imaginary part of the dielectric constant of the metal, where is the real part of surrounding dielectric medium. This loss can be overcome by gain medium such as dye at an optimum concentration. The gain γ_g needed to compensate for the SP loss is given by[47]

$$\gamma_g = \frac{2\pi}{\lambda_e} \frac{\varepsilon_d''}{(\varepsilon_d')^2} \left(\frac{\varepsilon_m' \varepsilon_d'}{\varepsilon_m' + \varepsilon_d'} \right)^{3/2} \qquad (15)$$

here ε_d'' is the imaginary part of the dielectric constant of dye medium which is the function of dye concentration (N) and emission cross section (σ_e). Hence the lowering in the reflection minimum of SP excitation could be due to dye assisted enhancement of SP.

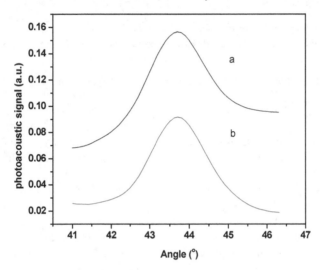

Figure 15. Photoacoustic investigation of (A) Bare Au film and (B) Au-Dielectric (porphyrin medium).

The photoacoustic investigation of SP assisted porphyrin excitation shows a decrease in PA signal (figure 15), which signal could be due to enhanced SP excitation in the presence porphyrin gain medium. The excited fluorophore in the gain medium emits plasmons instead of photons due to strong interaction between the excited dye molecules and electron vibration energy of SP. Any change in the property of the dye due to physio-adsorption and chemisorption will influence the excited SP, which will be reflected in reflective and photoacoustic measurements.

5. Conclusion

Photoacoustic investigation of surface plasmons was thus realized, establishing the relationship between SPR and photoacoustics. An in house developed photoacoustic technique was designed and demonstrated to be feasible as a real time sensor based on plasmon resonance spectroscopy through the repeatability of photoacoustic results using different modulation frequencies. Theoretical simulations done verify the experimental results for SPR, thereby providing a basis of comparison with the results of the photoacoustic experiment.

There is ample scope for future investigation on realizing photoacoustic bio-sensor of various types. The main aim is the identification of a target molecule by having a suitable biological recognition element attached to the surface of the sensor (be it metal, polymer or glass).

Author details

K. Sathiyamoorthy and V.M. Murukeshan
Nanyang Technological University, School of Mechanical and Aerospace Engineering, Singapore

C. Vijayan
Department of Physics Indian Institute of Technology Madras Chennai, India

6. References

[1]   G. Klenkar and B. Liedberg, Analytical and Bioanalytical Chemistry 391, 1679 (2008).

[2]   J. Borejdo, Z. Gryczynski, N. Calander, P. Muthu, and I. Gryczynski, Biophysical journal 91, 2626 (2006).

[3]   X. Chen, M. C. Davies, C. J. Roberts, K. M. Shakesheff, S. J. B. Tendler, and P. M. Williams, Analytical Chemistry 68, 1451 (1996).

[4]   S. Y. Wu, H. P. Ho, W. C. Law, C. Lin, and S. K. Kong, Opt. Lett. 29, 2378 (2004).

[5]   Y. H. Joo, S. H. Song, and R. Magnusson, Opt. Express 17, 10606 (2009).

[6]   J. D. Wright, C. von Bultzingslowen, T. J. N. Carter, F. Colin, P. D. Shepherd, J. V. Oliver, S. J. Holder, and R. J. M. Nolte, Journal of Materials Chemistry 10, 175 (2000).

[7]   L. Pang, G. M. Hwang, B. Slutsky, and Y. Fainman, APPLIED PHYSICS LETTERS 91, 123112 (2007).

[8]   Z. Genchev, N. Nedelchev, E. Mateev, and H. Stoyanov, Plasmonics 3, 21 (2008).

[9]   H.-S. Leong, J. Guo, R. G. Lindquist, and Q. H. Liu, Journal of Applied Physics 106, 124314 (2009).

[10] S. T. Persijn, E. Santosa, and F. J. M. Harren, Applied Physics B: Lasers and Optics 75, 335 (2002).

[11] J. Li, X. Gao, W. Li, Z. Cao, L. Deng, W. Zhao, M. Huang, and W. Zhang, Spectrochimica Acta Part A: Molecular and Biomolecular Spectroscopy 64, 338 (2006).

[12] M. Nägele, D. Hofstetter, J. Faist, and M. W. Sigrist, Analytical Sciences 17, 497 (2001).

[13] F. Müller, A. Popp, F. Kühnemann, and S. Schiller, Opt. Express 11, 2820 (2003).

[14] L.-y. Hao, J.-x. Han, Q. Shi, J.-h. Zhang, J.-j. Zheng, and Q.-s. Zhu, Review of Scientific Instruments 71, 1975 (2000).

[15] J. A. Balderas-Lopez, Review of Scientific Instruments 77, 064902 (2006).

[16] J. A. Balderas-Lopez, Review of Scientific Instruments 77, 086104 (2006).

[17] A. Rosencwaig, in *Advances in Electronics and Electron Physics*, edited by L. Marton (Academic Press, 1978), Vol. Volume 46, p. 207.

[18] C. K. N. Patel and A. C. Tam, Reviews of Modern Physics 53, 517 (1981).

[19] A. C. Tam and C. K. N. Patel, Appl. Opt. 18, 3348 (1979).

[20] C. K. N. Patel and A. C. Tam, Nature 280, 302 (1979).

[21] C. K. N. Patel and A. C. Tam, Applied Physics Letters 34, 467 (1979).

[22] A. Rosencwaig and A. Gersho, Journal of Applied Physics 47, 64 (1976).

[23] K. A. Azez, Journal of Alloys and Compounds 424, 4 (2006).

[24] Q. Shen, M. Inoguchi, and T. Toyoda, Thin Solid Films 499, 161 (2006).

[25] T. S. Silva, A. S. Alves, I. Pepe, H. Tsuzuki, O. Nakamura, M. M. F. d. A. Neto, A. F. da Silva, N. Veissid, and C. Y. An, Journal of Applied Physics 83, 6193 (1998).

[26] N. C. Fernelius, Appl. Opt. 18, 1784 (1979).

[27] L. C. Aamodt, J. C. Murphy, and J. G. Parker, Journal of Applied Physics 48, 927 (1977).

[28] J. W. G. Ferrell and Y. Haven, Journal of Applied Physics 48, 3984 (1977).

[29] B. C. Mohanty and S. Kasiviswanathan, Review of Scientific Instruments 76, 033103 (2005).

[30] K. Sathiyamoorthy, P. A. Kurian, C. Vijayan, and M. P. Kothiyal, edited by R. Guo, S. S. Yin and F. T. S. Yu (SPIE, San Diego, CA, USA, 2007), p. 669818.

[31] M. Yamamoto, Department of Energy and Hydrocarbon Chemistry, Kyoto University, Kyoto-Daigaku-Katsura, Nishikyo-ku, 615-8510, JAPAN 48, 1 (2008).

[32] N. A. George, C. P. G. Vallabhan, V. P. N. Nampoori, and P. Radhakrishnan, Optical Engineering 41, 251 (2002).

[33] F.-P. Montforts, Angewandte Chemie International Edition 43, 5431 (2004).

[34] M. L. Merlau, W. J. Grande, S. T. Nguyen, and J. T. Hupp, Journal of Molecular Catalysis A: Chemical 156, 79 (2000).

[35] N. A. Rakow and K. S. Suslick, Nature 406, 710 (2000).

[36] R. Ulrich and R. Torge, Appl. Opt. 12, 2901 (1973).

[37] K. Sathiyamoorthy, C. Vijayan, and M. P. Kothiyal, Review of Scientific Instruments 78, 043102 (2007).

[38] K. Sathiyamoorthy, J. Joseph, C. J. Hon, and M. V. Matham, edited by M. F. Costa (SPIE, Braga, Portugal, 2011), p. 80010K.

[39] A. Rosencwaig, Optics Communications 7, 305 (1973).

[40] V. Lepentsiotis and R. van Eldik, Journal of the Chemical Society, Dalton Transactions, 999 (1998).

[41] K. S. Suslick, F. V. Acholla, and B. R. Cook, Journal of the American Chemical Society 109, 2818 (1987).

[42] J. W. Lin and L. P. Dudek, Analytical Chemistry 51, 1627 (1979).

[43] T. Inagaki, K. Kagami, and E. T. Arakawa, Appl. Opt. 21, 949 (1982).

[44] T. A. El-Brolossy, T. Abdallah, M. B. Mohamed, S. Abdallah, K. Easawi, S. Negm, and H. Talaat, The European Physical Journal - Special Topics 153, 361 (2008).

[45] S. Negm and H. Talaat, Solid State Communications 84, 133 (1992).

[46] Q. Shen and T. Toyoda, Jpn. J. Appl. Phys. 39, 511 (2000).

[47] J. Seidel, S. Grafström, and L. Eng, Physical Review Letters 94, 177401 (2005).

Chapter 6

Electrically-Driven Active Plasmonic Devices

Young Chul Jun

Additional information is available at the end of the chapter

1. Introduction

Enhanced light-matter interactions in light-confining structures (such as optical cavities) have been extensively investigated for both fundamental studies and practical applications. Plasmonic nanostructures, which can confine and manipulate light down to the nanometer scale, are becoming increasingly important (Atwater 2007, Brongersma 2009). Plasmonic resonators and antennas can convert free-space light into intense, localized fields or enable coupling into deep-subwavelength-guided modes. Wherever subwavelength control over light is required, plasmonic structures are likely to play a vital role. Strong field enhancement in those structures can also alter light-matter interactions at a very fundamental level. Many areas of optical physics and devices can benefit from such extreme light concentration and manipulation (Schuller 2010). Moreover, these metal nanostructures can be simultaneously used as electrical contacts for current injection or application of electric fields (MacDonald 2010, Cai 2012). Thus, plasmonic structures are naturally suited for electrically driven devices. These active plasmonic devices can enable a broad range of new opto-electronic applications.

In this chapter, we review device concepts for electrically-driven active plasmonic devices, including the author's own work. We divide the following sections based on the physical mechanism leading to electrical control. In each section, the device operation and physics are explained. Finally, the conclusions and future prospects are presented.

2. Electric field induced liquid crystal reorientation

One possible way to obtain electrical control of plasmonic devices is to manipulate the refractive index of the dielectric medium adjacent to the metal surface. Liquid crystals (LCs) are an excellent example of such a tunable medium. The large broadband optical anisotropy of LCs makes them ideal for electrically tunable plasmonic structures (Kossyrev 2005, Chu 2006). For example, a E7 LC layer was employed in contact with metallic hole arrays, and electrical tuning of surface plasmon dispersion and optical transmission properties has been recently demonstrated (Dickson 2008). Plastic spacers (~ 60 um thick) were used to create the LC cell between the perforated gold film and the indium tin oxide (ITO) glass used as a top electrode. The static electric field controlled the LC's orientation in the cell. When the electric field was applied, the LC preferentially aligned with the field which was perpendicular to the interface. The maximum refractive index variation corresponding to a 90° reorientation of the LC was $n_e - n_o = 0.21$. (Ordinary and extraordinary refractive indexes, perpendicular and along the LC orientation, are $n_o = 1.525$ and $n_e = 1.734$.) This, in turn, induced a change in the SPP effective index at the Au/LC interface (given by Eq. 1) (Raether 1988) and optical response of the plasmonic crystals.

$$n_{SPP} = \sqrt{\varepsilon_{Au}\varepsilon_{LC} / (\varepsilon_{Au} + \varepsilon_{LC})} \tag{1}$$

Changes in both transmission amplitude and resonance frequency were observed. The changes in the transmission near the band-edges of the plasmonic crystals were most prominent, as expected.

3. Electrically induced thermal heating or phase transition

Another approach for electrical control is based on electrically induced thermal heating or phase transition. Figure 1a shows a 'thermo-plasmonic' device based on electrically-driven thermal heating and refractive index changes in a thermo-optic polymer (Nikolajsen 2004). Long-range (LR) surface plasmon polariton (SPP) waveguides for a wavelength of 1.55 μm were utilized to implement Mach-Zender interferometric modulators (MZIMs) and directional-coupler switches (DCSs) (Fig. 1a). First, 15 nm thin and 8 μm wide gold stripes were sandwiched between 15 μm thick layers of a thermo-optic polymer (BCB, benzocyclobutene) and heated by electrical signal currents. The LRSPP mode was excited by end-fire coupling with a single mode fiber. Gold stripes worked simultaneously as plasmonic waveguides and electrical heating elements. This configuration maximizes the influence of applied electrical signals. Low driving powers (< 10 mW for modulators and < 100 mW for switches), high extinction ratios (> 30 dB), and moderate response times (~ 1 ms) were demonstrated.

The dissipated power in the device for the temperature increase ΔT can be estimated as $P \sim 2\kappa \Delta T L w/d$, where $\kappa \sim 0.2$ W/mK is the polymer thermal conductivity, $w = 8$ μm is the stripe width, and $d = 15$ μm is the cladding thickness. The temperature increase needed for complete extinction at the modulator output is given as $\Delta T = (\partial n/\partial T)^{-1}(\lambda/2L)$, where $\partial n/\partial T \sim -2.5 \times 10^{-5}$ °C^{-1} is the polymer thermo-optic coefficient. Thus, the estimate for the modulator driving power is obtained as $P \sim 2\kappa \Delta T L w/d = (\partial n/\partial T)^{-1}(\kappa w \lambda/d) \approx 7$ mW, which is close to the measured value.

More recently, a semiconductor based electrical tuning of extraordinary optical transmission (EOT) through a metal hole array was reported (Shaner 2007). The gold aperture array, designed to operate in the mid-infrared spectral range (~ 1200 cm^{-1}), was fabricated on a doped GaAs epilayer (Fig. 1b). When the current flowed under the metal grating layer, the epilayer was resistively heated, and frequency tuning of the EOT spectrum (over 25 cm[-1]) was achieved (Fig. 1c).

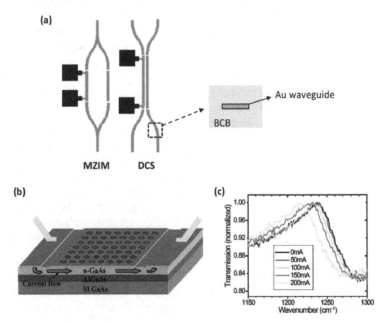

Figure 1. Active plasmonic devices based on electrically induced thermal heating (Nikolajsen 2004, Shaner 2007). (a) Schematic of LRSPP-based plasmonic modulator and switch using an electrically-driven refractive index change in a thermo-optic polymer (BCB). (b) Schematic of a tunable EOT device. The current path through n-doped GaAs epilayer is shown. (c) Transmission spectra through the EOT device. A redshift is observed as the current level increases.

Light transmission through such a structure relies on surface plasmon (SP) excitations at the air/metal and metal/semiconductor interfaces. The SP resonance condition for the metal/semiconductor interface is approximately given by

$$\sqrt{i^2 + j^2}\,\lambda = a_0\,\mathrm{Re}\!\left(\sqrt{\frac{\varepsilon_s\varepsilon_m}{\varepsilon_s + \varepsilon_m}}\right) \tag{2}$$

Here, λ is the free space wavelength, a_0 is the lattice constant, i and j are integers (related to reciprocal lattice vectors), and ε_s and ε_m are the permittivities of the semiconductor and metal. Thermal heating of the semiconductor layer affects both a_0 and ε_s, and thus induces the spectral tuning of the transmission peak. Thermal expansion of the heated sample can change a_0, which would lead to a red-shift in the SP resonance (Eq. 2). However, this is a relatively small effect. The semiconductor dielectric constant ε_s can be affected by temperature in several ways. The temperature dependence of the dielectric function can be modelled using the Drude approximation:

$$\varepsilon(\omega,T,N) = \varepsilon_\infty(T)\left(1 - \frac{\omega_p(T,N)^2}{\omega^2 + i\omega\Gamma(T,N)}\right) \quad \text{where} \quad \omega_p(T,N)^2 = \frac{N(T)q^2}{\varepsilon_0\varepsilon_\infty(T)m^*(T,N)} \qquad (3)$$

Here, ε_∞ is the background high frequency dielectric constant, ω_p is the plasma frequency, N is the electron density, m^* is the electron effective mass, and Γ is the damping term (= $1/\tau$, τ is the electron scattering time). The sample temperature T can be determined by monitoring the GaAs bandgap change (e.g. measuring the band edge PL as a function of current). High sample temperature can increase thermal generation of free carriers and affect the dielectric constant. This also affects the electron scattering time and effective mass.

Electrically induced domain switching or phase transition can modify the substrate refractive index too. Recently, domain switching in the ferroelectric barium-titanate (BaTiO$_2$) thin film was employed for active plasmonic devices (Dicken 2008). BaTiO$_2$ is a perovskite ferroelectric material that exhibits a large electrooptic coefficient (on the order of r ~ 100 pm/V) and large birefringence $\Delta n = 0.05$ (ordinary and extraordinary indexes are $n_o = 2.412$ and $n_e = 2.36$, respectively). The application of a bias across an Ag/BaTiO$_2$ SPP waveguide induced domain switching in the BaTiO$_2$ film. This modified the refractive index in the ferroelectric layer and modulated the optical output of a plasmonic interferometric device (working at $\lambda_0 = 688$ nm) by up to 15%.

Related works on tunable terahertz metamaterials (MMs) use phase transition in vanadium dioxide (VO$_2$) (Driscoll 2009). VO$_2$ is a correlated electron material that exhibits an insulator-to-metal (IMT) phase transition which can be controlled thermally, electrically, or optically. Local heating by electrical current through the VO$_2$ film induced the IMT and a change in the refractive index of the film. With increased temperature, the refractive index of VO$_2$ increased and the MM resonance frequency red-shifted. This electrically controlled frequency tuning is highly hysteretic and persistent; thus, this MM/VO$_2$ structure can also be used as a memory device. Similar electro-optic switching of plasmonic metamaterials was demonstrated in the near-infrared region, using the transition between crystalline and amorphous phases in a chalcogenide glass layer (Samson 2010).

4. Electrical carrier injection or depletion in semiconductor structures

We have so far considered electrical control based on liquid crystal reorientation, thermal heating, phase transition, etc. These are attractive enough for some applications, but they are rather slow processes. Therefore, for other applications, faster electrical control based on conventional semiconductor device technology may be preferred. Electrical tuning via carrier injection or depletion in semiconductor structures has been widely studied for dielectric photonics structures, e.g. for high speed, compact silicon photonic modulators (Soref 1987, Reed 2010). Carrier injection or depletion modifies the refractive index of a medium and can induce a change in the behavior of dielectric resonators.

A similar concept has been applied to active tuning of plasmonic MMs at terahertz frequencies (Chen 2006, Chan 2009). The free carrier absorption in a doped-GaAs substrate was dynamically controlled with an electric bias, by changing the carrier concentration in the substrate; this causes a strong amplitude modulation in the MM transmission. However, at higher frequencies (such as mid-infrared or mid-IR), the free carrier absorption is much smaller, so the transmission amplitude cannot be modulated in this way.

Nevertheless, using highly-doped semiconductor layers, we can still induce spectral tuning of mid-IR resonances by changing the refractive index of the substrate. Recently, electrically tunable mid-IR MMs based on depletion-type semiconductor devices was demonstrated (Jun 2012-1). Gold split ring resonator (SRR) arrays work simultaneously as an optical MM layer and electrical metal gate. With a reverse bias applied to the metal gate, the refractive index of the substrate directly underneath the plasmonic resonators varies through changes in the depletion width in a highly doped semiconductor. This results in frequency tuning of MM resonances. The mid-IR spectral range is technologically important for a number of applications, including chemical sensing and thermal imaging.

The images and schematics of the device are shown in Figs. 2a and 2b. The gold SRRs are connected to a metal gate via electrical bus lines and the whole MM layer works as an electrical gate. The MM layer is placed on top of an n-doped GaAs epilayer to form a metal-semiconductor junction. We need a highly doped n+ layer in order to induce a large dielectric constant change. An insulating barrier (e.g. undoped $Al_{0.3}Ga_{0.7}As$ layer) is also included to reduce leakage current. A modified SRR geometry was chosen because of its strong field enhancement in the two gaps and its compatibility with electrical connectivity, but other MM geometries can be used too.

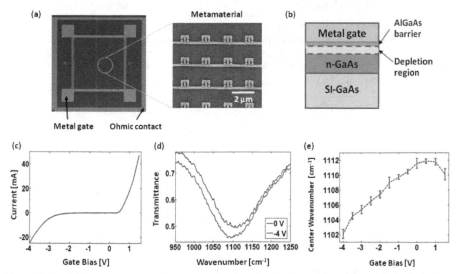

Figure 2. Fig. 2. Device structure and Measurement data (Jun 2012-1). (a) Image showing the device structure. The metamaterial layer (1 mm x 1 mm in size) is connected to the metal gate. These are surrounded by the outer Ohmic contact pad which contacts the n-doped layer. (b) Schematic of the substrate. It is composed of the barrier (30 nm $Al_{0.3}Ga_{0.7}As$) and doped semiconductor (n-GaAs, $N_D = 5$ x

10^{18} cm^{-3}) regions. The depletion region width in n-GaAs is varied by applying an external voltage. (c) IV curve exhibits a diode contact behavior. (d) Fourier transform infrared spectroscopy (FTIR) transmission measurement at room temperature for gate biases $V_G = 0$ V and -4 V. (e) The center frequency of the metamaterial resonance as a function of gate bias.

The dielectric constant ε of a semiconductor substrate can be modeled using the Drude approximation (Eq. 3). In the high doping regime (> 10^{18} cm^{-3}), the dielectric constant decreases rapidly with doping density (Fig. 3a). So, starting from an n+ GaAs doped layer, we can remove carriers with a reverse bias by increasing the depletion region to obtain a large dielectric constant change (e.g. $\Delta\varepsilon \sim 5.5$ for $\Delta N_D = 5 \times 10^{18}$ cm^{-3}, $\lambda_0 = 10$ μm).

(a)   **(b)**

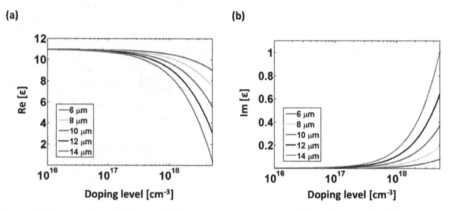

Figure 3. Drude model calculations of dielectric constants in n-doped GaAs for several mid-IR wavelengths (Jun 2012-1). (a) is the real part and (b) is the imaginary part of the dielectric constant.

The depletion width changes in an n-doped GaAs layer can be estimated using a textbook formula for a MIS (metal-insulator-semiconductor) capacitor:

$$W = \left[\frac{2\varepsilon_{GaAs}\varepsilon_0}{qN_D}(-\phi_s) \right]^{1/2} \tag{4}$$

where the surface potential ϕ_s is related to the gate voltage V_G as follows:

$$V_G = \phi_s - \frac{\varepsilon_{GaAs}}{\varepsilon_{AlGaAs}}W_{barrier}\left[\frac{2qN_D}{\varepsilon_{GaAs}\varepsilon_0}|\phi_s| \right]^{1/2} + \phi_{MS} \tag{5}$$

Here, ϕ_{MS} is the flat-band voltage. Note that the dielectric constants here are static values, which are different from the high frequency ones in Eq. 1. The surface potential ϕ_s is introduced to consider a voltage drop across a barrier layer and obtain the actual voltage bias applied in the doped semiconductor region. The depletion width gradually increases with a reverse bias, and the width change depends on the doping density, barrier material/thickness, etc.

We can model the MM resonance shifts using electromagnetic simulations. Here, finite difference time domain (FDTD) simulations are used to simulate the transmission spectra. We utilize dielectric constants and depletion widths obtained by previous equations. The wavelength-dependent dielectric constants from the Drude model are used for a doped semiconductor layer. We employ a different depletion width for each bias and repeat the numerical simulation. A broadband light pulse is incident from the top (normal to the MM plane) and polarized orthogonal to the SRR gap to excite the lowest MM resonance. The transmission is measured on the substrate side.

Figure 4 shows the calculated transmission spectrum. The transmission dip shows a spectral shift upon the application of a bias (Fig. 4a) and gradually red-shifts with increasing reverse bias (Fig. 4b) in agreement with our measurements. This red-shift can be easily understood from a LC resonator model. From $\omega_0 = 1/\sqrt{LC}$, we have $\omega_0 \propto 1/\sqrt{\varepsilon} = 1/n$, where n is the refractive index of a substrate. When the depletion width increases, the substrate refractive index increases (Fig. 3a) and the resonance red-shifts.

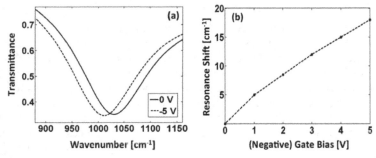

Figure 4. FDTD simulations of MM transmission spectra (Jun 2012-1). (a) Transmission spectra for the gate bias $V_G = 0$ V and -5 V. (b) Resonance red-shift as a function of reverse bias to the gate.

Electrical tunability can be further increased using asymmetric metamaterial designs (Jun 2012-2), which can be more sensitive to a substrate permittivity change. Recently, a classical analog of electromagnetically induced transparency (EIT) was demonstrated in several plasmonic and metamaterial structures. One such structure consists of two metal arcs with different lengths in order to have resonances at slightly different frequencies (Fig. 5, inset). The two metal arms are out-of-phase within a narrow frequency range, and the transmission spectrum has a transparency window due to the destructive interference from two resonator arms. Because this transparency window is caused by interference from two coupled resonators, it can be more sensitive to a substrate refractive index change. Figure 5a shows the simulated transmission spectrum for several different gate biases. We find that the interference part has larger tuning than other frequencies, as expected.

Depletion-type devices are also good for local index control. We can electrically address each resonator arm separately (Figs. 5a-5c). Depending on the bias scheme, we obtain quite different tuning behaviors. Here, we show a two-coupled resonator case. However, more sophisticated structures can be designed to enable a tailored spectrum change with a voltage bias. Electrically tunable metamaterials can be used for novel active infrared devices, such as optical filters, switches, modulators, and phase shifters.

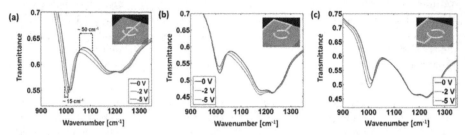

Figure 5. Asymetric metamaterial design based on coupled resonators (Jun 2012-2). Simulated transmission spectrum for several different gate biases, where (a) both resonator arms are connected to an electrical bus line – the straight gold line in the inset – and electrically biased, (b) only the smaller arc is connected to the electrical bus line, (c) only the larger arc is connected. The inset shows the unit cell of the FDTD simulation.

Other semiconductor structures or materials can be also adopted. For example, a metamaterial layer can be patterned on quantum well or quantum dot substrates to induce optical coupling between them (Gabbay 2011). The frequency and strength of this optical coupling can be tailored by designing proper metamaterial/substrate structures. Furthermore, this coupling can be electrically tuned by the stark shift of the quantum well or quantum dot transitions (Gabbay 2012), for instance. Graphene can be also employed for tunable plasmonic metamaterials. Due to its high polarizability, a single atomic layer of graphene can induce a noticeable change in the optical response of metamaterials (Papasimakis 2010). Electrostatic gating can be used to actively tune the optical response of graphene (Wang 2008), thus inducing a change in metamaterial responses. ITO can be also used for electrically carrier injection to induce a change in plasmonic mode indexes at visible and near-infrared wavelengths (Feigenbaum 2010).

A conventional metal-oxide-semiconductor (MOS) capacitor can be utilized as a plasmonic waveguide. Recently, a MOS field-effect plasmonic modulator was reported, termed a 'plasMostor' (Dionne 2009). That employs a metal-MOS-metal (Ag-SiO2-Si-Ag) waveguide structure, operating at 1.55 μm. It supports both photonic and plasmonic modes and has a transmission coefficient determined by interference between them; e.g., application of a gate bias (> 0.7 V) drives the MOS into accumulation and changes the Si index, which cut off the photonic mode and induces the transmission change. This plasmonic modulator can achieve modulation ratios approaching 10 dB in device volumes of half a cubic wavelength with femtojoule switching energies and the potential for gigahertz modulation frequencies.

5. Electrically pumped plasmon-emitting diode

The excitation of SPs with free-space light requires a special momentum matching technique (such as prism coupling, gratings, or scatterers). However, if the metal surface is very close to a light emitter (e.g. fluorescent molecules or quantum dots), the SPs can be excited directly via near-field coupling. Metal layers that support and guide the SP mode can also serve as contact electrodes. This brings an interesting opportunity for electrically pumped SP sources. The development of such electrical SP sources is important for miniaturized photonic circuitry and integrated sensing platforms.

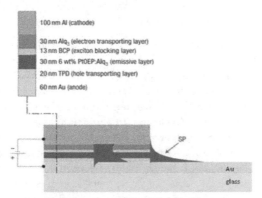

Figure 6. Schematic of an organic plasmon-emitting diode (Koller 2008). SPs are excited electrically through exciton recombination inside the MDM waveguide and later outcoupled to single surface SP wave as shown above.

Figure 6 shows an electrically pumped SP source using organic light-emitting diode (OLED) structures (Koller 2008). The organic active layers are placed in between two gold layers (i.e., a metal-dielectric-metal structure, or MDM). Metal layers work simultaneously as the SP waveguide and metal electrode. Electrically injected electrons and holes recombine in the active organic layer and generate the SP wave. The excited SP mode in the MDM waveguide can extend further into a single metal surface as shown in Fig. 6. The strong electromagnetic field intensity of the SP mode can enable a compact sensing platform without external illumination optics.

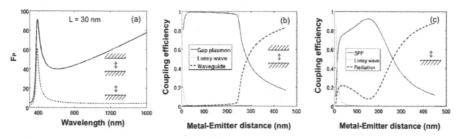

Figure 7. Light emission of a dipolar emitter inside an Ag/SiO$_2$/Ag MDM slab waveguide or near an Ag/SiO$_2$ interface (Jun 2008). We assume that the emitter has unit quantum efficiency. (a) Spontaneous emission enhancement as a function of free-space wavelength. In the MDM case, the emitter is in the center of the gap. The gap size L = 30 nm. (b) Fraction of dissipated energy into different decay pathways as a function of the metal-emitter distance. The emission wavelength $\lambda_0 = 800$ nm. (c) same as (b) but for the single metal/dielectric interface case.

A similar electrical SP source based on silicon CMOS processes has been reported (Walters 2010): gold cladding layers surround a semi-insulating layer of alumina that contains silicon quantum dots and form a MDM plasmonic waveguide. When a sufficient voltage is applied across the insulator layer, tunnelling electrons can excite the embedded quantum dots through impact ionization processes, producing the SP wave. The device fabrication and materials are compatible with well-established silicon microelectronics processes.

Because of their small optical mode volume, MDM structures can achieve large spontaneous emission enhancement and high coupling ratio into well-defined SPP waveguide modes (Jun 2008). Figure 7 explains this for a single dipolar emitter which is oriented normal to the interfaces. The spontaneous emission enhancement factor, F_P, is obtained by calculating the work done on the dipole by its own reflected field from the metal surfaces. Figure 7a shows F_P as a function of the free-space wavelength. In addition to a resonant enhancement peak around the SP resonance wavelength of silver, there is a strong non-resonant enhancement due to the tight confinement of modes between the two metallic films. In the non-resonant region, propagation lengths of SP modes become longer, which is desirable for many applications. For comparison, the case of a single metal surface is also plotted (dashed line), which clearly lacks such non-resonant enhancement. The light emission in MDM waveguides is also preferentially directed into plasmonic waveguide modes (Fig. 7b) over a large range of gap sizes. This behavior is in stark contrast to that observed for an emitter near a single metal surface, which exhibits relatively stronger coupling to lossy surface waves and free-space modes (Fig. 7c).

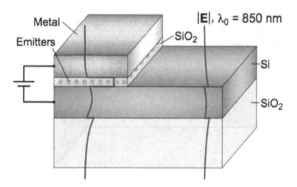

Figure 8. Schematic of a CMOS-compatible light-emitting device (Hryciw 2009). Typical electric field profiles for a TM mode are shown in red. Electrically pumped light emission in the MDS waveguide excites hybrid plasmonic modes and later can be outcoupled to SOI waveguide modes.

A high-index silicon slab supports guided dielectric modes. When this silicon slab is brought closer to a metal surface which supports SPP modes, a new hybrid plasmonic mode can form. A metal-dielectric-semiconductor (MDS) structure resembles MOS layers in conventional semiconductor devices and thus can be used as CMOS-compatible light sources. We can still achieve large field enhancement in the low-index dielectric region of such hybrid structures. Spontaneous emission enhancement in MDS waveguides is still significantly larger than those from a single metal surface. MDS waveguides exhibit broadband emission enhancement and high waveguide coupling ratio, similar to MDM waveguides (Hryciw 2009, Hryciw 2010). Furthermore, this MDS structure is more suitable for electrical carrier injection. The plasmonic modes in MDS waveguides can be efficiently converted into silicon slab waveguide modes for a longer distance transport (Fig. 8).

In addition to these plasmon-emitting diodes, there have been demonstrations on nanocavity lasers based on electrically pumped metal-coated cavities, MDM waveguides, and optically pumped hybrid plasmonic cavities (Martin 2010). In a 'spaser' (surface plasmon amplification by stimulated emission of radiation), the lasing mode may even be non-radiating, i.e. a near-field laser. Such coherent nanoscale light sources can lead to new applications in sensing and lithography.

6. Electrical control of light emission in plasmonic nanogaps

It was recently demonstrated that nanopatterned metal films such as metal slit-grooves can collimate and direct quantum dot (QD) emission in a designed fashion, working effectively as optical antennas (Jun 2011). The patterned metal films next to a slit can be used as electrical contacts to apply a field across the slit. In this section, we consider active electrical control of QD emission in such plasmonic nanogaps. We show how the metals defining the plasmonic cavity can be utilized to electrically control the emission intensity and wavelength. The unmatched, combined directional and electrical control over the emission of a large number of quantum emitters opens up a broad range of new opto-electronic applications for plasmonic antennas, facilitating the realization of a new class of active optical antennas.

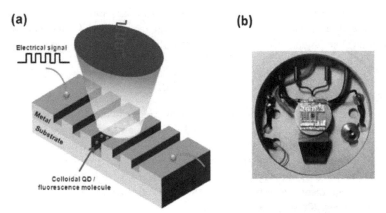

Figure 9. Electrical control of QD photoluminescence (Jun 2011). (a) Schematic of a plasmonic antenna device, in which colloidal CdSe/ZnS core/shell quantum dots are placed within the nanslit. An optical pump at 514 nm was directed from the below, and QD emission was collected from the top. (b) Top view of the mounted sample on the Linkam thermal stage.

First, two large metal contact pads were defined by standard optical lithography. Two metal pads were connected with a narrow metal stripe. A slit-groove structure was patterned later on this stripe with focused ion beam (FIB) milling (Fig. 9a). After FIB milling, the two metal pads were electrically disconnected. Then, a thin layer of colloidal CdSe/ZnS QDs mixed with PMMA (poly-methylmethacrylate) was spin-coated on top of the patterned metal structures. Electrical measurements were conducted at a low temperature using the Linkam thermal stage. A sample was mounted on the silver chuck of the Linkam thermal stage (Fig. 9b), which was cooled with liquid nitrogen flowing. Two metal contacts in the sample were wire-bonded for electric biasing. Because the chuck had a hole in the center, we could direct laser light (CW laser light at 532 nm) from the bottom and collect QD emission from the top. We applied external voltages across the metal gap to see the voltage-dependent QD emission spectrum.

Figure 10a shows the voltage-dependent QD emission spectrum measured from this slit-groove sample. As the applied DC voltage increases, the QD emission spectrum exhibited a red-shift and luminescence quenching. In most cases, the spectral changes were reversible by alternating external voltages, although sometimes QDs or the metal nanoslits were damaged under the strong field. The spectrum shift can be understood as the Stark effect of emission. The large electric field pulls apart electron-hole pairs inside QDs and induces the red-shift and broadening of the whole emission spectrum. Note that despite the intensity drop at the spectrum center, the QD emission intensity increased at longer wavelengths with applied voltages. However, the Stark effect alone does not explain the observed, large luminescence quenching. We believe this is possibly due to enhanced nonradiative recombination. A very large electric field in the gap can 'dissociate' electron-hole pairs, and those carriers can induce luminescence quenching by participating in nonradiative recombination processes such as Auger recombination.

Figure 10b shows the measured emission frequency shift as a function of applied voltage. The shift initially increases quadratically with voltage (as expected for the Stark shift) and then saturates at larger voltages (shaded region). According to quantum mechanical perturbation theory, the stark shift is given by

$$\Delta E = \mu F + \alpha F^2 + \dots \qquad (6)$$

where F is the electric field magnitude, μ permanent dipole moment, and α polarizability. It is known that a CdSe QD has a Wurtzite crystal structure and a permanent dipole moment. Thus, depending on the orientation of a QD, we can observe either a red- or blue-shift with the applied external field. However, averaging over many QDs in the gap, the linear term cancels out, and we obtain only a quadratic stark shift (i.e. red-shift). Thus, for our QD ensemble measurement, we expect to see the quadratic red-shift with external field, in agreement with Fig. 10b.

We can deduce the polarizability of QDs from this Stark shift data: we first obtain the electric field magnitude inside a QD as a function of applied voltage. Assuming that a QD is a sphere dielectric, we can analytically calculate the electric field F_{in} inside the QD:

$$F_{in} = -\frac{d\varphi_{in}}{dr} = \frac{3}{\varepsilon_{QD}/\varepsilon_{matrix}+2}F_0 = \frac{3}{\varepsilon_{QD}/\varepsilon_{matrix}+2}\frac{V_a}{W} \qquad (7)$$

where ε_{QD} and ε_{matrix} are the dielectric constants of QD and surrounding matrix medium, and F_0 is the incident field magnitude. The numbers 3 and 2 originate from the geometry (i.e. sphere).

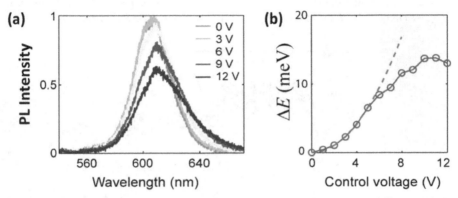

Figure 10. Bias-dependent QD emission (Jun 2011). (a) Emission spectra measured at different applied DC bias voltages across the nanogap. (b) The Stark shift of the emission as a function of applied voltage. The dashed line corresponds to a quadratic fitting curve, and the shaded region indicates a saturation effect for large voltages.

We have $F_0 = V_a/W$, considering that we have a voltage drop V_a across the gap W. In this experiment, we have W = 60 nm, $\varepsilon_{CdSe} = (2\varepsilon_\perp + \varepsilon_\parallel)/3 = 2/3 \times 9.23 + 1/3 \times 10.16 \approx 9.58$, $\varepsilon_{matrix} = \varepsilon_{PMMA} = 2.22$. By fitting data in Fig. 10b to the following equation (i.e. blue dotted curve in Fig. 10b), we obtain the QD polarizability.

$$\Delta E = \alpha F^2; \quad \alpha = 4\times10^{-5}\,meV\cdot cm^2\,/\,(kV)^2 \qquad (8)$$

Furthermore, we can apply a time-varying voltage signal and modulate the QD emission dynamically (Fig. 9a). A square-wave voltage signal was applied across the gap with a function generator. A 100 kHz voltage signal modulated between 0V and 10V with a 50% duty cycle. The QD emission was detected with a single photon avalanche detector (SPAD). We used the sync signal of the function generator as a start signal and the SPAD output as a stop signal in the photon counting module (Picoharp 300, Picoquant) to obtain the accumulated histogram of the modulated QD emission intensity (Fig. 11). A clear step appears in the time-resolved histogram: this means that the QD emission intensity was modulated with the external voltage signal. We accumulated the SPAD output for a few minutes to obtain a clear modulation step. This improved the signal-to-noise ratio but did not change the actual modulation depth. A modulation depth of M ~ 24 % was achieved, where $M = (I_{max} - I_{min})\,/\,I_{max}$.

This step appears only when we see QD emission spectrum change with voltage. For example, if incident laser light illuminated QDs on the bare quartz region (i.e. outside the metal stripe), we did not observe such a modulation step. The histogram just appears flat, i.e. no voltage dependent change. We also note that the observed modulation depth also depends on the detection wavelength. When there is a large change in QD emission spectrum, we observed larger modulation depth. This result presents new opportunities for electrically modulated optical antenna devices. In the above experiment, the modulation speed was rather moderate (100 kHz), but we can increase the speed to well above 100 MHz.

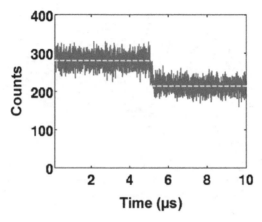

Figure 11. The time-resolved histogram of electrically modulated QD emission intensity (Jun 2011). The modulated square-wave signal (0-10 V, 100 kHz, 50% duty cycle) is applied. The step means that QD emission intensity is dynamically modulated with the external voltage signal.

Plasmonic modes in small metal nanogaps (< 100 nm) can further contribute to this electrical modulation. Spontaneous emission can be enhanced greatly in a small metal nanogap. The reduced lifetime can improve the responsivity of devices and increase the upper limit of the modulation speed. Furthermore, optimized side grooves can enable optical beaming. This electrically driven slit-groove structure acts as an active optical antenna, which can collimate and modulate QD emission in one device. The unmatched, combined directional and electrical control over the emission of a large number of quantum emitters can support a broad range of new opto-electronic applications.

More recently, electrically controlled nonlinear harmonic generation of light in plasmonic nanogaps was demonstrated (Cai 2011). The device structure is similar to the previous one (Fig. 9a). A commonly used polymer (PMMA, poly-methyl methacrylate) was inserted into the metal slit. The frequency doubled light originated from a third-order nonlinear susceptibility $\chi(2\omega; \omega, \omega, 0)$. The change in the second-harmonic signal as a function of applied DC voltage indicated a linear dependence of the frequency doubled output on the driving voltage, with the magnitude of the normalized change being over 7% per volt. Unlike the conventional second harmonic generation, which imposes a rigorous requirement on the lattice asymmetry of the nonlinear medium, third-order nonlinear responses are present in all materials. The subwavelength device size also eases the strict phase-matching condition and makes them suitable for chipscale, high density integration.

7. Conclusions

Plasmonics is emerging as a new chip-scale device technology that can bridge the size mismatch between nanoscale electronics and microscale photonics. It can combine the size of electronics and the speed/bandwidth of photonics in one device. It exploits the unique optical properties of metallic nanostructures to confine and manipulate light at the nanoscale. Strong optical field enhancement in plasmonic nanostructures can also modify light-matter interactions to unprecedented levels. Furthermore, plasmonic structures can serve as electrodes and thus perform optical and electrical functions simultaneously. This brings about many new opportunities for opto-electronic devices.

This chapter reviewed novel active plasmonic device concepts explored in the literature, including the author's own work. Electrically driven plasmonic device structures were introduced, and their device operation and physics were explained. Various physical mechanisms can be utilized for electrical control of optical properties, such as liquid crystal reorientation, thermal heating, phase transition, carrier injection/depletion, Stark shift of emitters, etc. The next step would be to develop device structures that are more efficient, compact, and scalable for large area, dense integration.

Author details

Young Chul Jun
Center for Integrated Nanotechnologies (CINT), Sandia National Laboratories, NM, USA

Acknowledgement

This work was performed, in part, at the Center for Integrated Nanotechnologies, a U.S. Department of Energy, Office of Basic Energy Sciences user facility. Sandia National Laboratories is a multi-program laboratory managed and operated by Sandia Corporation, a wholly owned subsidiary of Lockheed Martin Corporation, for the U.S. Department of Energy's National Nuclear Security Administration under contract DE-AC04-94AL85000. The author wishes to thank Dr. Igal Brener for his support in the preparation of this article.

8. References

Atwater, H.A. (2007), The promise of plasmonics, *Scientific American*, (April 2007), pp. 56-63

Brongersma, M.L. (2009), Schuller, J.A., White, J., Jun, Y.C., Bozhevolnyi, S.I., Sondergaard, T., Zia, R. Nanoplasmonics : components, devices, and circuits in *Plasmonic Nanoguides and Circuits*, (Pan Stanford Publishing, 2009), edited by Bozhevolnyi, S.I.

Cai, W. (2011), Vasudev, A.P., Brongersma, M.L. Electrically controlled nonlinear generation of light with plasmonics. *Science*, Vol.333, No.6050, (2011), pp. 1720-1723

Cai, W. (2012), Jun, Y.C., Brongersma, M.L. Electrical control of plasmonic nanodevices. SPIE Newsroom, (2012), doi: 10.1117/2.1201112.004060

Chan, W.L. (2009), Chen, H.-T., Taylor, A.J., Brener, I., Cich, M.J., Mittleman, D.M. A spatial light modulator for terahertz beams. *Appl. Phys. Lett.*, Vol.94, No.21, (2009), p. 213511

Chen, H.-T. (2006), Padilla, W.J., Zide, J.M.O., Gossard, A.C., Taylor, A.J., Averitt, R.D. Active terahertz metamaterial devices. *Nature*, Vol.444, (2006), pp. 597-600

Chu, K.C. (2006), Chao, C.Y., Chen, Y.F., Wu, Y.C., Chen, C.C. Electrically controlled surface plasmon resonance frequency of gold nanorods. Appl. Phys. Lett., Vol.89, No.10, (2006), p. 103107

Dicken, M.J. (2008), Sweatlock, L.A., Pacifici, D., Lezec, H.J., Bhattacharya, K., Atwater, H.A. Electrooptic modulation in thin film barium titanate plasmonic interferometers. *Nano Lett.*, Vol.8, No.11, (2008), pp. 4048-4052

Dickson, W. (2008), Wurtz, G.A., Evans, P.R., Pollard, R.J., Zayats, A.V. Electronically controlled surface plasmon dispersion and optical transmission through metallic hole arrays using liquid crystal. *Nano Lett.*, Vol.8, No.1, (2008), pp. 281-286

Dionne, J.A. (2009), Diest, K., Sweatlock, L.A., Atwater, H.A. PlasMOStor : a Metal-Oxide-Si field effect plasmonic modulator. *Nano Lett.*, Vol.9, No.2, (2009), pp. 897-902

Driscoll, T. (2009), Kim, H.-T., Chae, B.-G., Kim, B.-J., Lee, Y.-W., Jokerst, N.M., Palit, S., Smith, D.R., Ventra, M.D., Basov, D.N. Memory Metamaterials. *Science*, Vol.325, No.5947, (2009), pp. 1518-1521

Feigenbaum, E. (2010), Diest, K., Atwater, H.A. Unity-order index change in transparent conduction oxides at visible frequencies. *Nano Lett.*, Vol. 10, No. 6, (2010), pp. 2111-2116

Gabbay, A. (2011), Reno, J., Wendt, J.R., Gin, A., Wanke, M.C., Sinclair, M.B., Shaner, E., Brener, I. Interaction between metamaterial resonators and intersubband transitions in semiconductor quantum wells. *Appl. Phys. Lett.*, Vol. 98, No. 20, (2011), p. 203103

Gabbay, A. (2012), Brener, I. Theory and modeling of electrically tunable metamaterial devices using intersubband transitions in semiconductor quantum wells. Opt. Express, Vol.20, No.6, (2012), pp. 6584-6597

Hryciw, A. (2009), Jun, Y.C., Brongersma, M.L. Plasmon-enhanced emission from optically-doped MOS light sources. *Opt. Express*, Vol.17, No.1, (2009), pp. 185-192

Hryciw, A. (2010), Jun, Y.C., Brongersma, M.L. Electrifying plasmonics on silicon. *Nature Mater.*, Vol.9, (2010), pp. 3-4

Jun, Y.C. (2008), Kekatpure, R.D., White, J.S., Brongersma, M.L. Nonresonant enhancement of spotaneous emission in metal-dielectric-metal plasmon waveguide structures. *Phys. Rev. B*, Vol.78, No.15, (2008), p. 153111

Jun, Y.C. (2011), Huang, K.C.Y., Brongersma, M.L. Plasmonic beaming and active control over fluorescent emission. *Nature Comm.*, Vol.2, (2011), p. 283

Jun, Y.C. (2012-1), Gonzales, E., Reno, J.L., Shaner, E.A., Gabbay, A., Brener, I. Active tuning of mid-infrared metamaterials by electric control of carrier densities. *Opt. Express*, Vol.20, No.2, (2012), pp. 1903-1911

Jun, Y.C. (2012-2), Brener, I. Electrically tunable infrared metamaterials based on depletion-type semiconductor devices. *submitted* (2012)

Koller, D.M. (2008), Hohenau, A., Ditlbacher, H., Galler, N., Reil, F., Aussenegg, F.R., Leitner, A., List, E.J.W., Krenn, J.R. Organic plasmon-emitting diode. *Nature Photon.*, Vol.2, (2008), pp. 684-687

Kossyrev, P.A. (2005), Yin, A., Cloutier, S.G., Cardimona, D.A., Huang, D., Alsing, P.M., Xu, J.M. Electric field tuning of plasmonic response of nanodot array in liquid crystal matrix. Nano Lett., Vol.5, No.10, (2005), pp. 1978-1981

MacDonald, K.F. (2010), Zheludev, N.I. Active plasmonics : current status. *Laser Photonics Rev.*, Vol.4, No.4, (2010), pp. 562-567

Martin, T.H. (2010), Status and prospects for metallic and plasmonic nano-lasers. J. *Opt. Soc. Am. B*, Vol.27, No.11, pp. B36-B44.

Nikolajsen, T. (2004), Leosson, K., Bozhevolnyi, S.I. Surface plasmon polariton based modulators and switches operating at telecom wavelengths. *Appl. Phys. Lett.*, Vol.85, No.24, (2004), pp. 5833-5835

Papasimakis, N. (2010), Luo, Z., Shen, Z.X., Angelis, F.D., Fabrizio, E.D., Nikolaenko, A.E., Zheludev, N.I. Graphene in a photonic metamaterial, *Opt. Express*, Vo.18, No.8, (2010), pp. 8353-8359

Raether, H. (1988), *Surface Plasmons on Smooth and Rough Surfaces and on Gratings*, Springer-Verlag.

Reed, G.T. (2010), Mashanovich, G., Gardes, F.Y., Thomson, D.J. Silicon optical modulators. *Nature Photon.*, Vol.4, (2010), pp. 518-526

Samson, Z.L., MacDonald, K.F., Angelis, F.D, Gholipour, B., Knight, K., Huang, C.C., Fabrizio, E.D., Hewak, D.W., Zheludev, N.I. Metamaterial electro-optic switch of nanoscale thickness. *Appl. Phys. Lett.*, Vol.96, (2010), p. 143105

Schuller, J.A., Barnard, E.S., Cai, W., Jun, Y.C., White, J.S., Brongersma, M.L. Plasmonics for extreme light concentration and manipulation. *Nature Mater.*, Vol.9, (2010), pp. 193-204

Shaner, E.A. (2007), Cederberg, J.G., Wasserman, D. Electrically tunable extraordinary optical transmission gratings. *Appl. Phys. Lett.*, Vol.91, (2007), p. 181110

Soref, R.A. (1987), Bennett, B.R. Electrooptical effects in silicon. *IEEE J. Quantum Electron.*, Vol.QE-23, No.1, (1987), pp. 123-129

Walters, R.J. (2010), van Loon, R.V.A., Brunets, I., Schmitz, J., Polman, A. A silicon-based electrical source of surface plasmon polaritons. *Nature Mater.*, Vol.9, (2010), pp. 21-25

Wang, F. (2008), Zhang, Y., Tian, C.S., Girit, C., Zettl, A., Crommie, M.F., Shen, Y.R. Gate-variable optical transitions in graphene. Science, Vol.320, No.5873, (2008), pp. 206-209

Plasmonics Applications

Innovative Exploitation of Grating-Coupled Surface Plasmon Resonance for Sensing

G. Ruffato, G. Zacco and F. Romanato

Additional information is available at the end of the chapter

1. Introduction

Surface Plasmon Polaritons (SPPs) are confined solutions of Maxwell's equations which propagate at the interface between a metal and a dielectric medium and have origin from the coupling of the electromagnetic field with electron-plasma density oscillations inside the metal [1]. SPPs are localized in the direction perpendicular to the interface: field intensity decays exponentially from the surface with an extension length of the same order of the wavelength inside the dielectric and almost one order shorter in the metal [2]. These features make SPPs extremely sensitive to optical and geometrical properties of the supporting interface, such as shape, roughness and refractive indices of the facing media. Since these modes have a non-radiative nature, the excitation by means of a wave illuminating the metallic surface is possible only in the configurations providing the wavevector-matching between the incident light and SPP dispersion law (Surface Plasmon Resonance – SPR, see Figure 1). Prism-Coupling SPR (PCSPR) exploits a prism in order to properly increase incident light momentum and achieve SPP excitation, however this setup suffers from cumbersome prism alignment and it is not suitable for miniaturization and integration. A more amenable and cheaper solution consists in Grating-Coupling SPR (GCSPR), where the metal surface is modulated with a periodic corrugation. The plasmonic behaviour of these modulated metallic surfaces had been discovered since the early years of the last century by R.W. Wood [3] and the connection between Wood's anomalies and surface plasmons was finally established by J.J. Cowan and E.T. Arakawa [4]. A plane-wave illuminating the patterned area is diffracted by the periodic structure and it is possible for at least one of the diffracted orders to couple with SPP modes.

SPR has known an increasing interest in the design and realization of miniaturized label-free sensing devices based on plasmonic platforms. Surface plasmon modes in fact are extremely sensitive to changes in the refractive index of the facing dielectric medium: a thin coating film or the flowing of a liquid solution alter SPP dispersion curve and cause resonance conditions to change (Figure 2). Thus it is possible to detect refractive index variations by simply analyzing the resonance shift: it is the basis of modern SPR-sensing devices [5]. SPR reveals itself as a useful instrument for the study of surface optical properties and it is a highly suitable candidate as application with sensing purposes in a large variety of fields: environmental protection, biotechnology, medical diagnostics, drug screening, food safety and security.

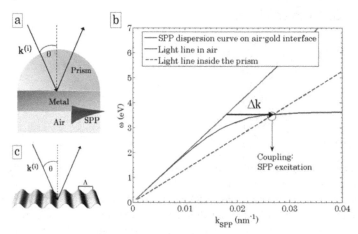

Figure 1. a) Kretschmann SPR configuration. b) SPP excitation with Prism-Coupling. The same effect of momentum-supply can be achieved by corrugating the metallic surface in the so-called Grating-Coupled SPR (c).

Currently, several groups are using different SPR approaches to detect the change of refractive index. Sensors using Prism-Coupled SPR (PCSPR) with Kretschmann configuration [6] can be readily combined with any type of interrogation: angular, wavelength, intensity or phase modulation [7]. PCSPR typically show refractive index sensitivity for angular interrogation architecture that ranges between 50-150°/RIU [8], with higher sensitivity at shorter wavelengths [9], and refractive index resolutions in the orders 10^{-6}-10^{-7} Refractive Index Units (RIU). However, PCSPR sensors suffer from cumbersome optical alignment and are not amenable to miniaturization and integration [10].

Grating-Coupled SPR (GCSPR) sensors instead, with either wavelength or angular interrogation, have been demonstrated to have sensitivity 2-3 times lower than PCSPR [11]. However, GCSPR has the intrinsic possibility to be used with different sensing architectures and interrogation systems. A parallel SPR angular detection was shown by Unfrict et al. to have the possibility for multi-detection for proteomic multi-array [12]. Homola's group demonstrated how a miniaturized GCSPR sensor implemented with a CCD allowed detection sensitivity of 50°/RIU and resolution of $5 \cdot 10^{-6}$ over 200 sensing parallel channels [13]. Alleyne has exploited the generation of an optical band gap by using prism-coupled to achieve sensitivity up to 680°/RIU by bandgap-assisted GCSPR [14]. A recent approach was reported by Telezhnikova and Homola [15] with the development of a sensor based on spectroscopy of SPPs down to $5 \cdot 10^{-7}$ RIU. The strong compatibility of gratings with mass production makes SPP couplers extremely attractive for fabrication of low-cost SPR platforms .

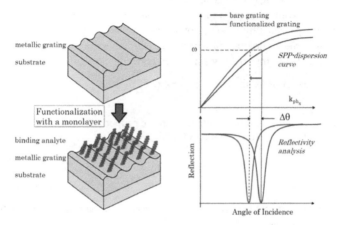

Figure 2. Surface Plasmon Resonance for sensing: the deposition of the analyte on the SPP-supporting surface results in a change of the dispersion curve and a consequent shift of the resonance dip.

Recently we have experimentally and theoretically described the effects of grating azimuthal rotation on surface plasmon excitation and propagation. More SPPs can be supported with the same illuminating wavelength [16] and sensitivity up to 1000°/RIU is achievable for the second dip in angular interrogation [17], which is one order of magnitude greater than that in a conventional configuration. On the top of that, the symmetry breaking after grating rotation makes polarization have a fundamental role on surface plasmon polaritons excitation [18] and the incident polarization must be properly tuned in order to optimize the coupling strength. This result leads to the possibility of exploiting the dependence of the optimal polarization on the resonance conditions, in order to design an innovative GCSPR configuration based on polarization interrogation in the conical mounting.

2. Nanofabrication of plasmonic gratings

2.1. Interferential lithography

Interferential Lithography (IL) is the preferred method for the fabrication of periodic patterns that must be spatially coherent over large areas [19]. It is a conceptually simple process where two coherent beams interfere in order to produce a standing wave which can be recorded over a sensitive substrate. The spatial period of the pattern can be as low as half the wavelength of the interfering light, allowing for structures down to about 100 nm using UV radiation. In Lloyd's Mirror Interferometer, one source is replaced by a mirror. Lloyd's mirror is rigidly fixed perpendicular to the surface and used to reflect a portion of incident wavefront back to the other half. The angle of interference and thus the grating periodicity are set just by rotating the mirror/substrate assembly around the point of intersection between the mirror and the substrate. Simple trigonometry guarantees that the light reflected off the mirror is always incident at the same angle as the original beam. Moreover since the mirror is in a rigid connection with the wafer chuck, vibration of the setup or wandering of the incoming beam do not affect the exposure: this results in a more stable configuration and prevents the need of phase locking systems. Furthermore, it allows to change and control the grating period without any alignment or critical adjustment between two different sources and thus the system is very convenient to calibrate and tune.

The basic principles in order to record a pattern of desired period and amplitude are the following:

- *Grating period*: the incident beam angle has to be properly adjusted in order to obtain the desired periodicity according to the following law:

$$\Lambda = \frac{\lambda}{2\sin\theta} \tag{1}$$

where Λ is the resulting spatial periodicity of the grating, λ the beam wavelength and θ the incident beam angle (see Figure 3).

- *Grating amplitude*: amplitude is controlled by varying the exposure dose. Keeping fixed the beam intensity the exposure dose can be varied by changing the exposure time. The higher the dose on the resist, the higher the amplitude that is obtained after resist development.

In our system a 50 mW Helium-Cadmium (HeCd) laser emitting TEM00 single mode at 325 nm was used as light source. After a 2 m long free-space propagation, the expanded laser beam illuminates both the sample and the perpendicular mirror. The designed sample holder offers translational and rotational degrees of freedom, while a rigid mechanical connection between the mirror and the sample-chuck prevents phase distortion. The possibility to translate the sample stage in two directions allows a fine positioning of the system in the zone where the Gaussian beam distribution reaches its maximum and the best conditions in terms of beam intensity, uniformity and spatial coherence are achieved. On the other hand the sample stage rotation around vertical axis, with 8 mrad resolution, allows a fine setting of fringes periodicity.

Metallic gratings can be processed through a sequence of steps [20]: resist spinning, IL exposure, resist devolpment, master replication, methal evaporation over the replica. Exposures can be performed over silicon samples of 2 x 2 cm² surface area. Silicon wafers have to be cleaned and pre-baked for 25 minutes at 120°C. A bottom anti-reflection layer can be also spun by spin coating in order to improve the exposure quality. Thus, a 120 nm thick film of photoresist S1805 (Microposit, Shipley European Limited, U.K.) and Propylene glycol monomethyl ether-1,2-acetate (PGMEA) solution (ratio 2:3) can be spun at the spinning rate of 3000 RPM for 30 seconds. After the exposure, a developing solution of MICROPOSIT MF-321 or MF-319 can be used.

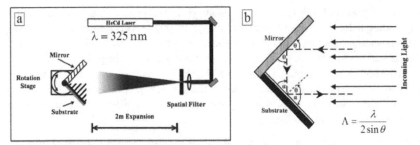

Figure 3. a) Scheme of the experimental IL setup. b) Detail of the sample holder configuration during the exposure.

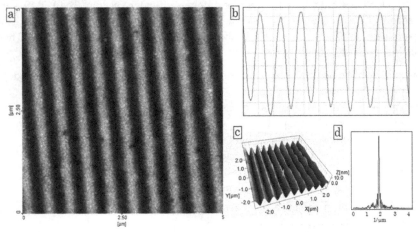

Figure 4. AFM analysis of a gold sinusoidal grating fabricated by interferential lithography of resist S1805 (period 505 nm, amplitude ~30 nm): top view (a), profile along the grating vector direction (b), 3D reconstruction (c), Fourier transform (d).

After the exposed resist has been developed, the result is a photonic crystal: a dielectric periodic surface which has no plasmonic feature yet. In order to realize a plasmonic crystal, the dielectric grating must be coated with a metallic layer of proper thickness, usually noble metals (silver or gold), for example by thermal evaporation. Common recipes consists in a silver layer with optimized thickness coated by a gold layer in order to prevent oxidation, or as an alternative an optimized gold layer over an adhesion layer of chrumium.

2.2. Soft-lithography replica

Soft-lithography is a useful technique in order to replicate pre-fabricated patterns with a nanometric resolution. It consists in making at first the negative replica of a pattern master onto a siliconic polymer and then in imprinting this pattern onto a photopolymeric substrate that cures when exposed to UV light.

In principle, a plasmonic grating could be obtained, as explained in the previous section, by simply coating a developed resist surface with specific noble metals, such as silver and gold over few nanometers of chromium or titanium as adhesion layers. However, many bio-molecular surface functionalization methods involve the use of organic solvents that may attack the photoresist pattern. To avoid this problem, a replica molding approach can be adopted in order to produce thiolene copies of the gratings [21]. Commercial thiolene resin (Norland Optical Adhesive – NOA61 [22]) can be chosen in order to exploit its relatively good resistance to organic solvents.

Resist patterns are usually replicated onto the thiolene resist film supported on microscope glass slides using polydimethylsiloxane (PDMS) molds. The latter are obtained by replicating the resist grating masters using RTV615 silicone. Base and catalyst of the two component silicone are mixed 10:1 ratio and degassed under vacuum. The PDMS is then cast against the resin masters and cured at 60°C, well below the resist post-exposure bake temperature (115° C) in order to prevent the resist pattern from distortion and after 2 hours the PDMS is peeled off from the resist master.

In order to obtain rigid and stable supports, the PDMS molds are bound to glass slides by exposing the flat backside of the PDMS mold and the glass slide to oxygen plasma before contacting them. The PDMS mold is then used to UV imprint the initial pattern onto a drop of NOA61 resin dripped on top of a glass slide by just slightly pressing the mold onto the liquid resin and exposing it to the UV light (365 nm) of a Hg vapor flood lamp at the proper distance and for enough time. After removing the PDMS mold, the replicated NOA61 grating can be coated with metal layers by thermal evaporation.

3. Grating-coupled Surface Plasmon Resonance in the conical mounting: simulation and results

3.1. Chandezon's method

A rigorous approach is necessary in order to exactly solve the problem of a monochromatic plane wave incident on a patterned surface and to simulate the optical response of such multilayered patterned structures. In the past decades, several numerical methods have been developed in order to compute the optical response of periodically modulated multilayered stacks. Among these algorithms, Chandezon's method (hereafter the C-method) revealed itself as one of the most efficient techniques for a rigorous solution of smooth grating diffraction problem. The algorithm is a curvilinear coordinate modal method by Fourier expansion that has gone through many stages of extension and improvement. The original theory was formulated by J. Chandezon et al. [23][24], for uncoated perfectly conducting gratings in classical mountings. Various author extended the method to conical diffraction gratings [25]. G. Granet et al. [26], T.W. Preist et al. [27] and L. Li et al. [28] allowed the various profile of a stack of gratings to be different from each other, although keeping the same periodicity. Solving the vertical faces case in a simple manner, J.P. Plumey et al. [29] have showed that the method can be applied to overhanging gratings.

In the numerical context, L. Li improved the numerical stability and efficiency of the C-method [30][31].

3.2. The coordinate transformation

The setup is as it follows: the grating profile is described by a differential curve $y = s(x)$, periodic in the x-direction with periodicity Λ. The basic feature of the C-method consists in the choice of a non-orthogonal coordinate system that maps the interfaces between different media to parallel planes:

$$u = y - s(x)$$
$$v = x \tag{2}$$
$$w = z$$

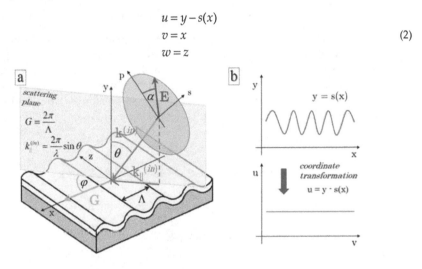

Figure 5. a) Scheme of the reference frame x-y-z and incidence angles θ (polar angle), φ (azimuth) and α (polarization). b) Pictorial description of the coordinate change effect under Chandezon's method.

Since it is a global coordinates transformation, Maxwell's equations covariant formalism is necessary. In a source-free medium the time-harmonic Maxwell equations in term of the covariant field component can be written as [32]

$$\partial_i \sqrt{g} g^{ij} E_j = 0 \tag{3}$$

$$\frac{1}{\sqrt{g}} \varepsilon^{ijk} \partial_j E_k = i\mu \frac{\omega}{c} g^{ij} H_j \tag{4}$$

$$\partial_i \sqrt{g} g^{ij} H_j = 0 \tag{5}$$

$$\frac{1}{\sqrt{g}} \varepsilon^{ijk} \partial_j H_k = -i\varepsilon \frac{\omega}{c} g^{ij} E_j \tag{6}$$

ε^{ijk} being the completely antisymmetric Levi-Civita tensor[1]. The parameter g is the determinant of the contravariant metric tensor g^{ij}, ε and μ are respectively the dielectric permittivity and the magnetic permeability of the medium. The covariant form[2] g_{ij} is the following:

$$g = \begin{pmatrix} 1+s'^2 & s' & 0 \\ s' & 1 & 0 \\ 0 & 0 & 1 \end{pmatrix} \tag{7}$$

In this case $g = 1$, since the coordinate transformation conserves the volume.

The reference frame is the following: the y-axis is perpendicular to the grating plane and the grating vector G is oriented along the x-axis positive direction. The scattering plane is perpendicular to the grating plane and forms an azimuth angle φ with the grating symmetry plane (see Figure 5.a). This reference frame has been chosen in order to simplify the coordinate transformation and the form of the metric tensor after an azimuthal rotation: under this convention, the grating is kept fixed and the scattering plane azimuthally rotates.

The incident wavevector $k^{(in)}$ is given by:

$$k^{(i)} = \frac{\omega}{c}\left(\sin\theta\cos\varphi, -\cos\theta, \sin\theta\sin\varphi\right) = \left(\alpha_0, -\beta_0, \gamma\right) \tag{8}$$

and the wavevector of the n-th diffraction order has the following form:

$$k^{(n)} = \left(\alpha_n, \beta_n, \gamma\right) \tag{9}$$

where

$$\begin{aligned} \alpha_n &= \alpha_0 + n \cdot G \\ \beta_n &= \sqrt{k^{(i)2} - \alpha_n^2 - \gamma^2} \end{aligned} \tag{10}$$

With these definitions and after some algebraic manipulations, curl equations can be rearranged into the following system of differential equations in the tangential components (H_x, H_z, E_x, E_z) unknown:

$$\frac{\partial H_x}{\partial u} = i\omega\varepsilon E_z + \frac{\partial}{\partial x}\left(\frac{s'}{1+s'^2}H_x\right) + \frac{\partial}{\partial x}\left(\frac{1}{1+s'^2}\frac{i}{\omega\mu}\frac{\partial E_z}{\partial x} + \frac{1}{1+s'^2}\frac{\gamma}{\omega\mu}E_x\right) \tag{11}$$

---

[1] ε^{ijk} equals 1 if (i,j,k) is an *even* permutation of (1,2,3), -1 if it is an *odd* permutation, 0 if any index is repeated.

[2] $g_{ij} = \Lambda^{i'}_i\Lambda^{j'}_j\delta_j$ where δ_j is Kronecker's delta, Λ^i_j is the Jacobi's matrix associated to the coordinate transformation: $\Lambda^i_j = \partial x^i/\partial x^j$. The metric tensor g_{ij} allows calculating the covariant components v_i of a vector v from the contravariant ones v^j and vice-versa: $v_i = g_{ij}v^j$, $v^i = g^{ij}v_j$.

$$\frac{\partial E_z}{\partial u} = \frac{s'}{1+s'^2}\frac{\partial E_z}{\partial x} + \frac{1}{1+s'^2}\frac{\gamma}{\omega\varepsilon}\frac{\partial H_z}{\partial x} + \frac{1}{1+s'^2}\left(i\omega\mu - i\frac{\gamma^2}{\omega\varepsilon}\right)H_x \qquad (12)$$

$$\frac{\partial H_z}{\partial u} = \frac{s'}{1+s'^2}\frac{\partial H_z}{\partial x} - \frac{1}{1+s'^2}\frac{\gamma}{\omega\mu}\frac{\partial E_z}{\partial x} - \frac{1}{1+s'^2}\left(i\omega\varepsilon - i\frac{\gamma^2}{\omega\mu}\right)E_x \qquad (13)$$

$$\frac{\partial E_x}{\partial u} = -i\omega\mu H_z + \frac{\partial}{\partial x}\left(\frac{s'}{1+s'^2}E_x\right) - \frac{\partial}{\partial x}\left(\frac{1}{1+s'^2}\frac{i}{\omega\varepsilon}\frac{\partial H_z}{\partial x} + \frac{1}{1+s'^2}\frac{\gamma}{\omega\varepsilon}H_x\right) \qquad (14)$$

where we used the following relation:

$$\frac{\partial F_i}{\partial z} = i\gamma F_i \qquad (15)$$

where F_i may stand for (H_x, H_z, E_x, E_z). In fact, since the grating vector has no components in the z-direction, the z-component γ of the incident momentum is conserved and is a constant parameter of the problem.

3.3. Numerical solution and boundary conditions

As suggested by Bloch-Floquet's theorem [33], thanks to the periodicity of the media in the x-direction, a generic field can be expanded in pseudo-Fourier series:

$$F(x,u,z) = e^{i\gamma z}\sum_{m=-\infty}^{+\infty} F_m(u)e^{i\alpha_m x} \qquad (16)$$

where $F_m(u)$ is a function periodic in x. Likewise, also the profile functions $C(x)$ and $D(x)$ are periodic in x and can be expanded as well. Laurent's rule can been applied for the Fourier factorization, assuming the continuity of profile function derivative [34]. In the case of sharp edges instead, equations should be rearranged in order to make the inverse-rule factorization applicable [35]. Since in our case of interest the grating profile is a regular function, we can apply Laurent's factorization. After truncation to a finite order N [36], the problem consists in the numerical solution of a system of $8N+4$ first order differential equations in each medium:

$$-i\frac{\partial U}{\partial u} = T U \qquad (17)$$

The problem is led to the diagonalization problem of the matrix T and the solution U^j in the j-th medium can be expressed as a function of eigenvectors V_q^j and eigenvalues λ_q^j of the matrix T:

$$U^j(u) = \sum_q b_q^j V_q^j e^{i\lambda_q^j u} \qquad (18)$$

where b_j^q are the weights of the corresponding eigenmodes in the expansion. Thereafter boundary conditions in each medium must be imposed: continuity of the tangential components at each interfaces and outgoing-wave conditions in the first (air) and last (substrate) media [24]. This leads to a system of 8N+4 equations in 8N+4 unknowns that is now possible to be solved numerically. Once the mathematical problem has been solved, the electromagnetic fields can be computed inside each j-th medium:

$$F^j\left(x,u,z\right)=e^{i\gamma z}\sum_{m=-N}^{+N}\sum_{n=-N}^{+N}F_{mn}^j e^{i\lambda_n^j u}e^{i\alpha_m x} \tag{19}$$

where F^j stands for (H_x, H_z, E_x, E_z), and F_{mn} are the j-field mn-Fourier weight. Transversal components (H_y, E_y) are calculated from Maxwell's equations as a function of the tangential components. By applying the metric tensor to covariant components F_j, the contravariant components F^j, which represent the physical fields, can be obtained. For incident wavelength λ, polar and azimuth angles (θ, φ) and polarization α, the implemented algorithm yields the spatial dependency of the diffracted fields everywhere in space for the modelled grating stack. An estimation of reflection and transmission coefficients for the different diffraction orders, in particular transmittance and reflectivity values (0-diffraction orders) can be obtained. By analysing both real and imaginary parts of eigenvalues λ_j, it is possible to distinguish between propagating and evanescent modes, which respectively contribute to the far-field and the near-field parts of the solutions. Thus by selecting only the evanescent contributions in the configurations where SPPs are excited, it is possible to describe the localized plasmonic fields of the excited modes.

3.4. Numerical results and comparison with experimental data

3.4.1. Grating optimization

C-Method has been implemented in MATLAB® code in order to compute the optical response of sinusoidal metallic gratings and simulations have been performed [37] with truncation order $N = 6$. The stack reproduces the typical multilayer bi-metallic gratings fabricated in laboratory by laser interference lithography in the so called Lloyd's configuration (section 2.1). Simulation code provides a fundamental tool in order to select the grating geometries that exhibit the best optical features and provide to nanofabrication the optimal windows of process for the realization of optimized optical components. Silver thickness must be properly chosen in order to optimize the plasmonic response of the structure and the optimal thickness strictly depends on the amplitude of grating modulation. Simulations have been performed at the incident wavelength $\lambda = 632$ nm for several values of the profile amplitude in the range 20-60 nm and for varying silver thickness in the range 10-80 nm, in the case of a sinusoidal grating with fixed period $\Lambda = 500$ nm, gold thickness 7 nm and a fixed chromium adhesion layer of 5 nm over a NOA61 substrate.

As Figure 6.a shows, for each amplitude value a silver thickness exists that optimizes the coupling of incident light with SPP modes, i.e. that minimizes the depth of the reflectivity dip. Figure 6.b shows some examples of reflectivity curves in angular scan for optimal combinations of profile amplitude and silver thickness. For increasing amplitude of the grating profile, the optimal silver thickness decreases (Figure 6.a) and the corresponding reflectivity curve becomes broader (Figure 6.b). This result seems to suggest the choice of a shallow grating modulation with the evaporation of the corresponding optimal thickness of silver: for an amplitude $A = 30$ nm the optimal silver thickness is around 80 nm. On the other hand, the coating with a great quantity of metal could affect the preservation of the original pattern and cause lack in accuracy of the final profile. The evaporation of about 40 nm of silver instead, could assure the control of the grating profile and at the same time a reasonable value for the reflectivity-dip FWHM (Full Width at Half Maximum).

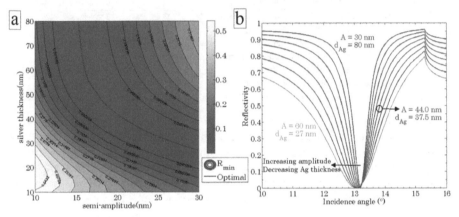

Figure 6. a) Reflectivity minimum as a function of profile semi-amplitude and silver-film thickness for fixed period 500 nm, fixed gold-film thickness 7 nm, incident wavelength 632 nm (sinusoidal profile). Superimposed blue line: optimal configurations. b) Simulated reflectivity in angle scan for values of amplitude and silver thickness along the optimal configuration line in (a): amplitude range 30-60 nm, step 4 nm.

3.4.2. Comparison with experimental data

Numerical results have been compared with experimental data from reflectivity analysis of the corresponding fabricated sample. The simulated stack reproduces the multilayer structure of the considered grating: air (upper medium), Au (8 nm), Ag (35 nm), Cr (9 nm), photoresist (70 nm), Si (substrate). For each layer the optical constants (refractive index n, extinction coefficient k) have been extrapolated from ellipsometric analysis and have been inserted into the code. From AFM analysis, the grating profile results sinusoidal with period 505 nm and peak-to-valley amplitude 26 nm.

As figure 7 shows, numerical estimation of grating reflectivity well fits experimental data within instrumental errors (~2%). Reflectivity measurements have been performed by means of the monochromatized 75 W Xe-Ne lamp of a spectroscopic ellipsometer VASE (J. A. Woollam), with angular and spectroscopic resolution respectively 0.001° and 0.3 nm

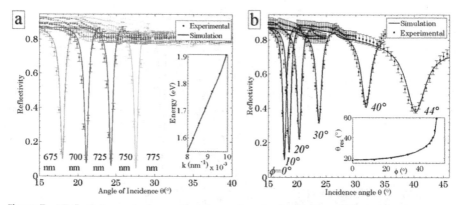

Figure 7. a) Reflectivity spectra for angular interrogation at null azimuth $\varphi = 0°$ for incident wavelengths λ in the range 675-775 nm with step 25 nm, p-polarization ($\alpha = 0°$). Experimental data points and simulation results (solid lines). In the inset graph: SPP dispersion curve $\omega-k$. b) Reflectivity spectra for angular interrogation, variable azimuth φ for the $\lambda = 675$ nm, p-polarization. In the inset graph: resonance angle θ_{res} as a function of the azimuth angle φ: experimental points and simulation curve (solid line).

Figure 7.a shows reflectivity spectra for angular interrogation at null azimuth $\varphi = 0°$ for incident wavelengths λ in the range 675-775 nm with step size 25 nm, p-polarization ($\alpha = 0°$). In angular interrogation, the reflectivity dip shifts towards greater resonance angles for increasing wavelength. Numerical results perfectly reproduce experimental data trends and from dip position it is possible to reconstruct SPP dispersion (Figure 7.a, inset graph).

Figure 7.b exhibits reflectivity spectra in angular interrogation for the azimuth values $\varphi = 0°$, 10°, 20°, 30°, 40°, 44°, incident wavelength $\lambda = 675$ nm and p-polarization ($\alpha = 0°$).

4. Sensitivity enhancement by azimuthal control

The analysis of the wavevector components allows a description of double SPP excitation using the schematic shown in Figure 8. The excitation of SPPs on a grating is achieved when the on-plane component $k_{||}^{(in)}$ of the incident light wavevector and the diffracted SPP wavevector k_{SPP} satisfy the momentum conservation condition:

$$k_{SPP} = k_{||}^{(in)} + n \cdot G$$
$$k_{SPP} = \frac{2\pi}{\lambda} \sin\theta_{res} \cdot (1,0) - \frac{2\pi}{\Lambda}(\cos\varphi, \sin\varphi) \tag{20}$$

where θ_{res} is the resonance polar angle, φ is the azimuth angle, Λ is the grating pitch, λ is the illuminating wavelength. Only the first diffraction order ($n = -1$) is considered because in our cases of interest, grating momentum G is always greater than k_{SPP}.

All quadrants of the circle in Figure 8 can be explored for SPP excitation as long as momentum conservation is satisfied. For symmetry reason, only k_y positive half space is considered. Soving eq. (20) in the resonance angle unknown, the following expression is obtained as a function of the azimuth angle φ.

$$\theta_{\mp} = \arcsin\left(\frac{\lambda}{\Lambda}\cos\varphi \mp \sqrt{M(\lambda)^2 - \left(\frac{\lambda}{\Lambda}\sin\varphi\right)^2}\right) \tag{21}$$

where $M(\lambda) = k_{SPP}/(2\pi/\lambda)$.

The largest circle in the k-space represents equi-magnitude G vectors at different azimuthal orientation. The two smaller circles represent all possible k_{SPP} vectors with equal magnitude respectively before and after surface functionalization and whose modulus, for shallow gratings, can be approximated by [2]:

$$k_{SPP} = \frac{2\pi}{\lambda}M(\lambda) = \frac{2\pi}{\lambda}\sqrt{\frac{\varepsilon_m\varepsilon_{eff}}{\varepsilon_m + \varepsilon_{eff}}} \tag{22}$$

where ε_m and ε_{eff} are the dielectric permittivity of the metal and the dielectric side respectively. After functionalization, SPP modulus increases because of the small increase in ε_{eff} due to the surface coating.

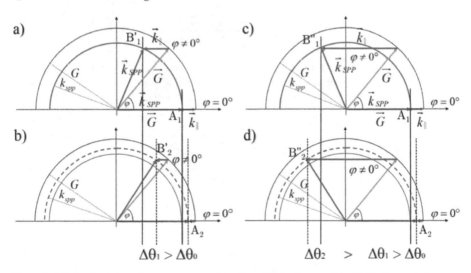

Figure 8. Schematic picture of wave-vector combination at SPP resonance ($n = -1$). The large circle represents equi-magnitude G vectors. The smaller circles represent equi-magnitude k_{SPP} vectors before (a, c), and after (b, d) surface functionalization for the first (a, b) and the second (c, d) SPP excitation. The blue arrows represent the photon on-plane wavevector and the red arrows represent the SPP propagation direction. The letters A and B represent the vector with azimuthal rotation $\varphi = 0°$ and $\varphi \neq 0°$ respectively.

The dashed line at the tip of the circle of radius G represents the x-component of the photon wave-vector $k_{\parallel}^{(in)}$, the only component that participates in SPP excitation. The line is scaled linearly in $\sin\theta$ so that the full length of the line at the incident angle θ of 90° corresponds to the maximum value of $k_{\parallel}^{(in)}$. The intersections of the $k_{\parallel}^{(in)}$ dashed horizontal line with the smaller k_{SPP} circle determine the conditions for which eq. (20) is satisfied and allows the identification of both incident angle θ_{res} for SPP resonance excitation and SPP propagation direction β . We consider first the case of the uncoated sample - the smallest of the semicircles. For example, point B on the G circle is identified by the azimuthal angle φ and allows the excitation of SPP at two possible conditions: B_1' and B_1'', with β^- and β^+ respectively. Within the double SPP range (point B), a small increment in wavelength makes the points B_1' and B_1'' merge to form a very broad. On the contrary, at $\varphi = 0°$ (point A_1) it is clear that the photon wave-vector can intersect the SPP circle only in the first quadrant but not the second, thus exciting only a single SPP for each wavelength. The same argument is applicable for light exciting SPP on the functionalized sample. Due to the larger k_{SPP}, a different excitation condition is expected. The intersection points changes from B_1' and B_1'' to B_2' and B_2'' and from A_1 to A_2. The sensitivity of the GCSPP is higher at high azimuthal angles because the condition for double SPP excitation around the circumference of the k_{SPP} circle generates a shifts in k-space between points B_1' and B_2', which is much larger than that between points A_1 and A_2 provided by a single SPP excitation condition for $\varphi = 0°$. The estimated refractive index sensitivity S of this configuration can be defined as:

$$S = \frac{\partial\theta}{\partial n} = \frac{\partial\theta}{\partial k_{\parallel}^{(in)}} \frac{\partial k_{\parallel}^{(in)}}{\partial k_{SPP}} \frac{\partial k_{SPP}}{\partial n} \tag{23}$$

In order to calculate S, we assumed the rippling amplitude A of the grating is so shallow ($A/\Lambda \sim 0.05$ in our case of interest) that the dispersion curve of SPPs traveling at the metal-dielectric interface of a grating can be approximated by the case of a flat sample (eq. (22)). The analytical expression for the sensitivity in angular interrogation can be found as:

$$S = -\frac{1}{\cos\theta_{res}} \left(\frac{M}{n_0}\right)^3 \frac{\sqrt{\dfrac{1}{\Lambda^2} + \dfrac{\sin^2\theta_{res}}{\lambda^2} - \dfrac{2\cos\varphi\sin\theta_{res}}{\Lambda\lambda}}}{\dfrac{\cos\varphi}{\Lambda} - \dfrac{\sin\theta_{res}}{\lambda}} \tag{24}$$

where M is defined in eq. (22), n_0 is the refractive index of the surrounding dielectric medium, the resonance angle θ_{res} is given by eq. (21) for fixed λ.

The functional behavior of sensitivity for the first and second SPP dip is shown in Figure 9 for a typical wavelength of $\lambda = 606$ nm. Both sensitivities diverge when φ approaches its critical value φ_{MAX}, i.e. the maximum φ angle that supports SPP resonances. In this configuration, incident photon momentum is tangential to the k_{SPP} circle and its length equals the x-component of the grating momentum so that the denominator in eq. (24) becomes null. Another condition for second dip sensitivity singularity is when φ approaches the critical azimuthal angle, φ_c, necessary to excite double SPP resonances, namely when the full length of the incident photon momentum is required to intersect the edge of the k_{SPP} circle. Since the incident angle $\theta = 90°$, $\cos\theta_{res}$ in the denominator of the first term eq. (24) approaches 0 and S diverges.

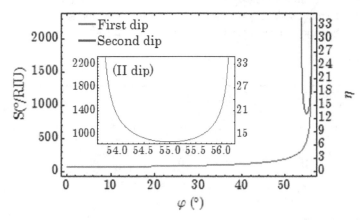

Figure 9. Sensitivity S_θ as a function of grating azimuthal angle φ for the two dips in SPR sensor with angular interrogation. The right y-scale refers to sensitivity values normalized to the first dip sensitivity at $\varphi = 0°$: $\eta = S_\theta(\varphi)/ S_\theta(\varphi = 0°)$.

Although azimuth φ values close to the critical values provides a great enhancement in S (up to 2400°/RIU, 35 times higher than the case $\varphi = 0°$), these configurations should be avoided because of experimental limits. For $\varphi \cong \varphi_c$, θ_{res} becomes large (>70-80°) and broad, becoming impossible to resolve the SPP minimum. In addition, when $\varphi \cong \varphi_{MAX}$ the two resonance dips merge into a single broad dip which makes the two minima hardly distinguishable. Thus only limited parts of the azimuthal angular range are suitable to enhance sensitivity significantly. The best conditions correspond to the middle of the "U-shape" of second dip functional behavior (Figure 9) where the sensitivity ranges from 900°/RIU to 1100°/RIU, about 15 times higher than $\varphi = 0°$ whose value is 67°/RIU. The sensitivity computed for the first SPP dip is smaller all over this range but it still provides values of the order of 500°/RIU.

4.1. Application to functionalization detection

Sensitivity enhancement with azimuthal rotation has been tested with a C12-functionalization of a bimetallic (37Ag/7Au) grating, period 487 nm, amplitude 25 nm. Reflectivity analyses performed before and after C12-coating are collected in Figure 10.a-b. The figure shows the experimental evidence of the azimuth rotation, where reflectivity spectra are reported at different wavelengths. As a reference Figure 10.a reports the reflectivity spectra using the conventional GCSPR configuration with $\varphi = 0°$. In this configuration, the difference in the reflectivity minima resonance angle θ_{res} before and after

C12 is typically less than 0.05°. On the contrary after an azimuthal rotation of the grating to an angle φ of about 60°, larger angular differences can be observed between the reflectivity dips (Figure 10.b). The resonance shift can reach values up to 3.1° as for the incident wavelength λ = 618 nm. By increasing wavelength from 606 to 618 nm, the two resonance dips in the reflectivity spectra get closer while resonance shift increases from 1.8° to 3.1°, until the two resonances merge together into a single broad dip at 620 nm.

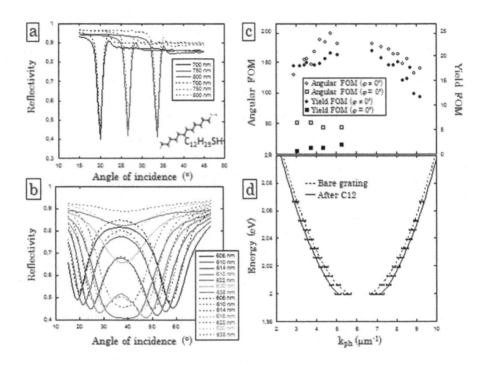

Figure 10. Comparison of SPR spectrum between uncoated (dashed line) and C12-coated (solid line) Au grating. (a) Grating with φ = 0°, for incident wavelength λ = 700-800 nm, step 50 nm, and (b) φ = 60°, (2 SPPs by single wavelength excitation condition), for incident wavelength λ = 606-638 nm, step 4 nm. c) Figure of merit for yield ($\cdot$, ■) and angular (o, ▢) interrogation of the SPR reflectivity minima. (d) Energy dispersion curve for the $k_{||}^{(in)}$ necessary for SPP excitation before (dashed line) and after (solid line) C12 SAM functionalization.

The experimental determinations of the wavevector $k_{||}^{(in)}$ of the incoming light necessary for the SPP excitation have been successfully fitted with the help of eq. (20), using as fitting parameters only the effective index of refraction and azimuthal angle (see Figure 10.d). After surface functionalization we have determined a total increment in index of refraction equal to Δn = 0.00357 ± 0.00007 RIU. This is a value that agrees with the estimation of RIU performed for effective refractive index change generated by a full surface coverage of close-packed self assembled C12 molecules, considering their length 1.46 nm with dielectric constant ε_{C12} = 2.12. Also the theoretical sensitivity determination (Figure 9) is in good agreement with the experimental determination obtained as a ratio between Δn, the angular differences before and after C12 coating and the Δn determination. The average experimental sensitivity is of the order of 520°/RIU for the first dip and reaches maximum values of 857°/RIU for the second dip. The final error of the refractive index determination has been estimated on the basis of chi-square minimization based on *a priori* determination of the SPP angular position. The reflectivity minima of this preliminary data set have been determined with a typical uncertainty of 0.07°. However the *a posteriori* determination of the SPP angular average deviation with respect to the dispersion curve best fit is much smaller, on the order of 0.015°, as can be confirmed by a simple graphical inspection. Taking into account this more realistic value for the final evaluation of the uncertainty, the value of 10^{-5} RIU. However, because our experimental system has an instrumental resolution of 0.001°, we believe it will be possible to greatly decrease the present angular uncertainty by increasing the statistical signal-to-noise ratio and using appropriate algorithms for data analysis. We expect that experimental uncertainties of Δn on the order of $5 \cdot 10^{-7}$ RIU is achievable. In order to better describe the detection improvement given by azimuthal rotation, we have also measured the typical figure of merit for angular and yield interrogation respectively defined as:

$$FOM_\theta = \frac{S_\theta}{\Delta\theta_{FWHM}}$$
$$FOM_Y = \frac{Y_{coat} - Y_{uncoat}}{Y_{uncoat}} \qquad (25)$$

where $\Delta\theta_{FWHM}$ is the angular full width at half maximum of the reflectivity minima, whereas Y_{coat} and Y_{uncoat} are the minimum yield of the reflectivity spectra collected before and after C12 functionalization at SPR resonances.

Figure 10.c shows the angular and yield *FOM* at zero and after the azimuthal rotation for all the reflectivity spectra. It clearly appears an enhancement of both the figures of merit after the azimuthal rotation that amounts up to a factor 4 and 10 for the angular and yield *FOM* respectively. This means that the distance between two dips, before and after the functionalization, scales with a factor greater than the enlargement of the reflectivity dip width. Moreover, it is worth to note that the reflectivity yield is even more sensitive than the angular position. It clearly appears from the reflectivity spectra of Figure 10.b that whereas the minimum yield between coated and uncoated are almost the same for zero azimuth, it changes dramatically after azimuthal rotation. Finally, we note that both the angular and the yield *FOM* have similar functional behavior: they increase approaching the condition of two dips merging when $\beta_+ = \beta_- = 90°$.

4.2. Application to solution analysis

A microfluidic cell was fabricated in PDMS by soft-lithography technique and embodied to the metallic grating. The considered metallic gratings exhibits a sensitivity $S_\phi(0°)$ = 64.9°/RIU for λ = 840 nm (Figure 11.a-b). As demonstrated in the previous section, angular sensitivity S_θ can be improved just with an azimuthal rotation of the grating plane. For a same fixed concentration of 10g(Nacl) in 200ml(water), corresponding to a refractive index variation with respect to distilled water Δn = 9.6 · 10^{-3} [38], reflectivity spectra have been collected for increasing azimuth angle φ up to 43° at the same incident wavelength. As Figure 11.c shows, resonance angle shift increases monotonically in modulus with increasing azimuth angle from $\Delta\theta(0°)$ = 0.59° to $\Delta\theta(43°)$ = 5.57° corresponding to a sensitivity enhancement $\Delta\theta(43°)/\Delta\theta(0°)$ = 9.5 of almost one order of magnitude: $S_\phi(43°)$ = 616.8°/RIU.

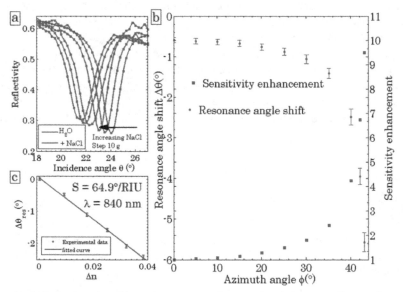

Figure 11. (a) Reflectivity in polar angular scan at incident wavelength 840 nm, null-azimuth, *p*-polarization, for increasing NaCl concentration in water solution: 0-50 g, step 10 g, in 200 ml of water. (b) Resonance angle shift as a function of refractive index variation of water solution with increasing NaCl concentration. (c) Resonance angle shift and sensitivity enhancement as a function of azimuth angle in the range 0-43°. NaCl concentration: 10g in 200ml of water (Δn = 9.6 · 10^{-3}).

This sensitivity-enhancement technique with azimuthal rotation has been exploited to reveal a lower sodium-chloride concentration than the previous one. A mass of 0.56 g has been dissolved into 200 ml of water and the resulting refractive index change is Δn = 5.4 · 10^{-4}. With this solution flowing through the microfluidic cell, reflectivity measurements have been performed for increasing grating rotation (Figure 12). The inset picture in Figure 12 shows the reflectivity spectra before and after salt dissolution for null azimuth and after grating azimuthal rotation up to 43°. While in the classical mounting, i.e. null azimuth, this variation is not detectable within the experimental error, in conical mounting instead grating azimuthal rotation increases SPR sensitivity and thus the system allows revealing the resonance angle shift. Resonance dips become broader when azimuthal angle increases, however the resonance angle shift $\Delta\theta$ scales with a factor greater than the enlargement of the dip full width half maximum $\Delta\theta_{FWHM}$. Thus the angular figure of merit FOM_θ increases and the detection improvement by azimuthal rotation is preserved. In this case we have $FOM_\theta(43°)/ FOM_\theta(0°)\cong4.3$.

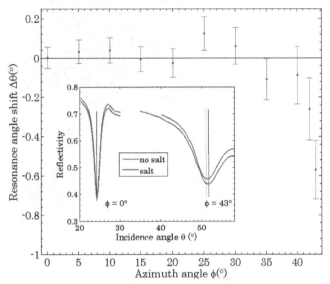

Figure 12. Resonance angle shift as a function of azimuth angle for sodium-chloride concentration: 0.56g/200 ml(water), $\Delta n = 5.4 \cdot 10^{-4}$. In the inset graph: reflectivity curve before (blue) and after (red) NaCl dissolution into water, angular scan for azimuth values $\varphi = 0°$ and 43°.

5. Innovative SPR with polarization modulation

In previous sections we experimentally and theoretically described the effects of grating azimuthal rotation on surface plasmon excitation and propagation. More SPPs can be supported with the same illuminating wavelength and a sensitivity enhancement of al least one order of magnitude is achievable than that in the conventional configurations. Here we consider how the symmetry breaking with grating rotation makes polarization have a fundamental role on surface plasmon polaritons excitation. As Figure 7.b shows, p-polarization becomes less and less effective for increasing azimuthal rotation and the reflectivity depth decreases. If the polar angle is fixed at resonance and a polarization scan is performed, the minimum of reflectivity R_{min} exhibits a harmonic dependency on the incidence polarization α with a periodicity of 180° (Figure 13):

$$R_{min} = f_0 - f_1 \cos(2\alpha + \alpha_0) \tag{26}$$

where f_0, f_1 and α_0 are fitting parameters that depend on the incidence angles θ and φ, incident wavelength λ and on the optical properties of the stack (thickness and dielectric permittivity of each layer). While for null azimuth $\alpha_0 = 0°$ and p-polarization is the most effective for SPP excitation, after grating rotation the phase term α_0 is non null and strictly depends on the incidence angles. By assuming that only the electric field component lying on the grating symmetry plane is effective for SPP excitation, on the basis of a vectorial approach, an analytical expression can be obtained [18] for the optimal polarization α_{opt} as a function of the azimuth angle φ and the resonance angle θ_{res}:

$$\tan \alpha_{min} = \tan \varphi \cdot \cos \theta_{res} \tag{27}$$

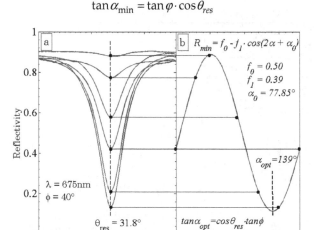

Figure 13. (a) Reflectivity for polar angle scan in the range 25°-38°, step 0.1°, at $\lambda = 675$ nm and azimuth $\varphi = 40°$ for varying incident polarization α in the range 0°-180°, step 30°. (b) Reflectivity minima as a function of polarization and fit curve.

If the grating surface is functionalized, the effective refractive index n_{eff} of the dielectric medium changes and resonance conditions are different. As a consequence of the shift in the resonance angle θ for a fixed azimuth φ, there is a shift $\Delta\alpha_0$ in the phase term α_0:

$$\Delta\alpha_0 = \frac{\partial\alpha_0}{\partial n} \Delta n \tag{28}$$

This result opens the route to a new GCSPR-configuration with polarization interrogation[39]. In this setup the grating is rotated of an azimuthal angle which is kept fixed. The illuminating wavelength is fixed and the incoming light impinges on the grating at the polar resonance angle. A rotating polarizer between source and sample-holder allows changing the polarization incident on the grating. Reflectivity data collected during a polarization scan can be fitted using eq. (26) and a variation of fitting parameters, e.g. amplitude f_1 or phase a_0, can be used in order to detect grating functionalization or for solution-concentration analysis, once the system has been properly calibrated.

A metallic grating with a period of 505 nm and amplitude of 26 nm, fabricated by interferential lithography (IL) followed by thermal evaporation of a gold (40 nm) metallic layer over 5 nm of chromium adhesion layer. Optical measurements have been performed in $\theta/2\theta$ symmetric reflectivity configuration, using the 75W Xe lamp of VASE Ellipsometer (J. A. Woollam). A self-assembled monolayer of dodecanethiol was deposited on the gold coated grating surfaces at room temperature [40].

First of all, reflectivity spectra have been collected in angular scan to identify the resonance angle position with a weighted centroid algorithm [41]. In order to exploit the shift enhancement, grating was azimuthally rotated and kept fixed at the value $\varphi = 57.8°$ wherein double SPP excitation is supported for the selected wavelength $\lambda = 625$ nm. In correspondence of the resonance angles, respectively $\theta = 30.8°$ and $\theta = 55.1°$ for the first and second dip, a polarization scan has been collected in the range 0-180°, step 10°, before and after C12 functionalization. The same analysis has been performed at the wavelength 635 nm, for the same azimuth, when the two dips merge into a single broad one centered in $\theta = 43°$.

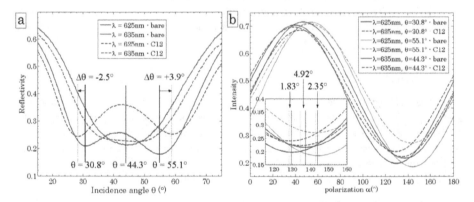

Figure 14. (a) Reflectivity for angular scan in the range 15°-75°, step 0.2°, at $\lambda = 625$ nm (blue lines) and $\lambda = 635$ nm (red lines), azimuth $\varphi = 57.8°$, incident polarization $\alpha = 140°$, before (solid lines) and after (dashed lines) functionalization with C12. (b) Polarization scan for $\lambda = 625$ nm (blue lines: I dip ($\theta = 30.8°$), green lines: II dip ($\theta = 55.1°$)), $\lambda = 635$ nm (red lines: merged dips ($\theta = 44.3°$), azimuth $\varphi = 57.8°$, before (solid lines) and after (dashed lines) C12-coating. Inset picture: phase shifts Δa_0.

By modelling the effective refractive index change Δn_{eff} with an effective medium approximation [42], it is possible to estimate the corresponding phase sensitivity S_α and moreover to calculate the refractive index resolution:

$$\sigma_{n,\alpha} = \frac{\sigma_\alpha}{S_\alpha} \qquad (29)$$

C12 has been assumed to form a monolayer 1.46 nm-thick with refractive index n=1.458 [43] and results in Δn_{eff} (625 nm) = 50.5 · 10^{-4} and Δn_{eff} (635 nm) = 48.1 · 10^{-4}. In the case of double SPP excitation (λ=625 nm), the phase shifts result $\Delta\alpha_0(I)$=1.834 ± 0.001° (first dip) and $\Delta\alpha_0(II)$= 2.353 ± 0.001° (second dip), corresponding respectively to sensitivity values $S_\alpha(I)$=363.2°/RIU and $S_\alpha(II)$= 465.9°/RIU, refractive index resolutions $\sigma_{n,\alpha}(I)$=2.7 · 10^{-6} RIU and $\sigma_{n,\alpha}(II)$=2.1 · 10^{-6} RIU. For λ=635 nm, we get S_α = 1022.7°/RIU and 2 · 10^{-6} RIU.

Thus refractive index changes of order 10^{-6} RIU are easily detectable and the resolution can be further improved to 10^{-7}-10^{-8} by reducing output noise σ or by increasing the number N of the points collected during the polarization scan.

Furthermore this technique provides a resolution at least two order greater than polar angle modulation with the same setup, which results around $\sigma_{n,\theta}$~10^{-4}-10^{-5} RIU. Moreover, while angle-modulation SPR becomes difficult near the merging dip condition, since dip position is hardly detectable, in the polarization-modulation case the analysis is still valid and it assures a greater sensitivity. On the top of that, since the output trend is a well-known function of polarization, this method assures a great accuracy on fitting parameters and their dependence on grating surface conditions provides a solution to detect and quantify surface functionalization or solution concentration. This method assures a competitive resolution down to 10^{-8} and limits the mechanical degrees of freedom just to the polarization control. The option of using an electronic modulator instead of a rotating polarizer, further assures the possibility to realize very compact, fast and low-cost high-resolution plasmonic sensors.

6. Conclusions

Plasmonic gratings have been demonstrated to assure a high-sensitive optical response to surface functionalization and to represent a promising and irreplaceable component for the realization of label-free devices for sensing purposes with considerable performance in refractive index sensitivity and resolution. The problem of designing and realizing metallic gratings for sensing applications has been studied and analysed through each step of the process-chain: simulation - nanofabrication - characterization.

Chandezon's method provides an algorithm for a rigorous and complete analysis of the diffraction problem of a multi-layered patterned. The numerical code provides a precise estimation of grating reflectivity and is an essential tool to design, for given geometry and material choice, the optimal profile that optimizes the coupling strength of incident light with surface plasmon polaritons. Simulation is essential to provide to nanofabrication the proper windows of process for the production of optimized supports.

As regards the nanofabrication of these components, interferential lithography is the preferred method to fabricate periodic pattern with a spatial coherence over large areas, while grating-replica process by soft-lithography assures the possibility of a fast and cheap throughput of perfectly replicated gratings. With respect to prism-coupling, grating nanofabrication technology assures the possibility to miniaturize and integrate the sensing components without a considerable increase of the total expenditure. Moreover, our results further highlight grating-coupling advantages rather than prism-coupling thanks to the sensitivity enhancement with azimuthal rotation. An enhancement at least one order greater than the conventional mounting has been theoretically and experimentally demonstrated.

If the azimuthal rotation of the grating support increases refractive index sensitivity, on the other hand incident polarization must be tuned in order to best couple incident light and optimize the optical response. The dependence of polarization angle on the resonance conditions suggests the exploitation of a new SPR configuration based on polarization modulation in the conical mounting. The phase term of a polarization scan, which is proportional to the optimal polarization, has been experimentally demonstrated to be a sensitive parameter for surface functionalization analysis. This innovative sensing configuration hugely simplifies the mechanical complexity of the device by limiting the degrees of freedom just to the rotating polarizer and assures the possibility of realizing very compact, fast and low-cost high-resolution plasmonic sensors based on polarization modulation.

Author details

G. Ruffato, G. Zacco and F. Romanato

University of Padova, Department of Physics "G. Galilei", Padova, Italy
Laboratory for Nanofabrication of Nanodevices (LaNN - Venetonanotech), Padova, Italy
Istituto Officina dei Materiali IOM-CNR National Laboratory, Trieste, Italy

Acknowledgement

This work has been supported by a grant from "Fondazione Cariparo" - Surface PLasmonics for Enhanced Nano Detectors and Innovative Devices (SPLENDID) – Progetto Eccellenza 2008 and from University of Padova – Progetto di Eccellenza "PLATFORM".

7. References

[1] Raether H., Surface Plasmons on Smooth and Rough Surfaces and on Gratings. Springer-Verlag, 1988.

[2] Maier S. A., editor. Plasmonics - Fundamentals and Applications. Springer, 2007.

[3] Wood R. W., On a remarkable case of uneven distribution of light in a diffraction grating spectrum, *Philos. Mag.* 4, 396-402 (1902).

[4] Cowan J. J., Arakawa E. T., Dispersion of surface plasmons in dielectric-metal coatings on concave diffraction gratings, *Zeitschrift fur Physik* 235, 97 (1970).

[5] Homola J., Yee S. S., Gauglitz G., Surface plasmon resonance sensors: review, *Sens. Actuators B* 54, 3-15 (1999).

[6] Kretschmann E., Z. *Phys.* 241, 313-324 (1971).

[7] Homola J., Surface Plasmon Resonance Based Sensors. Springer, 2006.

[8] Karlsson R., Stahlberg R., Surface plasmon resonance detection and multispot sensing for direct monitoring of interactions involving low-molecular-weight analytes and for determination of low affinities, *Anal. Biochem.* 228, 274-280 (1995).

[9] Homola J., Koudela I., Yee S. S., Surface plasmon resonance sensor based on diffraction gratings and prism couplers: sensitivity comparison, *Sens. Actuators B* 54, 16-24 (1999).

[10] Hoa X. D., Kirk A. G., and Tabrizian M., Towards integrated and sensitive surface plasmon resonance biosensors: a review of recent progress, *Biosens. Bioelectron.* 23, 151-160 (2007).

[11] Yoon K. H., Shuler M. L., and Kim S. J., Design and optimization of nano-grating surface plasmon resonance sensors, *Opt. Express* 14, 4842-4249 (2006).

[12] Unfricht D. W., Colpitts S. L., Fernandez S. M. and Lynes M. A., Grating-coupled surface plasmon resonance: a cell and protein microarray platform, *Proteomics* 5, 4432-4442 (2005).

[13] Dostalek J., Homola J., Miler M., Rich information format surface plasmon resonance biosensor based on array of diffraction gratings, *Sens. Actuators B* 107, 154-161 (2005).

[14] Alleyne C. J., Kirk A. G., McPhedran R. C., Nicorovici N. A. P., and Maystre D., Enhanced SPR sensitivity using periodic metallic nanostructures, *Opt. Express* 15, 8163-8169 (2007).

[15] Telezhnikova O., Homola J., New approach to spectroscopy of surface plasmons, *Opt. Lett.* 31, 3339-3341 (2006).

[16] Romanato F., Lee K. H., Kang H. K., Wong C. C., Zong Y. and Knoll W., Azimuthal dispersion and energy mode condensation of grating-coupled surface plasmon polaritons, *Phys. Rev. B* 77, 245435-245441 (2008).

[17] Romanato F., Lee K. H., Kang H. K., Ruffato G. and Wong C. C., Sensitivity enhancement in grating coupled surface plasmon resonance by azimuthal control, *Opt. Express* 17, 12145-12154 (2009).

[18] Romanato F., Lee K. H., Ruffato G. and Wong C. C., The role of polarization on surface plasmon polariton excitation on metallic gratings in the conical mounting, *Appl. Phys. Lett.* 96, 111103 (2010).

[19] Maldovan M., Thomas E. L., Periodic Materials and Interference Lithography. Wiley-VCH Verlag GmbH & Co. KGaA, 2008.

[20] Zacco G., Romanato F., Sonato A., Sammito D., Ruffato G., Morpurgo M., Silvestri D., Carli M., Schiavuta P. and Brusatin G., Sinusoidal plasmonic crystals for bio-detection sensors, *Microelectron. Eng.* 88, 1898-1901 (2011).

[21] Cabral J. T., Hudson S. D., Harrison C., and Douglas J. F., Frontal Photopolymerization for Microfluidic Applications, *Langmuir* 20, 10020-10029 (2004).

[22] Technical data sheet for NOA optical adhesives; Norland Products, Inc.: New Brunswick, NJ.

[23] Chandezon J., Maystre D. and Raoult G., A new theoretical method for diffraction gratings and its numerical application, *J. Optics (Paris)* 11, No. 4, 235-241 (1980).

[24] Chandezon J., Dupuis M. T. and Cornet G., Multicoated gratings: a differential formalism applicable in the entire optical region, *J. Opt. Soc. Am.* 72, No. 7, 839-846 (1982).

[25] Elston S. J., Bryan-Brown G. P. and Sambles J. R., Polarization conversion from diffraction gratings, *Phys. Rev. B* 44, No.12, 6393-6400 (1991).

[26] Granet G., Plumey J. P. and Chandezon J., Scattering by a periodically corrugated dielectric layer with non identical faces, *Pure Appl. Opt.* 4, No.1, 1-5 (1995).

[27] Preist T. W., Cotter N. P. K. and Sambles J. R., Periodic multilayer gratings of arbitrary shape, *J. Opt. Soc. Am. A* 12, No. 8, 1740-1749 (1995).

[28] Li L., Granet G., Plumey J. P. and Chandezon J., Some topics in extending the C-method to multilayer-coated gratings of different profiles, *Pure Appl. Opt.* 5, No.2, 141-156 (1996).

[29] Plumey J. P., Guizal B. and Candezon J., Coordinate transformation method as applied to asymmetric gratings with vertical facets, *J. Opt. Soc. Am. A* 14, No. 3, 610-617 (1997).

[30] Li L., Multilayer-coated diffraction gratings: differential method of Chandezon et al. revisited, *J. Opt. Soc. Am. A* 11, 2816-2828 (1994).

[31] Li L., Using symmetries of grating groove profiles to reduce computation cost of the C-method, *J. Opt. Soc. Am. A* 24, 1085-1096 (2007).

[32] Post E. J., Formal Structure of Electromagnetics. North-Holland, Amsterdam, 1962.

[33] Ashcroft N. W. and Mermin N. D., Solid State Physics. Saunders College Publishing, 1976.

[34] Zygmund A., Trigonometric Series. Cambridge U. Press, Cambridge, 1977.

[35] Li L., Use of Fourier series in the analysis of discontinuous periodic structures, *J. Opt. Soc. Am. A* 13, No. 9 (1996).

[36] Li L., Justification of matrix truncation in the modal methods of diffraction gratings, *J. Opt. A: Pure Appl. Opt.* 1, 531-536 (1999).

[37] Ruffato G., Zacco G. and Romanato F., Surface Plasmon Polaritons Excitation and Propagation on Metallic Gratings: Far-Field and Near-Field Numerical Simulations, *Journal of Materials Science and Engineering A* 1, No. 6, 768-777 (2011).

[38] Dorsey E., Properties of Ordinary Water-Substance, Reinhold Publishing Corporation 1940.

[39] Ruffato G. and Romanato F., Grating-Coupled Surface Plasmon Resonance in conical mounting with Polarization Modulation, *Opt. Lett.* 37 (2012).

[40] Ang X. F., Li F. Y., Tan W. L., Chen Z. and Wong C. C., Self-assembled monolayer for reduced temperature direct metal thermocompression bonding, *Appl. Phys. Lett.* 91, 061913 (2007).

[41] Johansen K., Stalberg R., Lundstrom I., Liedberg B., Surface plasmon resonance: instrumental resolution using photo diode arrays, *Meas. Sci. Technol.* 11, 1630-1638 (2000).

[42] Jung L. S., Campbell C. T., Chinowsky T. M., Mar M. N. and Yee S. S., Quantitative interpretation of the response of surface plasmon resonance sensors to adsorbed films, *Langmuir* 14, 5636-5648 (1998).

[43] Bain C. D., Troughton E. B., Tao Y. T., Evall J., Whitesides G. M., Nuzzo R. G., Formation of monolayer films by the spontaneous assembly organic thiols from solution onto gold, *J. Am. Chem. Soc.* 111, 321-335 (1989).

Chapter 8

Plasmon Mediated Energy Transport in PV Systems with Photo-Active Surface Modified Metallically in Nano-Scale and in Metallic Nano-Chains

Witold Jacak

Additional information is available at the end of the chapter

1. Introduction

Experimental and theoretical investigations of plasmon excitations in metallic nano-crystals rapidly grew up mainly due to possible applications in photo-voltaics and microelectronics. A significant enhancement of absorption of incident light in photo-diode-systems with active surfaces covered with nano-size metallic particles (of Au, Ag or Cu) with planar density 10^8-10^{10}/cm^2 was observed [24, 25, 27, 30, 32, 34]. These findings are of practical importance for enhancement of solar cell efficiency, especially for development of thin film cell technology. On the other hand, hybridized states of surface plasmons and photons result in plasmon-polaritons [20, 36], which are of high importance for applications in photonics and microelectronics [10, 20], in particular, for sub-diffraction transportation of converted light energy and information in metallically modified structures in nano-scale [21, 36].

Surface plasmons in nano-particles have been widely investigated since their classical description by Mie [23]. Many particular studies, including numerical modeling of multi-electron clusters, have been carried out [4, 5]. They were mostly developments of Kohn-Sham attitude in form of LDA (Local Density Approximation) or TDLDA (Time Dependent LDA) for small metallic clusters only [4, 5, 8, 9, 17], up to ca 200 electrons (as limited by numerical calculation constraints that grow rapidly with the number of electrons). The random phase approximation (RPA) was formulated [28] for description of volume plasmons in bulk metals and utilized also for confined geometry mainly in a numerical or semi-numerical manner [4, 5, 17]. Usually, in these analyzes the *jellium* model was assumed for description of positive ion background in the metal and the dynamics was addressed to the electron system only [5, 8, 17]. Such an attitude is preferable for clusters of simple metals, including noble metals (also transition and alkali metals).

The classical Mie frequencies [23] are not dependent on the sphere radius, in contrast to the experimental observations for both small metallic clusters and larger spheres. The emerging of the Mie response from the more general behavior was presented [4, 5, 8]. Numerical analyzes (for clusters up to 200 electrons) revealed the red-shift of Mie resonance mainly due to so-called spill-out of the electron cloud beyond the jellium rim. This effect is, however, not important for particles with radii larger than 2 nm, and for nanospheres, of size of $10 - 60$ nm, the other effects are responsible for size-dependent shifts of resonances, pronounced in large particles and opposite (with respect to the dependence versus a) to the red shift for small and ultra-small clusters.

The main factor for plasmon oscillations in large nanospheres was identified [11, 12, 15] as radiation phenomena, which for radius larger than 10 nm dominate plasmon damping as growing with the radius a as a^3. The pronounced cross-over in the vicinity of 10 nm (for Au, Ag and Cu) has been predicted [15]—for smaller nanospheres, the scattering damping scaling with a as $\frac{1}{a}$ (including inter-particle and boundary scattering and so-called Landau damping [35]) is important, while for larger spheres, the radiation losses are overwhelming and rapidly grow with a as a^3. Thus rather large nanoparticles would play a role in phenomena linked with energy transport employing plasmon radiation, which well corresponds to experimental observations [26, 27, 30, 31, 33, 34].

Metallic nanospheres (or nanoparticles of other shape) can act as light converters, collecting energy of incident photons in surface plasmon oscillations. This energy can be next transferred to the semiconductor substrate in a more efficient manner in comparison to the direct photo-effect. Experimental observations [26, 27, 30, 31, 33, 34] suggest, that the near-field coupling between plasmons in nanospheres and band electrons in the semiconductor substrate allows for significant growth of selective light energy transformation into a photo-current in the diode system. This phenomenon is not described in detail as of yet and moreover, some competitive mechanisms apparently contribute. Nevertheless, one can argue generally, that due to the nano-scale size of the metallic components, the momentum is not conserved, which leads to the allowance of all indirect optical inter-band transitions in the semiconductor layer, resulting in enhancement of the photo-current in comparison to the ordinary photo-effect, when only direct inter-band transitions are admitted.

In the present paper we review the RPA (random phase approximation) semi-classical theory of plasmons, formulated [11] for large metallic nanospheres in analogy to RPA theory of plasmons in bulk metal [28, 29] (Pines-Bohm theory). The advantage of this description [11, 15] consists in full-analytic formulation allowing for application to more complicated physical situations, as e.g., plasmons interaction with surrounding medium, in order to describe plasmonic enhancement of photo-voltaic (PV) effect in the case of metallic particles deposited on the photo-active layer of the semiconductor, or for description of transport of plasmon-polaritons, with potential applications for sub-diffraction nano-photonics [6].

The coupling of surface plasmons in the near-field zone with another systems, like a semiconductor substrate or other metallic nanosphere in the chain, is the main topic of the present paper. An assessment of the energy transfer via this channel, as made in the present paper within the Fermi golden rule scheme, reveals a high efficiency of plasmon-mediated energy transport and explains the experimentally observed PV efficiency growth of solar cell setups with surface metallic nano-modifications. In seems to be of a particular significance for thin film semiconductor solar cell technology including conjugated polymer semiconductor photo-active matrices with a great potential commercial usage, providing an increase of their efficiency by relatively not-costly modifications, e.g., by sparse coverings of photo-active surface with noble metal nanoparticles, which is feasible in various techniques.

The paper is organized as follows. In the next section a short review of the model is given including results of analyzes both for volume and surface plasmon excitations in spherical geometry. Damping phenomena with particular role of radiation losses for large nanospheres is described in the following section, in terms of the Lorentz friction. The separate discussion is addressed to giant enhancement of photo-voltaic effect by plasmon mediation in near-field zone. The comparison with experimental data related to radiation of plasmons in metallic nanospheres is also given. In the last section, the undamped mode of collective surface plasmon excitation along the metallic nanosphere chain is demonstrated, including nonlinear effects.

2. Surface and volume plasmons in large metallic nanospheres

2.1. RPA equation for collective electron oscillations in confined geometry

The Hamiltonian for the metallic nanosphere has the form:

$$\hat{H} = -\sum_{v=1}^{N} \frac{\hbar^2 \nabla_v^2}{2M} + \frac{1}{2} \sum_{v \neq v'} u(\boldsymbol{R}_v - \boldsymbol{R}_{v'}) - \sum_{j=1}^{N_e} \frac{\hbar^2 \nabla_j^2}{2m} + \frac{1}{2} \sum_{j \neq j'} \frac{e^2}{|\boldsymbol{r}_j - \boldsymbol{r}_{j'}|} \\ + \sum_{v,j} w(\boldsymbol{R}_v - \boldsymbol{r}_j), \tag{1}$$

where $\boldsymbol{R}_v$, $\boldsymbol{r}_j$ and M, m are the positions and masses of the ions and electrons, respectively; N is the number of ions in the sphere, $N_e = ZN$ is the number of collective electrons, $u(\boldsymbol{R}_v - \boldsymbol{R}_{v'})$ is the interaction of ions (ion is treated as a nucleus with electron core of closed shells) and $w(\boldsymbol{R}_v - \boldsymbol{r}_j)$ is the local pseudopotential of electron-ion interaction. Assuming the jellium model [3, 5, 7] one can write for the background ion charge uniformly distributed over the sphere: $n_e(r) = n_e \Theta(a - r)$, where $n_e = N_e/V$ and $n_e|e|$ is the averaged positive charge density, $V = \frac{4\pi a^3}{3}$ is the sphere volume and Θ is the Heaviside step-function.

A local electron density can be written as follows [28, 29]:

$$\rho(\boldsymbol{r}, t) = < \Psi_e(t)| \sum_j \delta(\boldsymbol{r} - \boldsymbol{r}_j)|\Psi_e(t) >, \tag{2}$$

with the Fourier picture:

$$\tilde{\rho}(\boldsymbol{k}, t) = \int \rho(\boldsymbol{r}, t) e^{-i\boldsymbol{k}\cdot\boldsymbol{r}} d^3 r = < \Psi_e(t)|\hat{\rho}(\boldsymbol{k})|\Psi_e(t) >, \tag{3}$$

where the operator $\hat{\rho}(\boldsymbol{k}) = \sum_j e^{-i\boldsymbol{k}\cdot\boldsymbol{r}_j}$.

Using the above notation one can rewrite the electron part of the Hamiltonian (1), $\hat{H}_e$, in the following form [11]:

$$\hat{H}_e = \sum_{j=1}^{N_e} \left[-\frac{\hbar^2 \nabla_j^2}{2m} \right] - \frac{e^2}{(2\pi)^3} \int d^3 k \tilde{n}_e(k) \frac{2\pi}{k^2} \left(\hat{\rho}^+(\boldsymbol{k}) + \hat{\rho}(\boldsymbol{k}) \right) \\ + \frac{e^2}{(2\pi)^3} \int d^3 k \frac{2\pi}{k^2} \left[\hat{\rho}^+(\boldsymbol{k}) \hat{\rho}(\boldsymbol{k}) - N_e \right], \tag{4}$$

where: $\tilde{n}_e(\mathbf{k}) = \int d^3 r n_e(\mathbf{r}) e^{-i\mathbf{k}\cdot\mathbf{r}}$, $\frac{4\pi}{k^2} = \int d^3 r \frac{1}{r} e^{-i\mathbf{k}\cdot\mathbf{r}}$.

The motion equation has the form:

$$\frac{d^2 \hat{\rho}(\mathbf{k})}{dt^2} = \frac{1}{(i\hbar)^2} \left[[\hat{\rho}(\mathbf{k}), \hat{H}_e], \hat{H}_e \right]. \tag{5}$$

Within the RPA and for Thomas-Fermi averaged kinetic energy formula [28], Eq. (5) attains the form [11]:

$$
\begin{aligned}
\frac{\partial^2 \delta\tilde{\rho}(\mathbf{r},t)}{\partial t^2} &= \left[\frac{2}{3}\frac{\epsilon_F}{m}\nabla^2 \delta\tilde{\rho}(\mathbf{r},t) - \omega_p^2 \delta\tilde{\rho}(\mathbf{r},t) \right] \Theta(a-r) \\
&- \frac{2}{3m}\nabla \left\{ \left[\frac{3}{5}\epsilon_F n_e + \epsilon_F \delta\tilde{\rho}(\mathbf{r},t) \right] \frac{\mathbf{r}}{r}\delta(a-r) \right\} \\
&- \left[\frac{2}{3}\frac{\epsilon_F}{m}\frac{\mathbf{r}}{r}\nabla\delta\tilde{\rho}(\mathbf{r},t) + \frac{\omega_p^2}{4\pi}\frac{\mathbf{r}}{r}\nabla \int d^3 r_1 \frac{1}{|\mathbf{r}-\mathbf{r}_1|}\delta\tilde{\rho}(\mathbf{r}_1,t) \right] \delta(a-r).
\end{aligned}
\tag{6}
$$

In the above formula ω_p is the bulk plasmon frequency, $\omega_p^2 = \frac{4\pi n_e e^2}{m}$ (it was taken into account that $\nabla\Theta(a-r) = -\frac{\mathbf{r}}{r}\delta(a-r)$). The solution of Eq. (6) can be decomposed into two parts related to the distinct domains:

$$\delta\tilde{\rho}(\mathbf{r},t) = \begin{cases} \delta\tilde{\rho}_1(\mathbf{r},t), & \text{for } r < a, \\ \delta\tilde{\rho}_2(\mathbf{r},t), & \text{for } r \geq a, \ (r \to a+), \end{cases} \tag{7}$$

corresponding to the volume and surface excitations, respectively. These two parts of local electron density fluctuations satisfy the equations (according to Eq. (6)):

$$\frac{\partial^2 \delta\tilde{\rho}_1(\mathbf{r},t)}{\partial t^2} = \frac{2}{3}\frac{\epsilon_F}{m}\nabla^2 \delta\tilde{\rho}_1(\mathbf{r},t) - \omega_p^2 \delta\tilde{\rho}_1(\mathbf{r},t), \tag{8}$$

and (here $\epsilon = 0+$)

$$
\begin{aligned}
\frac{\partial^2 \delta\tilde{\rho}_2(\mathbf{r},t)}{\partial t^2} &= -\frac{2}{3m}\nabla \left\{ \left[\frac{3}{5}\epsilon_F n_e + \epsilon_F \delta\tilde{\rho}_2(\mathbf{r},t) \right] \frac{\mathbf{r}}{r}\delta(a+\epsilon-r) \right\} \\
&- \left[\frac{2}{3}\frac{\epsilon_F}{m}\frac{\mathbf{r}}{r}\nabla\delta\tilde{\rho}_2(\mathbf{r},t) + \frac{\omega_p^2}{4\pi}\frac{\mathbf{r}}{r}\nabla \int d^3 r_1 \frac{1}{|\mathbf{r}-\mathbf{r}_1|} \left(\delta\tilde{\rho}_1(\mathbf{r}_1,t)\Theta(a-r_1) \right. \right. \\
&\left. \left. + \delta\tilde{\rho}_2(\mathbf{r}_1,t)\Theta(r_1-a) \right) \right] \delta(a+\epsilon-r).
\end{aligned}
\tag{9}
$$

2.2. Solutions of plasmon RPA equations

Eqs (8) and (9) can be solved upon imposed boundary and symmetry conditions. Let us represent both parts of the electron fluctuation in the following manner:

$$
\begin{aligned}
\delta\tilde{\rho}_1(\mathbf{r},t) &= n_e \left[f_1(r) + F(\mathbf{r},t) \right], \text{ for } r < a, \\
\delta\tilde{\rho}_2(\mathbf{r},t) &= n_e f_2(r) + \sigma(\Omega,t)\delta(r+\epsilon-a), \ \epsilon = 0+, \text{ for } r \geq a, \ (r \to a+),
\end{aligned}
\tag{10}
$$

and now let us choose the convenient initial conditions, $F(\mathbf{r},t)|_{t=0} = 0$, $\sigma(\Omega,t)|_{t=0} = 0$, ($\Omega$ is the spherical angle), moreover $(1 + f_1(r))|_{r=a} = f_2(r)|_{r=a}$ (continuity condition), $F(\mathbf{r},t)|_{r\to a} = 0$, $\int \rho(\mathbf{r},t)d^3 r = N_e$ (neutrality condition).

We arrive [11, 15] thus with the explicit form of the solutions of Eqs (8, 9):

$$
\begin{aligned}
f_1(r) &= -\frac{k_T a+1}{2}e^{-k_T(a-r)}\frac{1-e^{-2k_T r}}{k_T r}, \text{ for } r < a, \\
f_2(r) &= \left[k_T a - \frac{k_T a+1}{2}\left(1 - e^{-2k_T a}\right) \right] \frac{e^{-k_T(r-a)}}{k_T r}, \text{ for } r \geq a,
\end{aligned}
\tag{11}
$$

where $k_T = \sqrt{\frac{6\pi n_e e^2}{\epsilon_F}} = \sqrt{\frac{3\omega_p^2}{v_F^2}}$ (Thomas-Fermi inverse radius). For the time-dependent parts
of the electron fluctuations we find [11, 12]:

$$F(r,t) = \sum_{l=1}^{\infty} \sum_{m=-l}^{l} \sum_{n=1}^{\infty} A_{lmn} j_l(k_{nl}r) Y_{lm}(\Omega) \sin(\omega_{nl}t), \tag{12}$$

and

$$\sigma(\Omega,t) = \sum_{l=1}^{\infty} \sum_{m=-l}^{l} \frac{B_{lm}}{a^2} Y_{lm}(\Omega) \sin(\omega_{0l}t)$$

$$+ \sum_{l=1}^{\infty} \sum_{m=-l}^{l} \sum_{n=1}^{\infty} A_{lmn} \frac{(l+1)\omega_p^2}{l\omega_p^2 - (2l+1)\omega_{nl}^2} Y_{lm}(\Omega) n_e \int_0^a dr_1 \frac{r_1^{l+2}}{a^{l+2}} j_l(k_{nl}r_1) \sin(\omega_{nl}t), \tag{13}$$

where $j_l(\xi) = \sqrt{\frac{\pi}{2\xi}} I_{l+1/2}(\xi)$ is the spherical Bessel function, $Y_{lm}(\Omega)$ is the spherical function,
$\omega_{nl} = \omega_p \sqrt{1 + \frac{x_{nl}^2}{k_T^2 a^2}}$ are the frequencies of electron volume self-oscillations (volume plasmon

frequencies), x_{nl} are the nodes of the Bessel function $j_l(\xi)$, $k_{nl} = x_{nl}/a$, $\omega_{0l} = \omega_p \sqrt{\frac{l}{2l+1}}$ are
the frequencies of electron surface self-oscillations (surface plasmon frequencies).

From the above equations it follows thus that the local electron density (within RPA attitude)
has the form:

$$\rho(r,t) = \rho_0(r) + \rho_{neq}(r,t), \tag{14}$$

where the RPA equilibrium electron distribution is (correcting the uniform distribution n_e):

$$\rho_0(r) = \begin{cases} n_e[1 + f_1(r)], & \text{for } r < a, \\ n_e f_2(r), & \text{for } r \geq a, \ r \to a+ \end{cases} \tag{15}$$

and the nonequilibrium, of plasmon oscillation type, is:

$$\rho_{neq}(r,t) = \begin{cases} n_e F(r,t), & \text{for } r < a, \\ \sigma(\Omega,t)\delta(a+\epsilon-r), \ \epsilon = 0+, & \text{for } r \geq a, \ r \to a+ \, . \end{cases} \tag{16}$$

The function $F(r,t)$ displays volume plasmon oscillations, while $\sigma(\Omega,t)$ describes the surface
plasmon oscillations. Let us emphasize that in the formula for $\sigma(\Omega,t)$, Eq. (13), the first
term corresponds to surface self-oscillations, while the second term describes the surface
oscillations induced by the volume plasmons. The frequencies of the surface self-oscillations
are

$$\omega_{0l} = \omega_p \sqrt{\frac{l}{2l+1}}, \tag{17}$$

which, for $l = 1$, agrees with the dipole type surface oscillations described originally by Mie
[23], $\omega_{01} = \omega_p/\sqrt{3}$ (for simplicity, denoted hereafter as $\omega_1 = \omega_{01}$).

3. Damping of plasmons in large nanospheres

One can phenomenologically include damping of plasmons in analogy to oscillator damping
via the additional term, $-\frac{2}{\tau_0}\frac{\partial \delta\rho(r,t)}{\partial t}$, to the right-hand-side of plasmon dynamic equations.
They attain the form:

$$\frac{\partial^2 \delta\rho_1(r,t)}{\partial t^2} + \frac{2}{\tau_0}\frac{\partial \delta\rho_1(r,t)}{\partial t} = \frac{2}{3}\frac{\epsilon_F}{m}\nabla^2 \delta\rho_1(r,t) - \omega_p^2 \delta\rho_1(r,t), \tag{18}$$

and

$$\frac{\partial^2 \delta\rho_2(r,t)}{\partial t^2} + \frac{2}{\tau_0}\frac{\partial \delta\rho_2(r,t)}{\partial t} = -\frac{2}{3m}\nabla\left\{\left[\frac{3}{5}\epsilon_F n_e + \epsilon_F \delta\rho_2(r,t)\right]\frac{r}{r}\delta(a+\epsilon-r)\right\}$$
$$-\left[\frac{2}{3}\frac{\epsilon_F}{m}\frac{r}{r}\nabla\delta\rho_2(r,t) + \frac{\omega_p^2}{4\pi}\frac{r}{r}\nabla\int d^3r_1\frac{1}{|r-r_1|}\left(\delta\rho_1(r_1,t)\Theta(a-r_1)\right.\right. \tag{19}$$
$$\left.\left.+\frac{1}{\varepsilon}\delta\rho_2(r_1,t)\Theta(r_1-a)\right) + \frac{en_e}{m}\frac{r}{r}\cdot E(t)\right]\delta(a+\epsilon-r).$$

For the homogeneous forcing field $E(t)$ (this corresponds to dipole approximation satisfied for $a \sim 10 - 50$ nm, when $\lambda \sim 500$ nm), only dipole surface mode can be excited and the electron dynamics resolves to the equation for a single dipole type mode, described by the function $Q_{1m}(t)$. The function $Q_{1m}(t)$ satisfies the equation:

$$\frac{\partial^2 Q_{1m}(t)}{\partial t^2} + \frac{2}{\tau_0}\frac{\partial Q_{1m}(t)}{\partial t} + \omega_1^2 Q_{1m}(t)$$
$$= \sqrt{\frac{4\pi}{3}}\frac{en_e}{m}\left[E_z(t)\delta_{m0} + \sqrt{2}\left(E_x(t)\delta_{m1} + E_y(t)\delta_{m-1}\right)\right], \tag{20}$$

where $\omega_1 = \omega_{01} = \frac{\omega_p}{\sqrt{3\varepsilon}}$ (it is a dipole-type surface plasmon Mie frequency [23]). Only this function contributes to the dynamical response to the homogeneous electric field. Thus for the homogeneous forcing field, electron density fluctuations:

$$\delta\rho(r,t) = \begin{cases} 0, \ for \ r < a, \\ \sum\limits_{m=-1}^{1} Q_{1m}(t)Y_{1m}(\Omega) \ for \ r \geq a, \ r \to a+. \end{cases} \tag{21}$$

For plasmon oscillations given by Eq. (21) one can calculate the corresponding dipole,

$$D(t) = e\int d^3r\, r\delta\rho(r,t) = \frac{4\pi}{3}eq(t)a^3, \tag{22}$$

where $Q_{11}(t) = \sqrt{\frac{8\pi}{3}}q_x(t)$, $Q_{1-1}(t) = \sqrt{\frac{8\pi}{3}}q_y(t)$, $Q_{10}(t) = \sqrt{\frac{4\pi}{3}}q_x(t)$ and $q(t)$ satisfies the equation (cf. Eq. (20)),

$$\left[\frac{\partial^2}{\partial t^2} + \frac{2}{\tau_0}\frac{\partial}{\partial t} + \omega_1^2\right]q(t) = \frac{en_e}{m}E(t). \tag{23}$$

There are various mechanisms of plasmon damping, which could be effectively accounted for via phenomenological oscillator type damping term. All types of scattering phenomena, including electron-electron and electron-phonon interactions, as well as contribution of boundary scattering effect [6] cause significant attenuation of plasmons, in particular, in small metal clusters. All these contributions to damping time ratio scale as $\frac{1}{a}$ and are of lowering significance with the radius growth. In the following subsection we argue that damping of plasmons caused by radiation losses scales conversely, as a^3, and for large nanospheres this channel dominates plasmon attenuation.

3.1. Lorentz friction for plasmons

The nanosphere surface plasmons can be induced by a homogeneous electric field [15], while the volume mode excitations need field inhomogeneity on the radius scale (and therefore the visible light cannot excite volume modes in the nanospheres with radii of $10 - 50$ nm). Plasmon oscillations are themselves a source of the e-m radiation. This radiation takes away the energy of plasmons resulting in their damping, which can be described as the

Lorentz friction [18]. This damping was not included in τ_0 in Eq. (23). The latter accounted for scattering of electrons on other electrons, on defects, on phonons and on nanoparticle boundary—all they lead to damping rate expressed by the simplified formula [6]:

$$\frac{1}{\tau_0} \simeq \frac{v_F}{2\lambda_B} + \frac{cv_F}{2a}, \tag{24}$$

where, C is the constant of unity order, a is the nanosphere radius, v_F is the Fermi velocity in metal, λ_B is the electron free path in bulk (including scattering of electrons on other electrons, on impurities and on phonons [6]); for Ag, $v_F = 1.4 \times 10^6$ m/s and $\lambda_B \simeq 57$ nm (at room temperature); the latter term in the formula (24) accounts for scattering of electrons on the boundary of the nanoparticle, while the former one corresponds to scattering processes similar as in bulk. The other effects, as the so-called Landau damping (especially important in small clusters [9, 35]), corresponding to decay of plasmon for high energy particle-hole pair, are of lowering significance for nanosphere radii larger than $2-3$ nm [35] and are completely negligible for radii larger than 10 nm. Note that the similarly lowering role with the radius growth plays also electron liquid spill-out effect [5, 8], though it was of primary importance for small clusters [5, 17].

The e-m wave emission caused electron friction can be described as the additional electric field [18],

$$E_L = \frac{2}{3\varepsilon^{3/2}v^3} \frac{\partial^3 D(t)}{\partial t^3}, \tag{25}$$

where $v = \frac{c}{\sqrt{\varepsilon}}$ is the light velocity in the dielectric medium, and $D(t)$ is the dipole of the nanosphere. According to Eq. (22) we arrive at the following:

$$E_L = \frac{2e}{3\varepsilon v^2} \frac{4\pi}{3} a^3 \frac{\partial^3 q(t)}{\partial t^3}. \tag{26}$$

Substituting this into Eq. (23), we get

$$\left[\frac{\partial^2}{\partial t^2} + \frac{2}{\tau_0}\frac{\partial}{\partial t} + \omega_1^2\right] q(t) = \frac{en_e}{m} E(t) + \frac{2}{3\omega_1}\left(\frac{\omega_1 a}{v}\right)^3 \frac{\partial^3 q(t)}{\partial t^3}. \tag{27}$$

If one rewrites the above equation (for $E=0$) in the form

$$\left[\frac{\partial^2}{\partial t^2} + \omega_1^2\right] q(t) = \frac{\partial}{\partial t}\left[-\frac{2}{\tau_0} + \frac{2}{3\omega_1}\left(\frac{\omega_1 a}{v}\right)^3 \frac{\partial^2 q(t)}{\partial t^2}\right], \tag{28}$$

thus, one notes that the zeroth order approximation (neglecting attenuation) corresponds to the equation:

$$\left[\frac{\partial^2}{\partial t^2} + \omega_1^2\right] q(t) = 0. \tag{29}$$

In order to solve Eq. (28) in the next step of perturbation iteration, one can substitute, in the right-hand-side of this equation, $\frac{\partial^2 q(t)}{\partial t^2}$ by $-\omega_1^2 q(t)$ (acc. to Eq. (29)).

Therefore, if one assumes the above estimation, $\frac{\partial^3 q(t)}{\partial t^3} \simeq -\omega_1^2 \frac{\partial q(t)}{\partial t}$, one can include the Lorentz friction in a renormalized damping term:

$$\left[\frac{\partial^2}{\partial t^2} + \frac{2}{\tau}\frac{\partial}{\partial t} + \omega_1^2\right] q(t) = \frac{en_e}{m} E(t), \tag{30}$$

where

$$\frac{1}{\tau} = \frac{1}{\tau_0} + \frac{\omega_1}{3}\left(\frac{\omega_1 a}{v}\right)^3 \simeq \frac{v_F}{2\lambda_B} + \frac{Cv_F}{2a} + \frac{\omega_1}{3}\left(\frac{\omega_1 a}{v}\right)^3, \tag{31}$$

where we used for $\frac{1}{\tau_0} \simeq \frac{v_F}{2\lambda_B} + \frac{Cv_F}{2a}$ [6]. Renormalized damping causes a change in the shift of self-frequencies of free surface plasmons, $\omega_1' = \sqrt{\omega_1^2 - \frac{1}{\tau^2}}$.

Note also, that one can verify [15] the above calculated Lorentz friction contribution to plasmon damping by estimation of the energy transfer in the far-field zone (which can be expressed by the Poynting vector) and via comparison with the energy loss of plasmon oscillation. We have arrived [11, 15] at the same formula for damping time rate as given by Eq. (30). The radius dependent shift of the resonance resulting due to strong irradiation-induced plasmon damping was verified also experimentally [15] by measurement of light extinction in colloidal solutions of nanoparticles with different size (it is done [15] for Au, $10 - 80$ nm, and Ag, $10 - 60$ nm). These measurements clearly support the a^3 plasmon damping behavior, as described above for the far-field zone radiation losses in a dielectric surroundings.

If, however, in the vicinity of the nanosphere the another system is located, the situation would change. For the case when the nanosphere is deposited on the semiconductor surface, the near-field coupling of plasmons with semiconductor band electrons must be included.

4. Mediating of light energy transfer by surface plasmons in near-field regime

Let us calculate the probability (per time unit) of inter-band transitions of electrons in substrate semiconductor covered with metallic nanospheres (with radius a), induced by photon-excited surface plasmons in metallic components. These plasmons excited in nanospheres deposited on the semiconductor layer, couple to the semiconductor band electrons in the near-field regime. The situation is similar to the ordinary photo-effect, though the perturbation of the electron system in semiconductor is not of a plane-wave form as it was in the case of direct illumination of semiconductor by incident photons, but attains the form of dipole-type near-field electric interaction [11, 18]. This causes a change of the matrix element in the relevant Fermi golden formula. The potential for the near-field interaction of the surface Mie type plasmons with the band electrons can be written as [18]:

$$w = e\psi(\mathbf{R}, t) = \frac{e}{\varepsilon_0 R^2}\hat{\mathbf{n}} \cdot \mathbf{D}_0 sin(\omega t + \alpha) = w^+ e^{i\omega t} + w^- e^{-i\omega t},$$
$$w^+ = \left(w^-\right)^* = \frac{e}{\varepsilon_0 R^2}\frac{e^{i\alpha}}{2i}\hat{\mathbf{n}} \cdot \mathbf{D}_0. \tag{32}$$

The terms w^+, w^- correspond to emission and absorption, respectively ($\hat{\mathbf{n}} = \frac{\mathbf{R}}{R}$, $\mathbf{D}_0$ is the dipole plasmon amplitude). We choose the first one in order to consider emission of energy from the plasmon oscillations and transfer of it to the electron system in the semiconductor substrate. The semiconductor band system we model in the simplest single-band parabolic effective mass approximation. The inter-band transition probability is given by the Fermi golden rule,

$$w(\mathbf{k}_1, \mathbf{k}_2) = \frac{2\pi}{\hbar}\left|< \mathbf{k}_1|w^+|\mathbf{k}_2 >\right|^2 \delta(E(\mathbf{k}_1) - E(\mathbf{k}_2) + \hbar\omega), \tag{33}$$

where Bloch states in the conduction and valence bands we assume as planar waves (for the sake of simplicity),

$$\Psi_{k_1} = \frac{1}{(2\pi)^{3/2}} e^{ik_1 \cdot R - iE(k_1)t/\hbar}, \Psi_{k_2} = \frac{1}{(2\pi)^{3/2}} e^{ik_2 \cdot R - iE(k_2)t/\hbar},$$
$$E(k_1) = -\frac{\hbar^2 k_1^2}{2m_p^*} - E_g, E(k_2) = \frac{\hbar^2 k_2^2}{2m_n^*}, \tag{34}$$

(indices n, p refer to electrons from the conduction and valence bands, respectively, E_g is the forbidden gap).

The electron wave functions are normalized to the Dirac delta, which corresponds to infinite movement and the continuous energy spectrum. The wave function modulus squares do not have, in this case, probability interpretation (which must be normalized to the unity), therefore the expression, $w(k_1, k_2) = \frac{2\pi}{\hbar} |< k_1|w^+|k_2 >|^2 \delta(E(k_1) - E(k_2) + \hbar\omega)$, has to be divided by the delta Dirac square, i.e., by the factor, $\left(\frac{V}{(2\pi)^3}\right)^2$. This factor corresponds to the probability proper normalization, because,

$$\frac{1}{(2\pi)^3} \int d^3r e^{ik \cdot r} = \delta(k) \simeq (for\ k = 0)\ \frac{V}{(2\pi)^3}\ (V \to \infty). \tag{35}$$

Note, that the same factor occurs also due to density of states, when one integrates over all initial and final states k_1, k_2. Two integrals will give the factor $\left(\frac{2V}{(2\pi)^3}\right)^2$, (2 is caused by spin degeneration). Thus, both renormalized factors cancel mutually themselves.

Taking into account the above described renormalization, one can find the matrix element,

$$< k_1|w^+|k_2 >= \frac{1}{(2\pi)^3} \int d^3R \frac{e}{\varepsilon_0 2i} e^{i\alpha} \hat{n} \cdot D_0 \frac{1}{R^2} e^{-i(k_1-k_2) \cdot R}. \tag{36}$$

Let us introduce the vector $q = k_2 - k_1$. One can choose the coordinate system in such a way, that the vector q is oriented along the z axis, and the vector D_0 lies in the plane zx (as is depicted in the Fig. 1). Then, $q = (0,0,q)$, $\hat{n} = \frac{R}{R} = (sin\Theta_1 cos\psi_1, sin\Theta_1 sin\psi_1, cos\Theta_1)$, $R = R(sin\Theta_1 cos\psi_1, sin\Theta_1 sin\psi_1, cos\Theta_1)$, $D_0 = D_0(sin\Theta, 0, cos\Theta)$ and

$$q \cdot R = qRcos\Theta_1, \tag{37}$$
$$\hat{n} \cdot D_0 = D_0(sin\Theta sin\Theta_1 cos\psi_1 + cos\Theta cos\Theta_1). \tag{38}$$

Hence,

$$< k_1|w^+|k_2 >$$
$$= \frac{1}{(2\pi)^3} \int d^3R \frac{e}{\varepsilon_0 2i} e^{i\alpha} \hat{n} \cdot D_0 \frac{1}{R^2} e^{-i(k_1-k_2) \cdot R}$$
$$= \frac{1}{(2\pi)^3} \frac{ee^{i\alpha}}{\varepsilon_0 2i} D_0 \int_a^\infty \frac{R^2}{R^2} dR \int_0^\pi$$
$$\times sin\Theta_1 d\Theta_1 \int_0^{2\pi} d\psi_1 \{cos\Theta cos\Theta_1 + sin\Theta sin\Theta_1 cos\psi_1\} e^{iqRcos\Theta_1}$$
$$= \frac{1}{(2\pi)^3} \frac{ee^{i\alpha}}{\varepsilon_0 2i} D_0 cos\Theta 2\pi \int_a^\infty dR \int_0^\pi cos\Theta_1 sin\Theta_1 d\Theta_1 e^{iqRcos\Theta_1}, \tag{39}$$

the integer over $d\psi_1$ vanishes the second term in the parenthesis and only the first term contributes with the factor 2π. Note that,

$$\int_0^\pi cos\Theta_1 sin\Theta_1 d\Theta_1 e^{ixcos\Theta_1} = -i\frac{d}{dx} \int_0^\pi sin\Theta_1 e^{ixcos\Theta_1} = -i\frac{d}{dx} 2\frac{sinx}{x}. \tag{40}$$

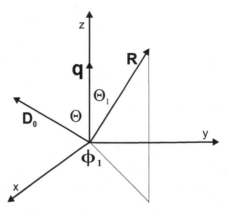

Figure 1. The reference frame is chosen in the way that the vector $\mathbf{q}$ is oriented along the axis z, while the vector $\mathbf{D}_0$ lies in the plane xz

Thus,

$$
\begin{aligned}
< \mathbf{k}_1|w^+|\mathbf{k}_2 > &= \frac{-1}{(2\pi)^3}\frac{ee^{i\alpha}}{\varepsilon_0}D_0cos\Theta(2\pi)\int_a^\infty dR\frac{1}{q}\frac{d}{dR}\frac{sinqR}{qR} \\
&= \frac{1}{(2\pi)^2}\frac{ee^{i\alpha}}{\varepsilon_0}D_0cos\Theta\frac{1}{q}\frac{sinqa}{qa} \to_{a\to 0} \frac{1}{(2\pi)^2}\frac{ee^{i\alpha}}{\varepsilon_0}D_0cos\Theta\frac{1}{q}.
\end{aligned}
\tag{41}
$$

The lower limit of integral with respect to R is taken as a (the nanosphere radius). In this manner one can limit the accessible space for planar waves in the semiconductor, if assumes, for a model, a completely embedded nanosphere in surrounding material. This is, however, a limiting approximation as tunneling effect certainly would allow an access to the inner space, to some extent, at least. In order to account for that the nanosphere is deposited on the semiconductor surface we will confine integration with respect to $d\Theta_1$ to the segment $[0, \pi/2]$, instead of $[0, \pi]$. We obtain in this case the following matrix element,

$$
\begin{aligned}
< \mathbf{k}_1|w^+|\mathbf{k}_2 > &= \frac{1}{(2\pi)^3}\frac{ee^{i\alpha}}{2i\varepsilon_0}D_0cos\Theta(2\pi)\int_a^\infty dR\frac{1}{q}\frac{d}{dR}\frac{1-e^{iqR}}{qR} \\
&= \frac{1}{(2\pi)^2}\frac{ee^{i\alpha}}{2i\varepsilon_0}D_0cos\Theta\frac{1}{q}\frac{1-e^{iqa}}{qa} \to_{a\to 0} \frac{1}{(2\pi)^2}\frac{ee^{i\alpha}}{2i\varepsilon_0}D_0cos\Theta\frac{1}{q},
\end{aligned}
\tag{42}
$$

because,

$$
\int_0^{\pi/2} cos\Theta_1 sin\Theta_1 d\Theta_1 e^{ixcos\Theta_1} = -i\frac{d}{dx}\int_0^{\pi/2} sin\Theta_1 e^{ixcos\Theta_1} = \frac{d}{dx}\frac{1-e^{ix}}{x}.
\tag{43}
$$

In the limit $a = 0$, the modulus of the matrix element will diminish twice in comparison to the case when integration over $d\Theta_1$ had been taken over the whole space.

Now we will integrate over all initial and final states of both bands. The related integration over $\mathbf{k}_1, \mathbf{k}_2$, one can substitute with integration over $\mathbf{q}, \mathbf{k}_2$. Scalar products are invariant against coordinate systems rotations, therefore the result of integration will be the same if $\mathbf{q}$ was along z axis or $\mathbf{D}_0$ is oriented now along the z axis – Θ gives, in the latter case, the deviation of $\mathbf{q}$ with respect to z direction – this choice of the reference frame is convenient for integration with respect to $d\mathbf{q}$.

Thus we arrive with the formula for transition probability in the following form,

$$
\delta w = \int d^3k_1 \int d^3k_2 \left[f_1(1-f_2)w(\mathbf{k}_1, \mathbf{k}_2) - f_2(1-f_1)w(\mathbf{k}_2, \mathbf{k}_1) \right],
\tag{44}
$$

where, f_1, f_2 assign the temperature dependent distribution functions (Fermi-Dirac distribution functions) for initial and final states, respectively. The emission and absorption were included, but for room temperatures one can assume, $f_2 \simeq 0$ and $f_1 \simeq 1$, which leads to,

$$\delta w = \int d^3 k_1 \int d^3 k_2 w(\mathbf{k}_1, \mathbf{k}_2). \tag{45}$$

In the above formula we avoided the density state factors canceled by probability renormalization, as was mentioned above.

We have to calculate the following integral,

$$\delta w = \int d^3 k_2 \int d^3 q \frac{e^2}{(2\pi)^3 \hbar \varepsilon_0} \frac{D_0^2 \cos^2 \Theta}{q^2} \frac{\sin^2(qa)}{(qa)^2} \delta \left(\frac{\hbar^2 k_1^2}{2m_n^*} + \frac{\hbar^2 k_2}{2m_p^*} - (\hbar\omega - E_g) \right), \tag{46}$$

with $\delta \left(\frac{\hbar^2 k_1^2}{2m_n^*} + \frac{\hbar^2 k_2}{2m_p^*} - (\hbar\omega - E_g) \right) = \frac{1}{\alpha+\beta} \frac{1}{2\beta' q k_2} \delta \left(\cos\Theta_2 - \frac{k_2^2 + \beta' q^2 - \gamma'}{2\beta' q k_2} \right)$, where, $\alpha = \frac{\hbar^2}{2m_n^*}$, $\beta = \frac{\hbar^2}{2m_p^*}$, $\gamma = \hbar\omega - E_g$, $\alpha' = \frac{\alpha}{\alpha+\beta}$, $\beta' = \frac{\beta}{\alpha+\beta}$, $\gamma' = \frac{\gamma}{\alpha+\beta}$. For each integral, with respect to k_2 and q the coordinate system can be rotated independently, and for integration over dk_2 we choose again orientation of $\mathbf{q}$ along the axis z, which leads to the spherical angle Θ_2 resulting form the product, $\mathbf{k}_2 \cdot \mathbf{q} = k_2 q \cos\Theta_2$.

The expression for δw attains thus the form,

$$\delta w = \frac{e^2 D_0^2}{(2\pi)^3 \hbar \varepsilon_0^2} \int d^3 q \frac{\sin^2(qa)}{q^2 a^2} \frac{\cos\Theta}{q^2} \\ \int_0^\infty dk_2 k_2^2 \int_0^\pi d\Theta_2 \sin\Theta_2 \int_0^{2\pi} d\psi_2 \frac{1}{\alpha+\beta} \frac{1}{2\beta' q k_2} \delta \left(\cos\Theta_2 - \frac{k_2^2 + \beta' q^2 - \gamma'}{2\beta' q k_2} \right). \tag{47}$$

Integration over $d\Theta_2$ employs the Dirac delta. The relevant nonzero contribution (due to integration over $d\cos\Theta_2$) is conditioned by the inequality,

$$-1 < \frac{k_2^2 + \beta' q^2 - \gamma'}{2\beta' q k_2} < 1, \tag{48}$$

which resolves itself to the condition,

$$|\beta' q - \sqrt{\gamma' - (1 - \beta')\beta' q^2}| < k_2 < \beta' q + \sqrt{\gamma' - (1 - \beta')\beta' q^2}. \tag{49}$$

Thus,

$$\delta w = \frac{1}{(\alpha+\beta) 2\beta'} \frac{e^2 D_0^2}{2\pi \hbar \varepsilon_0^2} \int_0^{\sqrt{\frac{\gamma'}{\beta'(1-\beta')}}} dq q^2 \frac{\sin^2 qa}{(qa)^2} \\ \times \int_0^\pi d\Theta \sin\Theta \cos^2\Theta \frac{1}{q^3} \int_{|\beta' q - \sqrt{\gamma' - (1-\beta')\beta' q^2}|}^{\beta' q + \sqrt{\gamma' - (1-\beta')\beta' q^2}} dk_2 k_2. \tag{50}$$

and hence,

$$\delta w = \frac{2}{3} \frac{\sqrt{\gamma'}}{\alpha+\beta} \frac{e^2 D_0^2}{2\pi \hbar \varepsilon_0^2} \frac{1}{\zeta} \int_0^{\frac{1}{\zeta}} dx \frac{\sin^2(x(a/\zeta))}{(x)^2 (a/\zeta)^2} \sqrt{1 - x)^2}, \tag{51}$$

where, $\zeta = \sqrt{\frac{(1-\beta')\beta'}{\gamma'}}$. Therefore,

$$\delta w = \frac{4}{3} \frac{\mu^2 (m_n^* + m_p^*) 2(\hbar\omega - E_g) e^2 D_0^2}{\sqrt{m_n^* m_p^*} 2\pi \hbar^5 \varepsilon_0^2} \int_0^1 dx \frac{\sin^2(xa/\zeta)}{(xa/\zeta)^2} \sqrt{1 - x^2} \\ = \frac{4}{3} \frac{\mu^2}{\sqrt{m_n^* m_p^*}} \frac{e^2 D_0^2}{2\pi \hbar^3 \varepsilon_0^2} \zeta^2 \int_0^1 dx \frac{\sin^2(xa\zeta)}{(xa\zeta)^2} \sqrt{1 - x^2}, \tag{52}$$

where, $\zeta = 1/\zeta = \frac{\sqrt{2(\hbar\omega - E_g)(m_n^* + m_p^*)}}{\hbar}$.

In limiting cases, we obtain finally,

$$\delta w = \begin{cases} \frac{1}{3}\frac{\mu^2(m_n^* + m_p^*)2(\hbar\omega - E_g)e^2 D_0^2}{\sqrt{m_n^* m_p^*}2\hbar^5\varepsilon_0^2}, & a\zeta \ll 1, \\ \frac{4}{3}\frac{\mu^2(m_n^* + m_p^*)2(\hbar\omega - E_g)e^2 D_0^2}{\sqrt{m_n^* m_p^*}4a\zeta\hbar^5\varepsilon_0^2}, & a\zeta \gg 1, \end{cases} \tag{53}$$

In the latter case the following approximation was applied,

$$\int_0^1 \frac{\sin^2(xa\zeta)}{(xa\zeta)^2} \approx (for\ a\zeta \gg 1)\frac{1}{a\zeta}\int_0^\infty d(xa\zeta)\frac{\sin^2(xa\zeta)}{(xa\zeta)^2} = \frac{\pi}{2a\zeta},$$

while in the former one, $\int_0^1 dx\sqrt{1-x^2} = \pi/4$.

The result must be multiplied by 4 (due to spin) and, in the case of half-space, divided by 4. In the considered limiting cases we obtain thus,

$$\delta w = \begin{cases} \frac{1(4)}{3}\frac{\mu\sqrt{m_n^* m_p^*}(\hbar\omega - E_g)e^2 D_0^2}{\hbar^5\varepsilon_0^2} & for\ a\zeta \ll 1, \\ \frac{1(4)}{3}\frac{\mu^{3/2}\sqrt{2}\sqrt{\hbar\omega - E_g}e^2 D_0^2}{a\hbar^4\varepsilon_0^2} & for\ a\zeta \gg 1, \end{cases} \tag{54}$$

where (4) is addressed to the completely embedded nanosphere.

One can notice that the above formulae are quite distinct in comparison to the ordinary photo-effect. In the latter case, the perturbation is given by the vector potential in the kinematic momentum (for gauge, $div\mathbf{A} = 0$),

$$\hat{H} = \frac{(-i\hbar\nabla - \frac{e}{c}\mathbf{A}(\mathbf{R},t))^2}{2m^*} \simeq \frac{(-i\hbar\nabla))^2}{2m^*} + \frac{i\hbar e}{m^* c}\mathbf{A}(\mathbf{R},t)\cdot\nabla. \tag{55}$$

For monochromatic plane wave, this perturbation has the form,

$$\begin{aligned} w(\mathbf{R},t) &= \frac{ie\hbar}{cm^*}(\mathbf{A}_0\cdot\nabla)cos(\omega t - \mathbf{k}\cdot\mathbf{R} + \alpha) \\ &= \frac{ie\hbar}{2cm^*}(\mathbf{A}_0\cdot\nabla)(e^{i(\omega t - \mathbf{k}\cdot\mathbf{R} + \alpha)} + e^{-i(\omega t - \mathbf{k}\cdot\mathbf{R} + \alpha)}). \end{aligned} \tag{56}$$

Because in this case, both states of band electrons and photon have the form of plane waves, the matrix element in the Fermi golden rule will be proportional to the Dirac delta with respect to the momentum sum. This expresses momentum conservation in translationally invariant system, what is a case for photon interacting with the semiconductor,

$$\begin{aligned} <\mathbf{k}_1|w(\mathbf{R},t)|\mathbf{k}_2> &= \frac{i\hbar e}{2cm^*(2\pi)^3}\int d^3 R e^{-i(\mathbf{k}_1 + \mathbf{k})\cdot\mathbf{R}}(\mathbf{A}_0\cdot\nabla)e^{i\mathbf{k}_2\cdot\mathbf{R}} \\ &= -\frac{e\hbar}{2cm^*}\mathbf{A}_0\cdot\mathbf{k}_2\delta(\mathbf{k}_1 + \mathbf{k} - \mathbf{k}_2). \end{aligned} \tag{57}$$

Because of high value of photon velocity c, only vertical transitions, $\mathbf{k}_1 = \mathbf{k}_2$, are admitted in the ordinary photo-effect.

Nevertheless, this is no case for nanosphere plasmon interacting in near-field regime with the semiconductor substrate. This system is not translationally invariant and indirect inter-band electron excitations are admitted resulting in enhancement of total transition probability.

To compare with the ordinary photo-effect, let us recall the appropriate calculus upon the Fermi golden rule scheme,

$$w(\mathbf{k_1}, \mathbf{k_2}) = \frac{\pi e^2 \hbar}{2c^2 m^{*2}} k_2^2 A_0^2 \cos^2 \Theta \delta^2 (\mathbf{k_1} - \mathbf{k_2}) \delta(E_1 - E_2 + \hbar\omega), \tag{58}$$

where, Θ is the angle between $\mathbf{k_2}$ i $\mathbf{A_0}$. One can use the following approximation in order to get rid the Dirac delta square,

$$\delta^2(\mathbf{k_1} - \mathbf{k_2}) = \delta(0)\delta(\mathbf{k_1} - \mathbf{k_2}) \simeq \frac{V}{(2\pi)^3} \delta(\mathbf{k_1} - \mathbf{k_2}), \tag{59}$$

Similarly as was previously discussed, the normalization of probability must be performed. Thus we get,

$$w(\mathbf{k_1}, \mathbf{k_2}) = \frac{\pi e^2 \hbar}{2c^2 m^{*2}} \frac{V}{(2\pi)^3} k_2^2 A_0^2 \cos^2 \Theta \delta(\mathbf{k_1} - \mathbf{k_2}) \delta(E_1 - E_2 + \hbar\omega). \tag{60}$$

The integration over all states in both electron bands results in,

$$\delta w = \int d^3 k_1 \int d^3 k_2 \frac{\pi e^2 \hbar}{2c^2 m^{*2}} \frac{V}{(2\pi)^3} k_2^2 A_0^2 \cos^2 \Theta \delta(\mathbf{k_1} - \mathbf{k_2}) \delta(E_1 - E_2 + \hbar\omega), \tag{61}$$

or

$$\delta w = \int d^3 k_2 \frac{\pi e^2 \hbar}{2c^2 m_p^{*2}} \frac{V}{(2\pi)^3} k_2^2 A_0^2 \cos^2 \Theta \delta(E_1(\mathbf{k_2}) - E_2(\mathbf{k_2}) + \hbar\omega), \tag{62}$$

where the effective mass in the initial valence state is m_p^* (moreover, $A_0 = \frac{c}{\omega} E_0$). We thus obtain,

$$\delta w = \frac{1}{12} \frac{e^2 \hbar}{m_p^{*2}} \frac{V}{\pi} \frac{E_0^2}{\omega^2} \frac{2^{3/2} \mu^{5/2}}{\hbar^5} (\hbar\omega - E_g)^{3/2}. \tag{63}$$

One can rewrite it in the form,

$$\delta w = \frac{4\sqrt{2}}{3} \frac{\mu^{5/2} e^2}{m^{*2} \omega \varepsilon_0 \hbar^3} \left(\frac{\varepsilon_0 E_0^2 V}{8\pi\hbar\omega} \right) (\hbar\omega - E_g)^{3/2}, \tag{64}$$

this expression should be multiplied by 4 (due to spin degeneracy, though for circular polarization of photons this factor is only 2, due to angular momentum selection rules).

Taking into account, that the number of photons in the volume V equals to, $\left(\frac{\varepsilon_0 E_0^2 V}{8\pi\hbar\omega} \right)$, then the probability of single photon attenuation by the semiconductor per time unit, attains the form,

$$q = \delta w \left(\frac{\varepsilon_0 E_0^2 V}{8\pi\hbar\omega} \right)^{-1} = \frac{4(4)\sqrt{2}}{3} \frac{\mu^{5/2} e^2}{m^{*2} \omega \varepsilon_0 \hbar^3} (\hbar\omega - E_g)^{3/2}, \tag{65}$$

(factor (4) corresponds here to spin degeneration of band electrons).

In order to assess efficiency of the plasmon near-field coupling channel one can estimate the ratio of the probability of energy absorption in the semiconductor via mediation of surface plasmons (per single photon incident on the metallic nanospheres, q_m) to the energy attenuation in the semiconductor directly from a planar wave illumination (also per single photon, q). The ratio $\frac{q_m}{q}$ turns out to be of order of $10^4 \frac{\beta 40}{H[nm]}$ (at a typical surface density of nanoparticles, $n_s \sim 10^8/\text{cm}^2$), which (including the phenomenological factor β [to account for proximity surface effects, not directly included], and H—the semiconductor photo-active layer depth) is sufficient to explain the scale of the experimentally observed strong enhancement of absorption and emission rates. The strong enhancement of this transition probability is linked with allowance of momentum-non-conserved transitions, which is, however, reducing with the radius a growth, according to formulae (54). For e.g., $a \sim 10 - 60$ nm, this gives reducing of the transition probability, which we have included by the effective phenomenological factor β (fitted from the experiment, $\beta \simeq 28 \times 10^{-3} \left(\frac{50}{a[nm]}\right)^2$) in the case of atomic limit, $a = 0$ [11]. The enhancement of the near-field induced inter-band transition, in the case of large nanospheres, is, however, still significant as the reducing role of size-related quenching of transitions is partly compensated by $\sim a^3$ growth of the dipole amplitude of plasmon oscillations.

High efficiency (even if decreased with growing a) of the near-field energy transfer from surface plasmons to the semiconductor substrate is caused mainly by a contribution of all inter-band transitions, not restricted here to the direct (vertical) ones as for ordinary photo-effect, due to the absence of the momentum conservation constraints for the nanosystem. The strengthening of the probability transition due to all indirect inter-band paths of excitations in the semiconductor is probably responsible for the observed experimentally strong enhancement of light absorption and emission in diode systems mediated by surface plasmons in nanoparticle surface coverings [26, 27, 30, 31, 33, 34].

In the balanced state, when the incoming energy of light is transferred to the semiconductor via plasmon near-field coupling, we deal with the stationary solution of driven and damped oscillator. The driving force is the electric field of the incident planar wave, and the damping force is the near-field energy transfer described by the $\frac{1}{\tau}$ (calculated using formulae (54) in the manner as described in [13]). The resulting red-shifted resonance with simultaneously reduced amplitude allows for the accommodation to the balance of energy transfer to the semiconductor with incident photon energy. Within this model, the amplitude of resonant plasmon oscillations $D_0(\omega)$ is thus shaped by $f(\omega) = \frac{1}{\sqrt{(\omega_1^2 - \omega^2)^2 + 4\omega^2/\tau^2}}$. The extremum of red-shifted resonance is attained at $\omega_m = \omega_1\sqrt{1 - 2(\omega_1\tau)^{-2}}$ with corresponding amplitude $\sim \tau/\left(2\sqrt{\omega_1^2 - \tau^{-2}}\right)$. This shift is proportional to $1/(\omega_1\tau^2)$ and scales with nanosphere radius a similarly (diminishes with decreasing a) as in the experimental observations [30] (note again that for scattering-induced $1/\tau_0$ the dependence on a is opposite [grows with decreasing a]).

In order to compare with the experiment we can estimate the photo-current in the case of a metallic modified surface. This photo-current is given by $I' = |e|N(q + q_m)A$, where N is the number of incident photons and q and q_m are the probabilities of single photon attenuation in the ordinary photo-effect [16] and in that one enhanced due to the presence of metallic nanospheres; $A = \frac{\tau_f^n}{t_n} + \frac{\tau_f^p}{t_p}$ is the amplification factor ($\tau_f^{n(p)}$ is the annihilation time of both sign carriers, $t_{n(p)}$ is the drive time for carriers [the time of traversing the distance between electrodes]). From the above formulae, it follows that (here $I = I'(q_m = 0)$, i.e., the photo-current without metallic modifications)

$$\frac{I'}{I} = 1 + 7.95 \cdot 10^5 c_0 \frac{m_p^*}{m_n^*} \left(\frac{2a}{100[1nm]} \sqrt{\frac{\hbar\omega_1[eV]}{x}} \left(\frac{m_p^*}{m} + \frac{m_n^*}{m}\right)\right)^3 \phi(x), \qquad (66)$$

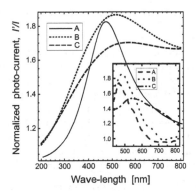

Figure 2. Dependence of the normalized photo-current $\frac{I'}{I}(\lambda)$–comparison with the experimental data [30] for A: $a = 25$ nm (better fitting for 19 nm), $n_s = 6.6 \times 10^8$ 1/cm^2, B: $a = 40$ nm, $n_s = 1.6 \times 10^8$ 1/cm^2, C: $a = 50$ nm, $n_s = 0.8 \times 10^8$ 1/cm^2 ($H = 3$ μm)

where $c_0 = \frac{4\pi a^3}{3}\beta\frac{n_s}{H}$, with n_s as the surface density of metallic nanospheres, H as the semiconductor layer depth, $\phi(x) = \frac{x^2}{(x^2-1)^2+4x^2/x_1^2}\frac{1}{\sqrt{x-x_g}}$, $x = \omega/\omega_1$, $x_1 = \tau\omega_1$, $x_g = E_g/(\hbar\omega_1)$, $\hbar\omega_1 = 2.72$ eV, $m_{n(p)}$ as the effective mass of conduction band and valence band carriers (for Si, $m_n^* = 0.19(0.98)$ m and $m_p^* = 0.16(0.52)$ m, for light (heavy) carriers, band gap $E_g = 1.14$ eV, $\varepsilon = 12$), m as the bare electron mass.

The results are summarized in Fig. 2, and reproduce well the experimental behavior [30]. Both channels of photon absorption resulting in photo-current in the semiconductor sample are included, the direct ordinary photo-effect absorption with probability of transitions given by q and the plasmon mediated absorption with probability q_m, respectively. Note also that some additional effects like reflection of the incident photons or destructive interference on metallic net would contribute and it was phenomenologically accounted for in the plasmon mediated channel by the experiment fitted factor β. These corrections are, however, rather not strong for the considered low densities of metallic coverings of order of 10^8/cm^2, and nanosphere sizes well lower than the resonant wave-length, though for larger concentrations and larger nanosphere sizes, would play a stronger reducing role [19, 32]. The resonance threshold was accounted for by the damped resonance envelope function $\phi(x)$ in Eq. (66) including also semiconductor band-gap limit.

As indicated in Fig. 2, the relatively high value of $\frac{q_m}{q} \sim 10^4\frac{\beta 40}{H[nm]}$ enables a significant growth of the efficiency of the photo-energy transfer to the semiconductor, mediated by surface plasmons in nanoparticles deposited on the active layer, by increasing β or reducing H (at constant n_s). However, because of the fact that an enhancement of β easily induces the overdamped regime, a greater perspective would be thus lowering H, the layer depth. The overall behavior of $I'/I(\omega) = 1 + q_m/q$ calculated according to the relation (66), and depicted in Fig. 2, agrees quite well with the experimental observations [30], in the position, height and shape of the photo-current curves for distinct samples (the strongest enhancement is achieved for $a = 40$ nm).

5. Nonlinear effect for undamped plasmon-polariton propagation along metallic nano-chain

In the present chapter we report the RPA description using a semiclassical approach for a large metallic nanosphere (with radius of several tens nm, and with 10^5-10^7 electrons), in an all-analytical calculus version [11]. The electron liquid oscillations of compressional and translational type result in excitations inside the sphere and on its surface, respectively. They are referred to as volume and surface plasmons. Damping of plasmons due to electron scattering and due to radiation losses (accounted for via the Lorentz friction force) is included. The shift of the resonance frequency of dipole-type surface plasmons (only such plasmons are induced by homogeneous time-dependent electric field), due to damping phenomena, well fits with the experimental data for various nanosphere radii [15].

Collective dipole-type surface plasmon oscillations in the linear chain of metallic nanospheres were then analyzed and wave-type plasmon propagation along the chain was described [13, 36]. A coupling in the near field regime between oscillating dipoles in neighboring nanospheres, together with retardation effects for energy irradiation, allowed for appearance of undamped propagation of plasmon waves (called plasmon-polaritons) along the chain in the experimentally realistic region of parameters (for separation of spheres in the chain and nanosphere radii). These effects are of a particular significance for plasmon arranged non-dissipative and sub-diffraction transport of converted light energy and information along metallic chains with promising expected applications in nano-electronics.

The undamped mode of plasmon-polaritons occurs, however, on the rim of stability of the linear approach [13]. It means that the zero damping rate separates the region with positive its value (corresponding to ordinary attenuation of plasmon-polaritons) and the region with negative damping rate (corresponding to instable plasmons-polariton modes). The latter indicates an unphysical behavior. i.e., the arte-fact of the linear approximation. In order to regularize this description the nonlinear corrections must be thus included. Inclusion of nonlinear corrections induced by the invariant form of the Lorentz friction results in such regularization. The instability region of linear approach is entirely covered by the region of undamped wave propagation with the amplitude accommodated to the nonlinearity scale and is independent of the initial condition (despite of its magnitude). This phenomenon, familiar in other nonlinear systems [2], seems to be of a particular significance for understanding of collective plasmon excitations with interesting possible applications.

5.1. Nonlinear corrections to plasmon dynamics in phenomenological approach

Even if the derivation of plasmon dynamics equation in the form of effective harmonic oscillator equation is rigid upon quantum approach of quasiclassical RPA method [11], the inclusion of plasmon attenuation of scattering type and of radiation losses type needs some phenomenological assumptions. They resolve themselves to extension of quantum RPA harmonic oscillator formulation to the damped oscillator equation form with attenuation described by heuristically assumed damping rates. It is proved [11, 14] that radiation losses in the case of the free far-field zone radiation (i.e., in the case of vacuum or dielectric surroundings of metallic nanosphere with oscillating plasmons) can be accounted for as the Lorentz friction [18]. Nevertheless, when in the near-field zone (closer than the wave length corresponding to plasmon frequency) the energy receiver (i.e., other charge system, like semiconductor with its band system or another metallic nanosphere, as in the chain) is located, the irradiation losses are dominated by energy transfer via this near-field zone coupling channel. Existence of charged system of energy receiver modifies retarded e-m potential of emitting system and this modifies the Lorentz friction formula, which was derived in the standard form for dielectric surroundings. The enhancement of radiation losses in the case when the nanoparticles with dipole Mie surface plasmons (excited by incident external light) are deposited on the semiconductor surface lies behind the observed PV efficiency growth in new generation of solar cells metallically modified [24, 25, 27, 30, 32, 34]. In this case the related attenuation rate can be also estimated by application of the Fermi golden rule to the semiconductor inter-band transitions induced by dipole near-field coupling with plasmons [11, 14], as was demonstrated above. As it was proved [11], the resulting attenuation rate scales with nano-sphere radius, a, in a different way in comparison to far-field radiation losses in dielectric medium, as is displayed by the formulae (53), additionally with a renormalization coefficient expressed in terms of the band system parameters [11]. One can expect the similar behavior in the case of the near-field coupling between nanospheres in the chain, but for the sake of effectiveness of modeling one can assume that related attenuation rate has the form as that for the Lorentz friction renormalized only by some coefficient phenomenologically assumed and accounting for a mentioned modification of e-m potential space distribution, caused by the receiver presence apart of the source.

Let us consider first a single metallic nanosphere with dipole type surface oscillations with the dipole,

$$D = eN_e a R, \qquad (67)$$

where a is the nanosphere radius (of order of $10 - 50$ nm), N_e is the number of electrons, e is the electron charge. Oscillating dipole emits radiation and related energy losses can be expressed by the Lorentz friction (eventually renormalized by some phenomenological factor, as commented above). The general relativistic invariant formula for Lorentz friction force is as follows [18],

$$
\begin{aligned}
g_i &= \frac{2}{3} \frac{(eN_e)^2}{c} \left(\frac{d^2 u_i}{ds^2} + u_i u_k \frac{d^2 u_k}{ds^2} \right), \\
ds &= cdt \sqrt{1 - \frac{v^2}{c^2}}, \\
u &= \frac{v}{c\sqrt{1 - \frac{v^2}{c^2}}}, u_4 = \frac{i}{\sqrt{1 - \frac{v^2}{c^2}}}, \\
v &= a\omega_1 \dot{R}, \dot{R} \equiv \frac{1}{\omega_1} \frac{dR}{dt},
\end{aligned}
\qquad (68)
$$

The latter expression slightly modifies the definition of dipole derivation in (68) as it incorporates plasmon frequency, for the sake of simplicity of further formulae. As the velocity can be expressed by the dipole derivation, the lowest order of power expansion of relativistic denominator gives the nonlinear corrections to the Lorentz friction, which can be accounted for in the formula for the Lorentz friction induced electric field,

$$E_l = \frac{2}{3} \frac{\sqrt{\epsilon}}{c^3} (eN_e) a\omega_1^3 \left\{ \dddot{R} + \frac{\epsilon a^2 \omega_1^2}{c^2} a \left(\frac{3}{2} R^{(3)} \dot{R}^2 + 3\ddot{R} (\dot{R} \cdot \ddot{R}) + \dot{R} (\dot{R} \cdot \dddot{R}) \right) \right\}, \qquad (69)$$

with nanosphere dipole plasmon frequency, $\omega_1 = \frac{\omega_p}{\sqrt{3\epsilon}}$ (in a dielectric surroundings). In the above expression for the Lorentz friction induced electric field, the first term with the third order derivation of the dipole is the usual linear part of the Lorentz friction, while the next contribution (in parenthesis) is the lowest order nonlinear correction.

In the framework of perturbation method of solution of dynamic equation of oscillatory type for the dipole, one can notice that in the zero order of perturbation (when dissipation is neglected), $\ddot{R} + R = 0$, from which it follows that $\dddot{R} = -\dot{R}$. Thus one can simplify the formula (69), which leads to dynamical equation for the plasmon dipole including, in the first order of perturbation, the damping of plasmons due to scattering $\left(\frac{1}{\tau_0}\right)$ and due to radiation losses, i.e., due to the Lorentz friction (expressed by the rest of the right-hand-side of the equation (70) with the first term of linear contribution and with the next term of nonlinear type; the overall renormalization coefficient due to near-field coupling as commented above, is assumed here 1 for the sake of simplicity, but it could be changed in the final result),

$$\ddot{R} + R = -\frac{2}{\tau_0 \omega_1}\dot{R} - \frac{2}{3}\left(\frac{\omega_p a}{\sqrt{3}c}\right)^3\dot{R} + \frac{2}{3}\left(\frac{\omega_p a}{\sqrt{3}c}\right)^5\left\{-\frac{5}{2}\dot{R}\left(\dot{R}\cdot\dot{R}\right) + 3R\left(\dot{R}\cdot R\right)\right\}. \quad (70)$$

The above equation is complicated in mathematical sense and advanced methods of solutions must be applied, as described in [2]. According to this asymptotical methods for solution of nonlinear differential equation (70), one can find the solution in the following form,

$$R(t) = \frac{A_0 e^{-\frac{t}{\tau}}}{\sqrt{1 + \frac{9}{8}\gamma A_0^2\left(1 - e^{-\frac{2t}{\tau}}\right)}}\cos\left(\omega_1 t + \Theta_0\right), R = R\hat{n}, \hat{n} = \frac{r}{r},$$

$$\frac{1}{\tau\omega_1} = \frac{1}{\tau_0\omega_1} + \frac{1}{3}\left(\frac{\omega_p a}{\sqrt{3}c}\right)^3 \approx \frac{1}{3}\left(\frac{\omega_p a}{\sqrt{3}c}\right)^3, \quad (71)$$

$$\gamma = \tau\omega_1\frac{1}{3}\left(\frac{\omega_p a}{\sqrt{3}c}\right)^5.$$

In the formulae (71) both coefficients $\frac{1}{\tau\omega_1}$ and γ can be renormalized eventually by the mentioned above phenomenological factor (still held here 1, for simplicity). From the form of equation (71) it follows that $\frac{1}{\tau\omega_1}$ is always positive. Note that the scattering term, $\frac{1}{\tau_0} = \frac{v_F}{2a} + \frac{C v_F}{2\lambda_B}$, where v_F is the Fermi velocity in metal material of the nanosphere, λ_B is the bulk mean free path and C is the constant of unity order [11, 14]. This scattering term is negligible (for nanosphere radius beyond 10 nm) in comparison with the linear contribution of the Lorentz friction, as it is demonstrated in Fig. (3). The nonlinear correction scale is given by $\gamma \approx 10^{-4}\left(\frac{a[nm]}{10}\right)^2$. As this coefficient is small, one can neglect the related contribution in the denominator for the dipole solution (71), which results in ordinary linear solution of damped oscillations, which means that the nonlinear corrections to the Lorentz friction have no significance in the case of plasmon oscillations of a single nano-sphere.

This situation changes, however, considerably in the case of collective plasmon excitation propagating along the metallic nano-chain, as it will be described in the following paragraph.

5.2. Collective plasmon wave-type propagation along the nano-chain in the nonlinear regime

In the case of the metallic nano-chain one has to take into account the mutual affecting of nanospheres in the chain. Assuming that in the sphere located in the point r we deal with the

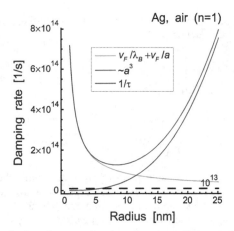

Figure 3. Contribution to the damping rate of surface plasmon oscillations in the nano-sphere versus the nano-sphere radius, including the scattering attenuation (green line) and the linear Lorentz friction damping (blue line); for radii greater than ca 10 nm the second channel dominates in the overall damping (red line)

dipole D, then in the other place r_0, the dipole type electric field attains the form as follows (including electro-magnetic retardation),

$$E\left(r, r_0, t\right) = \frac{1}{\varepsilon r_0^3} \left\{ 3n_0 \left(n_0 \cdot D\left(r, t - \frac{r_0}{v}\right)\right) - D\left(r, t - \frac{r_0}{v}\right) \right\},$$
$$n_0 = \frac{r_0}{r_0}, v = \frac{c}{\sqrt{\varepsilon}}. \tag{72}$$

This allows for writing of the dynamical equation for plasmons at each nanosphere of the chain which can be numbered by integer l (d is here the separation between nanospheres in the chain). The first term of the right-hand-side in the following formula (73) describes the dipole type coupling between nanospheres [13] and the other two terms correspond to contribution due to plasmon attenuation (in the latter term the Lorentz friction caused electric field accounts also for nonlinear corrections). The index α enumerates polarizations, longitudinal and transversal ones with respect to the chain orientation.

$$\ddot{R}_\alpha + R_\alpha\left(ld, t\right) = \sigma_\alpha \frac{a^3}{d^3} \sum_{m=-\infty, m \neq l}^{\infty} \frac{R_\alpha\left(md, t - \frac{d|l-m|}{v}\right)}{|l-m|^3} - \frac{2}{\tau_0 \omega_1} \dot{R}_\alpha\left(ld, t\right) + \frac{e}{ma\omega_1^2} E_\alpha\left(ld, t\right),$$
$$\sigma_\alpha = \begin{cases} -1, \alpha = x, y \\ 2, \alpha = z \end{cases} \tag{73}$$

The summation in the first term of the right-hand-side of the equation (73) can be explicitly performed in the manner as presented in [13], because it is the same as for the linear theory formulation. Similarly as in the linear theory framework, one can change to the Fourier picture taking advantage of the chain periodicity (in analogy to crystals with the reciprocal lattice), i.e.,

$$R_\alpha\left(ld, t\right) = \tilde{R}_\alpha\left(k, t\right) e^{-ikld},$$
$$0 \leq k \leq \frac{2\pi}{d}, \tilde{R}_\alpha\left(k\right) \sin\left(t\omega_1 + \beta\right). \tag{74}$$

Thus the equation (73) can be rewritten in the following form (the Lorentz friction term was represented similarly as in equation (70)),

$$\ddot{\tilde{R}}_\alpha(k,t) + \tilde{\omega}_\alpha^2 \tilde{R}_\alpha(k,t) \left\{ \frac{1}{\tau_\alpha \omega_1} + \frac{1}{3}\left(\frac{\omega_p}{\sqrt{3}c}\right)^5 \left(\frac{5}{2}|\dot{\tilde{R}}_\alpha(k,t)|^2 - 3|\tilde{R}_\alpha(k,t)|^2\right) \right\},$$

$$\tilde{\omega}_\alpha^2 = 1 - 2\sigma_\alpha \frac{a^3}{d^3}\cos(kd)\cos\left(\frac{d\omega_1}{v}\right),$$

$$\frac{1}{\tau_{x,y}\omega_1} = \frac{1}{\tau_0\omega_1} + \frac{1}{4}\left(\frac{\omega_1 d}{v}\right)\frac{a^3}{d^3}\left(\left(\frac{\omega_1 d}{v}\right)^2 - (kd - \pi)^2 + \frac{\pi^2}{3}\right),$$

$$\frac{1}{\tau_z\omega_1} = \frac{1}{\tau_0\omega_1} + \frac{1}{2}\left(\frac{\omega_1 d}{v}\right)\frac{a^3}{d^3}\left(\left(\frac{\omega_1 d}{v}\right)^2 - (kd - \pi)^2 + \frac{\pi^2}{3}\right),$$

(75)

In the above formulae the remarkable property is linked with the expressions for the attenuation rate for both polarizations. Two last expressions in equation (75) give these damping rates explicitly and one can notice that they may change their signs depending on values for d,a,k. In Fig. (4) the region of negative value for damping rates is depicted (for both polarizations).

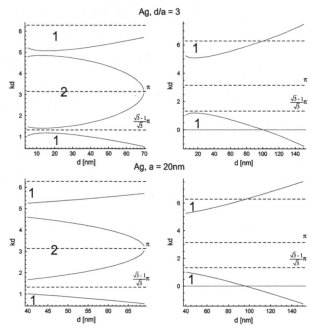

Figure 4. Regions for negative value of damping rates for plasmon-polaritons in the chain (1 for longitudinal polarization modes and 2 for transversal ones, in the nolinear formulation framework; for the linear theory (red lines gives the position of vanishing damping rate for longitudinal modes of plasmon-polaritons and blue lines the same for transversal modes)

Applying the same methods for solution of the nonlinear equation (75) as in the former paragraph using the asymptotic methods [2], one can find the corresponding solutions for both regions with positive and negative damping rate, respectively,

$$for \; \frac{1}{\tau_\alpha \omega_1} > 0,$$

$$\tilde{R}_\alpha\,(k,t) = \frac{A_{\alpha 0}e^{-\frac{t}{\tau_\alpha}}}{\sqrt{1+\gamma_\alpha A_{\alpha 0}^2\left(1-e^{-\frac{2t}{\tau_\alpha}}\right)}}\cos\left(\omega_\alpha t+\Theta_0\right)\to_{t\to\infty} 0,$$

$$\gamma_\alpha = |\tau_\alpha \omega_1|\left(\frac{\omega_1 a}{c}\right)^3\frac{1}{4}\left(\frac{5}{2}\tilde{\omega}_\alpha^2-1\right);$$

$$for \; \frac{1}{\tau_\alpha \omega_1} < 0,$$ \hfill (76)

$$\tilde{R}_\alpha\,(k,t) = \frac{A_{\alpha 0}e^{\frac{t}{|\tau_\alpha|}}}{\sqrt{1+\gamma_\alpha A_{\alpha 0}^2\left(e^{\frac{2t}{|\tau_\alpha|}}-1\right)}}\cos\left(\omega_\alpha t+\Theta_0\right)\to_{t\to\infty}\frac{1}{\sqrt{\gamma_\alpha}}\cos\left(t\omega_\alpha+\Theta_0\right),$$

$$D_\alpha = \frac{eN_e a}{\sqrt{\gamma_\alpha}}\frac{1}{2}\left\{e^{i(\omega_\alpha t - kld)}+e^{-i(\omega_\alpha + kld)}\right\}.$$

From the above formulae it follows that for positive attenuation rate we deal with ordinary damped plasmon-polariton propagation, not strongly modified in comparison to linear theory (due to small value of the factor γ). Nevertheless, in the case of negative damping rate the solution behaves differently, i.e., in longer time scale this solution stabilizes on the constant amplitude value independently of initial conditions. This remarkable property characterizes undamped propagation of plasmon-polariton along the chain (if one turns back to dipole explicit form (67), then typical planar wave formula with constant amplitude describes this undamped mode, as written in the last equation of (76)). The region of negative damping corresponds thus, within the nonlinear approach, to undamped modes with the fixed amplitude. Let us note that the same region was linked with instability of the linear theory (which was, however, the arte-fact of the linear approach).

Finally, one can calculate the group velocity of the undamped plasmon-polariton mode in the following form,

$$v_\alpha = \frac{d\omega_\alpha}{dk} = \omega_1 d\frac{\sigma_{alpha}a^3\sin\left(kd\right)\cos\left(\frac{d\omega_1}{c}\right)}{d^3\sqrt{1-2\sigma_\alpha\frac{a^3}{d^3}\cos\left(kd\right)\cos\left(\frac{d\omega_1}{c}\right)}}.$$ \hfill (77)

From this formula it follows that the group velocity of the undamped wave type collective plasmon excitation (called plasmon-polariton) may attain different values depending on a,d,k.

Indicated above undamped mode of propagation of collective surface plasmons seems to match with experimentally observed long range propagation of plasmon excitations along the metallic nano-chain [1, 6, 21, 22, 36]. The constant and fixed value of the amplitude for these oscillations (76) are independent of initial conditions, which means that these excitations will be present in the system even if are excited by arbitrarily small fluctuations. Thus one can conclude that these are self-exciting modes which are always present in the system provided that radii of spheres and their separation in the chain have values for which attenuation rates (75) are negative.

6. Conclusions

We have demonstrated the practical utilization of RPA semiclassical description of plasmon oscillations in metallic nanospheres. The oscillatory form of dynamics both for volume and surface plasmons, rigorously described upon the RPA semiclassical limit fits well with the large nanosphere case, when the nanosphere radius is greater than 10 nm and lower than 60 nm, (for Au, Ag or Cu material), what is confirmed by experimental observations, on the other hand. The most important property of plasmons on such large nanospheres is the very strong e-m irradiation caused by these excitations, which results in quick damping of oscillations. The attenuation effects for plasmons were not, however, included into the quantum RPA model. Nevertheless, they could be included in a phanomenological manner, taking advantage of the oscillatory form of dynamic equations. Some information on plasmon damping can be taken from microscopic analyzes of small metallic clusters (especially made by LDA and TDLDA methods of numerical simulations employing Kohn-Sham equation). For larger nanospheres, these effects, mainly of scattering type (also Landau damping), are, however, not specially important as diminishing with radius growth as $\frac{1}{a}$. The irradiation effects overwhelming the energy losses in the case of large nanospheres can be grasped in terms of the Lorentz friction, which reduces the charge movement. This approach has been analyzed in the present paper. Two distinct situations were indicated, the first one–of the free radiation to far-field zone in dielectric/vacuum surroundings of single nanoparticle and the second one, when in the near-field zone of plasmons on the nanosphere, an additional charged system is located. This additional charge system of the e-m energy receiver strongly modifies the e-m potentials of the source and in this way modifies energy emission. In particular, the Lorentz friction is modified in this case in comparison to simple free emission to far-field zone. The e-m energy receiver located close to emitting nanosphere, could be semiconductor (as in the case of metallically modified solar cells) or other metallic nanospheres (as in the case of metallic nano-chain). Both these situations have been analyzed in this chapter.

The energy transfer from surface plasmon oscillations in metallic nanospheres deposited on photo-active surface of semiconductor to the substrate electron band system has been described upon the scheme of Fermi golden rule. The dependence of the effect with respect to the nanosphere dimension has been analyzed in details and the comparison with experimental data is given. We have elucidated the reason of the experimentally observed giant enhancement of PV efficiency caused by plasmon effect. Surface plasmons in metallic nanocomponents deposited on the semiconductor surface act as solar energy converters. The surface plasmons excited in the metallic nanospheres by incident photos couple with band electrons in the semiconductor substrate in the near-field regime. This dipole type coupling is more effective in comparison to the ordinary photo-effect coupling of electrons with incident photons, due to allowance of indirect inter-band transitions in the metallically locally modified systems. The metallic nanosphere coupled with substrate electrons in near-field zone of plasmon oscillations is not translational invariant system and thus the momentum conservation does not constrain quantum transitions in this case. This leads to strong enhancement of the inter-band transition probability, quenched, however, with the nanosphere radius growth, in the manner as presented in this paper.

With regard to collective plasmon fluctuations in ordered metallic systems, we have shown previously that along the infinite nano-chain the collective plasmon-polaritons can propagate (being collective surface plasmons coupled by e-m field in near-field zone), which at certain values of nanosphere radii and separation in the chain, appear as undamped modes. Simultaneously, the instability regions of linear theory of plasmon-polariton dynamics occur, which shows that the nonlinear corrections must be included. In this paper we have developed the nonlinear theory of collective plasmon-polariton dynamics along the chain, including nonlinear corrections to Lorentz friction force. Even though the related nonlinearity is small, it suffices to regularize the instable linear approach. As the most important observation, we noted the presence of undamped excitations (instead of those instable within the linear approach), which have fixed amplitude independently how small or large the initial conditions were. This excitations, typical for nonlinear systems, would have some practical significance, e.g., to enhance sensitivity of antennas with coverings by plasmon systems offering self-induced collective plasmon-polaritons in wide range of frequencies, which would be excited by even very small signal (the energy to attain the stable level of plasmon-polariton amplitude would be supplied, in this case, by an external auxiliary source).

Acknowledgments

Supported by NCN project No: 166719 Sonata 2

Author details

Jacak Witold
Institute of Physics, Wrocław University of Technology, Wyb. Wyspiańskiego 27, 50-370 Wrocław, Poland

7. References

[1] Barnes, W. L., Dereux, A. & Ebbesen, T. W. [2003]. *Nature* 424: 824.

[2] Bogolubov, N. N. & Mitropolskyj, J. A. [2005]. *Asymptotical methods in theory of nonlinear oscillations*, Nauka, Moscow.

[3] Bohren, C. F. & Huffman, D. R. [1983]. *Absorption and Scattering of Light by Small Particles*, Wiley, New York.

[4] Brack, M. [1989]. *Phys. Rev. B* 39: 3533.

[5] Brack, M. [1993]. *Rev. of Mod. Phys.* 65: 667.

[6] Brongersma, M. L., Hartman, J. W. & Atwater, H. A. [2000]. *Phys. Rev. B* 62: R16356.

[7] Ekardt, W. [1984]. *Phys. Rev. Lett.* 52: 1925.

[8] Ekardt, W. [1985]. *Phys. Rev. B* 31: 6360.

[9] Ekardt, W. [1986]. *Phys. Rev. B* 33: 8803.

[10] Garcia de Abajo, F. J. [2010]. *Rev. Mod. Phys.* 82: 209.

[11] Jacak, J., Krasnyj, J., Jacak, W., Gonczarek, R., A.Chepok & Jacak, L. [2010]. *Phys. Rev. B* 82: 035418.

[12] Jacak, L., Krasnyj, J. & Chepok, A. [2009]. *Fiz. Nisk. Temp.* 35: 491.

[13] Jacak, W., Krasnyj, J., Jacak, J., Chepok, A., Jacak, L., Donderowicz, W., Hu, D. & Schaadt, D. [2010]. *J. Appl. Phys.* 108: 084304.

[14] Jacak, W., Krasnyj, J., Jacak, J., Donderowicz, W. & Jacak, L. [2011]. *J. Phys. D: Appl. Phys.* 44: 055301.

[15] Jacak, W., Krasnyj, J., Jacak, J., Gonczarek, R., Chepok, A., Jacak, L., Hu, D. & Schaadt, D. [2010]. *J. Appl. Phys.* 107: 124317.

[16] Kiriejew, P. S. [1969]. *Physics of Semiconductors*, PWN, Warsaw.

[17] Kresin, V. V. [1992]. *Phys. Rep.* 220: 1.

[18] Landau, L. D. & Lifshitz, E. M. [1973]. *Field Theory*, Nauka, Moscow.

[19] Losurdo, M., Giangregorio, M. M., Bianco, G. V., Sacchetti, A., Capezzuto, P. & Bruno, G. [2009]. *Solar Energy Mat. & Solar Cells* 93: 1749.

[20] Maier, S. A. [2007]. *Plasmonics: Fundamentals and Applications*, Springer, Berlin.

[21] Maier, S. A., Kik, P. G. & Atwater, H. A. [2003]. *Phys. Rev. B* 67: 205402.

[22] Markel, V. A. & Sarychev, A. K. [2007]. *Phys. Rev. B* 75: 085426.

[23] Mie, G. [1908]. *Ann. Phys.* 25: 337.

[24] Morfa, A. J., Rowlen, K. L., Reilly, T. H., Romero, M. J. & Lagemaat, J. [2008]. *Appl. Phys. Lett.* 92: 013504.

[25] Okamoto, K., Niki, I., Scherer, A., Narukawa, Y. & Kawakami, Y. [2005]. *Appl. Phys. Lett.* 87: 071102.

[26] Okamoto, K., Niki, I., Shvartser, A., Narukawa, Y., Mukai, T. & Scherer, A. [2004]. *Nature Mat.* 3: 601.

[27] Pillai, S., Catchpole, K. R., Trupke, T., Zhang, G., Zhao, J. & Green, M. A., [2006]. *Appl. Phys. Lett.* 88: 161102.

[28] Pines, D. [1999]. *Elementary Excitations in Solids*, ABP Perseus Books, Massachusetts.

[29] Pines, D. & Bohm, D. [1952]. *Phys. Rev.* 92: 609.

[30] Schaadt, D. M., Feng, B. & Yu, E. T. [2005]. *Appl. Phys. Lett.* 86: 063106.

[31] Stuart, H. R. & Hall, D. G. [1998]. *Appl. Phys. Lett.* 73: 3815.

[32] Sundararajan, S. P., Grandy, N. K., Mirin, N. & Halas, N. J. [2008]. *Nano Lett.* 8: 624.

[33] Wen, C., Ishikawa, K., Kishima, M. & Yamada, K. [2000]. *Sol. Cells* 61: 339.

[34] Westphalen, M., Kreibig, U., Rostalski, J., Lüth, H. & Meissner, D. [2003]. *Sol. Energy Mater. Sol. Cells* 61: 97.

[35] Yannouleas, C., Broglia, R. A., Brack, M. & Bortignon, P. F. [1989]. *Phys. Rev. Lett.* 63: 255.

[36] Zayats, A. V., Smolyaninov, I. I. & Maradudin, A. A. [2005]. *Phys. Rep.* 408: 131.

Chapter 9

Plasmonic Conducting Polymers for Heavy Metal Sensing

Mahnaz M. Abdi, Wan Mahmood Mat Yunus, Majid Reayi and Afarin Bahrami

Additional information is available at the end of the chapter

1. Introduction

Conducting polymers are a sub-group of organic and inorganic electrical conductors which can be considered as interdisciplinary science and technology. They are in the intersection of three disciplines: physics, chemistry and engineering. These macromolecular materials are unique combination of electronic and optical properties of metals and semiconductors with the processing advantages and mechanical properties of polymers [1].

Conducting polymers (CPs) such as polyaniline, polypyrrole and polythiophen have received great attention to the chemists and physicists during the last decade due to their potential applications in different fields such as mechanical and optical sensors, actuators, light-emitting diodes (LED), transistors, energy storage, supercapacitors. [2-5]. Conducting sensors based on PPy film with different dopant were developed for detecting volatile aromatic hydrocarbons [6] and volatile organic solvent sensors have been fabricated using PPy films on conducting glass substrates [7].

Different techniques including electrochemical quartz crystal microbalance [5], Fourier transform infrared spectroscopy (FTIR) [8], photoacoustic effect [9], and electromagnetic interference shielding effectiveness (EMI SE) [10-11] have been used to study the optical and electrical properties of conducting polymers. Surface plasmon resonance (SPR) is one of these methods that provided considerable information about electropolymerization, doping and dedoping processes [12].

Surface plasmon resonance (SPR) is a quantum optical-electrical phenomenon arising from the interaction of light with a metal surface. Surface plasmons are surface electromagnetic waves that propagate in a direction parallel to the metal/dielectric (or metal/vacuum) interface. The waves are on the boundary of the metal and the external medium (air or water for example); thereby these oscillations are very sensitive to any change of this boundary, such as the adsorption of molecules to the metal surface [13].

2. Basic of surface plasmon resonance

When a polarized light strikes the area between the glass and air, at a certain angle (θ), a total internal reflection (TIR) is generated. In this case the incident light is equal to the reflective light and an electric field intensity, known as evanescent wave is generated. When a metal is placed at the interface between the glass and air, at a specific angle, the plasmon on the metal surface couples with the energy from the light and cause a reduction in the intensity of the reflected light, achieve the resonance at the given wavelength and angle, where SPR is occurring. Any changes in the chemical composition of the environmental within the metal surface of the plasmon field cause a shift in the angle of the reflected light. Thus the magnitude of this change can be determined by measuring the angle of the reflected light by the metal surface using a sensogram.

SPR is an optical phenomenon related to charge density oscillation at the interface between two materials that their signs of dielectric constant are opposite [14]. The basic of SPR is the propagation of surface plasmon wave (SPW) along the interface of metal and dielectric which behaves like quasii-free electron plasma. The wave number of light should be matched with the wave number of surface plasmon given as:

$$n_p \sin\theta = \mathrm{Re}\left[\sqrt{\frac{\varepsilon_d \varepsilon_m}{\varepsilon_d + \varepsilon_m}}\right] \tag{1}$$

where n_p , θ are refractive index of the prism and angle of resonance, while ε_m and ε_d are the dielectric constants of metal and last medium, respectively [13]. The thickness and dielectric constant of thin films deposited on the gold surface can be measured by this optical technique.

The electrochemical surface plasmon resonance (ESPR) technique is good combination of surface plasmon resonance with electrochemical measurements, which can be used for simultaneous optical characterization and electrochemical polymerization of the electrode–electrolyte interface [5]. In this technique, the real ($\Delta\varepsilon_{real}$) and imaginary ($\Delta\varepsilon_{imaginary}$) parts of the dielectric constant are parameter of the average thickness of the adsorbed layer, electrode potential modulation (ΔV), as well as changes in the surface charge density ($\Delta\delta$) as follows [15-16].

$$\frac{\Delta\theta_{SPR}}{\Delta V} \approx C_1\left(\frac{\Delta\varepsilon_{real}}{\Delta V}\right) + C_2\left(\frac{\Delta\varepsilon_{imaginary}}{\Delta V}\right) + C_3\left(\frac{\Delta d}{\Delta V}\right) + C_4\left(\frac{\Delta\delta}{\Delta V}\right) \tag{2}$$

where C_1, C_2, C_3 and C_4 are constants, Δd is the change of the average thickness of the adsorbed layer, and $\Delta\delta$ represents the effects of the electron density on the surface plasmon resonance angle. Above mentioned formula shows a good combination of surface plasmon resonance with electrochemical measurements, which can be used for simultaneous optical characterization and electrochemical polymerization of the electrode–electrolyte interface [5].

If the dielectric constant experessed as:

$$\varepsilon = \varepsilon_{real}\left(\lambda\right) + i\left(\frac{2\sigma\lambda}{C}\right) \qquad (3)$$

It has been shown that the imaginary part of the dielectric constant can be changed by variation of conductivity of material layer and this relation is monitored as:

$$\varepsilon_{imaginary} = \frac{2\sigma\lambda}{C} \qquad (4)$$

where C is the speed of light with wavelength of λ incident on the material with conductivity of σ.

3. Application of SPR in Conducting polymers

In recent years, surface plasmon resonance (SPR) has been widely demonstrated as an effective optical technique for in situ investigation of optical and electrical properties of conducting polymers such as polypyrrole films [5, 15]. Damos and his coworker showed that during electropolymerization and doping/dedoping processes of thin polypyrrole films on flat gold surfaces a significant change in the surface plasmon angle position was produced. These changes in the electrochemical and optical properties of the polypyrrole films were due to the changes in the real and imaginary parts of the complex dielectric constant during doping/dedoping processes [5].

They prepared PPy films by in situ electropolymerization using potentiostatic, galvanostatic and potential cycling process (potentiodynamic) methods and showed that the shape of the reflectivity curves changed during polymerization process, indicating that the growth processes are also efficient in producing polymer films with high conductivity. In addition, they observed that during polymerization the resonance angle shifted toward higher values which can be related to the changes in the optical properties at the electrode–solution interface when polymer grow at the surface of the metal.

On the other hand, the changes of real and imaginary part of dielectric constant for polypyrrole films polymerized by potentiodynamic method was more identical and reversible compared with those polymerized by potentiostatic or galvanostatic conditions indicating higher ability for the ions exchange of the films prepared by potentiodynamic. It was shown that the real part of the dielectric constant for polypyrrole film shows a small decrease and the imaginary part increased by switching the potential between the reduced and oxidized states of the polypyrrole films. It was rsulted that the changes in the imaginary part of the dielectric constant are associated with changes of the electronic energy states of the polypyrrole according to the polaron/bipolaron model [5].

During the oxidation and doping process, intermediary energy levels are created between the valance band (VB) and conduction band (CB); causing a decrease in the energy required for the electronic transition. As a result, the polymers with higher conductivity exhibit high values of "$\varepsilon_{imaginary}$" which results in a decrease in the real part of the complex dielectric constant. The study on the optical and electrical properties of polypyrrole at different oxidation levels showed that there is a correlation between conductivity and formation of polaron and bipolaron species [17].

The electrical properties of the PPy-CHI composite films was investigated in our previous work [18] and it was revealed the enhanced electrical properties of composite films in the presence of a certain amount of CHI. It was observed that the wavelength of electron transition in Uv-vis spectrum increased with the increase in CHI content, indicating a decrease of the band gaps between valance band (VB) and conduction band (CB).

SPR technique was used to detect ochratoxin A (OTA) which is a toxic fungal metabolite and acts as a DNA damaging agent [19]. In this research, Jorn prepared sensing area by in situ electropolymerziation of pyrrole in the presence of chloride dopant. He observed that PPy prepared in the presence higher amount of NaCl presented lower conductivity and so that the SPR angle increased. On the other hand, they believed that adsorption of Cl⁻ on the gold layer affected double-layer charging, resulted in a decrease in SPR angle. The linearity range of 0.1-10 µg/ml (with $R^2 = 0.9689$) was expersed for this ochratoxin A (OTA) sensor.

Guedon et al, [20], reported a DNA hybridization sensor based on ppy- ODN matrix coupled with SPR imaging. They prepared SPR chip by electrospotting pyrrole and pyrrole linked to an oligonucleotide (ODN). The SPR imaging configuration is closely based on the same configuration with regular SPR, in which the metal surface is imaged on a CCD camera via an imaging lens. This kind of SPR has advantages of sensing on several areas of the gold surface at the same time [21-22]. Electrospotting method is used for the construction of biochips on individually addressable microelectrode arrays. In this work [20] they optimized hybridization response of the polypyrrole/DNA probes versus spot thickness and density of immobilized probe. The optimal thickness of the spot was found to be close to 11 nm and a surface density of polypyrrole/DNA probes of 130 fmol/ mm² (590 pg/mm²) was reported to optimize the hybridization signal that can be detected directly.

4. Heavy metals and SPR detection

Heavy metals such as mercury and lead ions have long been recognized as a harmful environmental pollutant. Lower levels of lead can affect on the central nervous system, kidney, and blood cells [23] and in severe cases can cause convulsions, coma, and even death. Lead interferes with the development of the nervous system and is therefore particularly toxic to children and unborn babies, causing potentially permanent learning and behavior disorders [24]. Overexposure to mercury can damage the central nervous system which can cause memory loss, loss of appetite, personality changes, and lack of coordination, reproductive problems, and possibly death [25].

There are many techniques for analysis trace metal e.g., atomic absorption, atomic emission, and fluorescence spectrometry [26], inductively coupled plasma-mass spectrometry, (ICP-MS), [27] and electrochemical techniques (such as ion-selective potentiometry and anodic stripping voltammetry) [8]. In contrast of their attractive analytical "figures of merit", each of these techniques suffers of some disadvantages. Applications of all these methods require knowledge of chemistry and instrumentation and need exactitude apparatuses (28-29)

A major disadvantage of ICP-MS is the high capital cost of the instrumentation. In addition, this instrument is bulky and not selective to different charge states of an element. The non linearity of the calibration curves is a disadvantage of the atomic absorption spectroscopy, (AAS), technique (in absorbance range higher than 0.5 to 1). Running and investment costs of inductively coupled plasma atomic emission spectroscopy, (ICP-AES), are high. Another disadvantage of ICP-AES is spectral interferences (many emission lines). Voltammetric methods are simple, inexpensive, and portable but they suffer of interferences inherent in complex sample matrix. Anodic stripping voltammetry can only measure amalgam-forming metal species. Thereby, the complementary methods have developed to overcome some of these shortcomings [29-30].

Surface plasmon resonance (SPR) is an effective optical technique to detect monolayer thicknesses of the material on the conducting surface and has been used to study the optical and electrical properties of conducting polymers such as polypyrrole films. This technique exhibited a good sensitivity, stability, and reproducibility and some times the changes in refractive index of approximately 10^{-5} can be detected by SPR technique [31]. Usually a thin layer of gold is used as the SPR excitation layer. This technique was used to detect and characterize deposited organic layers and functionalized polymer films on SPReeta sensors[32], and also in biosensor such as pseudomonas aeruginosa cells [33]. One of the important parameters used for evaluating the quality of biodiesel is copper corrosion and the standard test of copper strip is used for detection this ion. It has been shown [34] that surface Plasmon resonance based on the self-assembly of squarylium dye containing cystamine can be used as sensing area for determining copper concentration.

In our previous work [35], we prepared a sensing layer based on a composite of conducting polymer (polypyrrole) and chitosan for the detection of Cu^{2+}. Different corrosion levels classified as class 1a according to the standard copper strip test (CST) could be recognized in the sample using this optical sensor. In this work four types of biodiesel samples were used to detect different levels of corrosion and it was found that the sensor based on polypyyrole-chitosan can measure different levels of corrosion more precisely than the standard CST. In addition, the variation of reflectivity with thickness was studied in this research and it was observed that the reflectivity monotonically increased with increasing thickness of polymer film.

In addition, the refractive index and resonance angle of the composite film were determined by Feresenel theory and the results are presented in table 1. It was found that the real part of the refractive index decreased and the imaginary part increased with increasing thickness, as expected from the Kramers–Kroning relation [36-37].

Yu et al., [31] developed a SPReeta sensor for Hg(II) detection using polypyrrole (PPy) and 2-mercaptobenzothiazole (2-MBT). Binding interactions of mercury ion with polypyrrole (PPy) and (2-MBT) was monitored by surface plasmon resonance (SPR) technique. They used 2-MBT as chelating agent to forms complexes with Hg(II) in order to enhance the specificity of Hg²⁺ determination. The PPy-modified SPReeta surface area was prepared by in situ pyrrole electropolymerization on the gold surface at a constant potential of +0.85V for 100–150 s. A solution of 2-MBT in ethanol/water (1:9 v/v) was employed to serve as a sensitivity enhancement agent for detecting Hg²⁺ bound on the PPy surface. They reported the detection limit of 0.01 ppm with an increase of SPR angle of 20±10 RU.

Thickness t (nm)	Resonance angle θ_R (deg)	Real part ε_{real}	Imaginary Part ($\varepsilon_{imaginary}$)	Reflectivity R
20.8	58.224	1.6654	0.153	0.260
31.8	60.726	1.6321	0.169	0.381
40.2	60.896	1.6035	0.186	0.458
49.7	62.818	1.5812	0.200	0.520
58.6	63.336	1.5661	0.230	0.579

Table 1. Refractive index and resonance angle of PPy–CHI film. The refractive index and the thickness of the gold layer were 0.235+3.31i and 49 nm, respectively.

5. SPR Detection of Mercury and Lead Ions

5.1. Set up configuration and Formulation

The aim of this research was to fabricate a sensor based on surface plasmon resonance (SPR) using conducting polymer to detect trace amounts of mercury and lead ions. SPR characterisation of the thin films was performed by using a set up assembled according to the Kretschmann's prism configuration (Figure 1). Data analysis was done by Matlab software using Fresnel formula .The He-Ne laser (with wavelength of 632.8nm) was used as light source and silicon detector (Newport, 818-SL) and Lock-In-Amplifier were used for detection the light. The sensitivity of amplifier was 20 mV and the chopper frequency was 125 Hz.

The values of optical constants were calculated by comparing the experimental SPR data to the theoretical values provided by the Fresnel theory applied to the SPR optical configuration used for our study [38]. The values of the real and imaginary parts of the refractive indexes of PPy and PPy-CHI films were obtained by nonlinear least square fitting using Fresnel equations to the three-layer of SPR system.

In the Fresnel equation for p-polarized light the value of reflection coefficient can be written as:

$$r = \frac{E_i}{E_r} \tag{5}$$

where E$_i$ and E$_r$ are the incident and reflected electrical fields, respectively [38]. At the interfaces between N layers, the reflection coefficient is as follow:

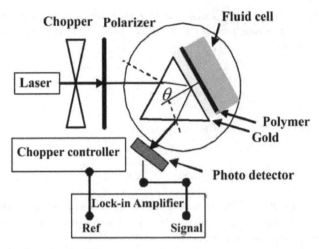

Figure 1. Set up configuration for SPR measurements.

$$r = \frac{m_{21} + m_{22}\tau_2 - m_{11}\tau_0 - m_{12}\tau_2\tau_0}{m_{21} + m_{22}\tau_2 + m_{11}\tau_0 + m_{12}\tau_2\tau_0} \qquad (6)$$

where m_{ij} are the matrix layer elements and for single layer with the thickness t, the matrix element is

$$M = \begin{bmatrix} \cos\delta & -i\dfrac{\sin\delta}{\tau_1} \\ -i\tau_1\sin\delta & \cos\delta \end{bmatrix} \qquad (7)$$

Where the δ (the phase shift due to the beam passing through different layer) and τ_1 can be written as:

$$\delta = \frac{2\pi}{\lambda} t n_1 \cos\theta_{t_1} \qquad (8)$$

$$\tau_1 = \frac{n_1}{\cos\theta_{t_N}}\sqrt{\varepsilon_0\mu_0} \qquad (9)$$

Where $\lambda\ \varepsilon_0\ \mu_0$ are the wavelength of light used during the measurements, dielectric constants, and permeability. If the refractive index of gold and the sample obey following relation, resonance takes place [13].

$$n_p\sin\theta_R = \sqrt{\frac{n_m^2 n_d^2}{n_m^2 + n_d^2}} \qquad (10)$$

Where θ_R, n_p, n_m, and n_d are the resonance angle, refractive indices of the prism, metal (gold layer) and dielectric medium (sample), respectively and the refractive index of the sample can be determined from:

$$n_d = \sqrt{\frac{n_m^2 n_p^2 \sin^2 \theta_R}{n_m^2 - n_p^2 \sin^2 \theta_R}} \tag{11}$$

The angle of incidence at the interface between the prism and the gold layer is obtained from following Equation [39]:

$$\theta_2 = A - \arcsin\left(\left(n_{air} / n_p\right)\sin\theta_1\right) \tag{12}$$

Where A is the angle of the prism, θ_1 is the angle of incidence of the light beam directed to the prism, and n_{air} is the refractive index of air.The refractive index and resonance angle of the sample will be found by minimizing the sum [40]

$$\Gamma = \sum_\theta \left[R_{Exp}\left(\theta_2, n_2\right) - R_{Theory}\left(\theta_2, n_2\right) \right] \tag{13}$$

Where R_{Exp} and R_{Theory} are the experimental and theoretical, reflectivity.

In the Fresnel equation the reflectivity is a function of θ (angle of incidence light at the interface between the prism and surface), wavelength of light, refractive index of sample as well as refractive index and thickness of gold layer [41].

5.2. Pyrrole electropolymerization

A thin layer of gold (49 nm) was deposited onto a glass microscope slide and electrochemical deposition of PPy-CHI and PPy films on the gold substrate were performed potentiostatically imposing a constant potential of 0.85 V (vs SCE) for 150 sec. In order to form a thin film on the gold substrate with a homogeneous surface, electrodeposition should be done at low potential in short time. The sample of PPy-CHI was prepared in a solution containing 0.015 M Pyrrole, 0.005 M P-TS and 0.07% (w/v) chitosan in acetic acid. The results from FT-IR, electrical conductivity, and XRD study confirmed the incorporation of chitosan in PPy structure [42].

The PPy film was prepared using the same composition of pyrrole and P-TS without chitosan. The surface of the polymer films was washed with distilled deionized water (DDW). The thin slides were attached to the prism with high index glass (ZF52, n=1.85) by using index-matching fluid (n=1.52 new port company). The precision of rotator stage was 0.003. The refractive indexes of PPy and PPy-CHI thin films (in contact with air) by SPR technique were obtained in our previous work [43]. Since they produced a sharp peak of resonance angle, it was shown these polymers are capable for using in sensitive optical sensors.

DDW was run on the surface of the films to establish a baseline. When a stable baseline of SPR angle was obtained, the PPy and PPy-CHI film exhibited a SPR angular profiles that was deep enough for precise definition of the resonance angle. The values of the real and imaginary parts of the refractive indexes of PPy and PPy-CHI films were obtained by nonlinear least square fitting using Fresnel equations to the three-layer of SPR system.

The resonance of the Au-PPy thin film occurred at larger angles compared to the Au, which can be ascribed to a change in the actual refractive index of the gold layer when the PPy film is deposited on it. Figure 2 shows the surface plasmon resonance (SPR) reflectivity curves for the glass/Au/DDW, (Au), and glass/Au/PPy/DDW, (Au-PPy), and glass/Au/PPy-CHI/DDW, (Au-PPy-CHI), systems.

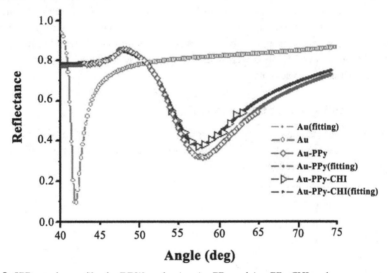

Figure 2. SPR angular profiles for DDW on the Au, Au-PPy and Au- PPy-CHI surface.

The resonance angle (θ_{spr}) and the optical constant for the PPy film were $(57.67\pm0.01)^0$ and $1.598+0.159i$, respectively. SPR angle for the PPy-CHI film appeared in lower value indicating formation of a thin film of polypyrrole with chitosan on the gold surface. The resonance angle (θ_{spr}) and the optical constant of the PPy-CHI film were $(56.75\pm0.01)^0$ and $1.601+0.165i$, respectively. The dip widths of the curves for the PPy-CHI and PPy films were larger than Au/DDW, which was proportional to the attenuation of the surface plasmon. The thickness of the thin films was calculated using the Fresnel equation and the values of 26 nm and 28 nm were obtained for the PPy-CHI and PPy thin films, respectively. When a stable baseline of SPR angle was obtained for PPy and PPy-CHI films, the surface of the films were ready for binding tests for Hg^{2+} and Pb^{2+} samples.

5.3. Binding of Hg²⁺ and Pb²⁺ on to PPy and PPy-CHI films

Different concentrations of Hg^{2+} and Pb^{2+} in aqueous solution were injected in to the fluid cell attached to the films, and the SPR angle (resonance units = RU) was monitored to detect any binding interactions. The increase in the SPR angle (ΔRU) for each Hg^{2+} and Pb^{2+} concentration was determined based on the resonance angle of the ions (after the ion binding event) and the DDW baselines before running ions samples over a time duration of Δt.

These ΔRU results indicated that Hg^{2+} and Pb^{2+} were binding successfully with the PPy and PPy-CHI. Each sample was run for 20 min. The concentrations of ions in the parts per million range produced the changes in the SPR angle minimum in the region of 0.03 to 0.07. Due to the high sensitivity of SPR, it also is possible to detect concentrations in the parts per billion ranges. A control experiment for the Hg^{2+} and Pb^{2+} ions with bare Au surface was done for 30 min, which produced no changes in the resonance angle indicating that the polymer was necessary to react with the ions. After exposure to each sample concentration, all of the PPy and PPy-CHI films were washed with DDW to establish a baseline, as the history of the films was irrelevant.

5.4. Sensitive optical sensor for Hg²⁺ and Pb²⁺detection

The increase in the SPR angle for each Hg^{2+} and Pb^{2+} concentration indicated that these ions were binding successfully with PPy. The typical sensograms for 0.5–12 ppm Hg^{2+} and Pb^{2+} binding on PPy are presented in Figure 3 and 4.

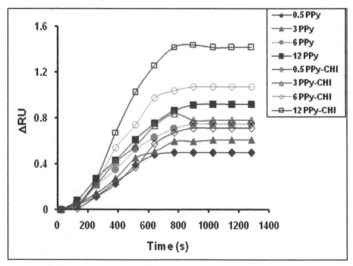

Figure 3. The sensograms for 0.5–12 ppm Hg^{2+} binding on PPy and PPy-CHI surface. Filled symbols (PPy surface), Open symbols (PPy-CHI surface).

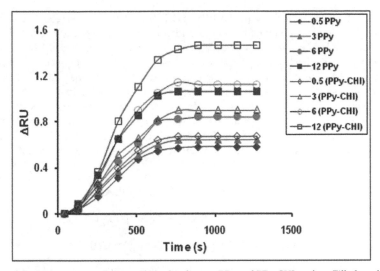

Figure 4. The sensograms for 0.5–12 ppm Pb^{2+} binding on PPy and PPy-CHI surface. Filled symbols (PPy surface), Open symbols (PPy-CHI surface).

It can be observed that, first, ΔRU increased over the time for each sample, but after almost 700 sec the graphs showed a constant value. This result can be explained by saturation of the binding sites available on the PPy surface. It seems that there is a chemical binding of ions in water with the PPy thin film that was immobilized on the gold surface. The interaction between Hg^{2+} and PPy was also reported by Yu et al. [31], as the natural preference for a combination of soft acid (Hg^{2+}) and soft base (PPy). On the other hand, there are some studies on the treatment of solutions of metal ions with redox polymers, such as polypyrroles, polyanilines and polythiophenes, to reduce the ions to a lower valence [44]. These literatures show that PPy can interact with certain heavy metals through an acid and base interaction or an oxidation reaction.

5.5. Sensitivity enhancement using CHI

Chitosan was used to prepare the PPy-CHI composite film to function as a sensitivity enhancement agent for detecting for heavy metals ions. DDW was run to establish a baseline. The decreasing SPR angle for the PPy-CHI film when compared to the PPy film indicated that chitosan incorporated inside the PPy structure. The sensograms for 0.5–12 ppm Hg^{2+} and Pb^{2+} binding on the PPy- CHI are also presented in Figure 3 and 4. The composite films of PPy- CHI showed the higher ΔRU compared to the PPy film for each time and concentration, indicating a higher sensitivity of the PPy in the presence of chitosan. In fact, the natural biopolymers have adsorption properties toward certain heavy metals. Of the many absorbents identified, chitosan has the highest sorption capacity for several metal ions [45]. The sorption proceeds by electrostatic attraction on protonated amine groups in acidic solutions resulted in an enhancement of the original ΔRU and improvement of sensitivity of sensor.

In order to show the sensitivity of the PPy and PPy-CHI films for metal ions, standard calibration curves were plotted for sensors and are shown in Figure 5 and 6. Different concentrations of ions were run for 20 min and all ΔRU were taken from the first 13 min of each binding curves. Each concentration was repeated 3 times and the standard deviation data were in the region of 0.06 to 0.18 and it is shown as error bar on the diagram of sensitivity. These figures clearly show the enhancement of sensitivity for sample determination in the presence of chitosan. In the both sensograms the SPR dip shift is directly correlated with the ions concentration ($[Hg^{2+}]$ or $[Pb^{2+}]$). Linear regression analysis of the sensors yielded the following equations with all correlation coefficients R^2 greater than 0.97, revealing good linearity for the relationship between the SPR dip shift and ions concentration:

$$\Delta\theta = 6.50 \times 10^{-1} + 6.41 \times 10^{-2}\ [Hg^{2+}]\ (PPy\text{-}CHI)$$

$$\Delta\theta = 4.97 \times 10^{-1} + 3.69 \times 10^{-2}\ [Hg^{2+}]\ (PPy)$$

$$\Delta\theta = 6.75 \times 10^{-1} + 6.77 \times 10^{-2}\ [Pb^{2+}]\ (PPy\text{-}CHI)$$

$$\Delta\theta = 5.50 \times 10^{-1} + 4.24 \times 10^{-2}\ [Pb^{2+}]\ (PPy)$$

From standard calibration curve, it can be seen that the sensitivity of PPy-CHI sensor was almost 2 times higher than that for the PPy sensor.

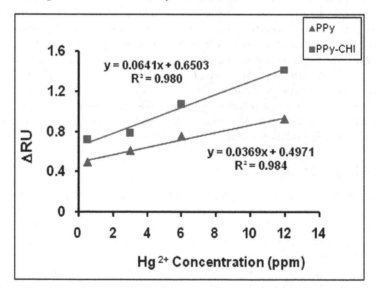

Figure 5. Standard calibration curves of ΔRU versus Hg^{2+} ion concentration for PPy and PPy-CHI films.

It can be seen that all the standard calibration curves showed a significant deviation from 0. This deviation was due to saturatation of the binding sites available on the polymer surface, a fundamental limitation of maximum uptake capacity [46].

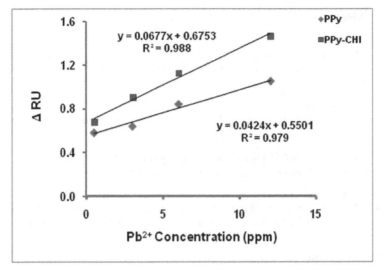

Figure 6. Standard calibration curves of ΔRU versus Pb²⁺ ion concentration for PPy and PPy-CHI films.

6. Conclusion

Application of biopolymers such as chitosan, as an electrical or optical material has rarely been reported. However, using of these materials in electrical devices are now of interest because of the current environmental threat and societal concern. Chitosan as a biodegradable, biocompatible, nontoxic and low-cost polymer was found to improve the electrical and mechanical properties of the PPy film. The changes in the electrochemical and optical properties of the polypyrrole films are due to the changes in the real and imaginary parts of the complex dielectric constant when is combined with chitosan.

Chitosan as natural biopolymers shows a high sorption capacity for several metal ions. Following the interaction of CHI and metal ions to enhance the original ΔRU, lower concentrations of ions could be readily detected with less interference. The aim of this study was to prepare a sensor area based on PPy-CHI composite films for detecting Mercury and Lead ions. The refractive indexes of PPy-CHI were successfully measured by surface plasmon resonance (SPR) technique. This technique was used to fabricate the PPy and PPy-CHI optical sensors for monitoring trace amount of Hg²⁺ and Pb²⁺. Specific binding of chitosan with Hg²⁺ and Pb²⁺ ions were presented as an increase in resonance angles. Pb²⁺ ion showed higher value for ΔRU due to stronger biding with polymer and the sensors based on PPy and PPy-CHI were more sensitive to Pb²⁺ compared to the Hg²⁺.

The research on the preparation and characterization of conducting polymers is still continuing in order to find conducting polymers with improved electrical and mechanical properties. Since chitosan could improve the electrical and optical properties of the PPy film, further research could be extended to prepare the composite films using other biopolymer such as Nano cellulose with other conducting polymers.

Author details

Mahnaz M. Abdi
Deprtemet of Chemistry, Faculty of Science, Universiti Putra Malaysia, Serdang, Selangor, Malaysia
Institute of Tropical Forestry and Forest Products (INTROP), Universiti Putra Malaysia, Serdang, Selangor, Malaysia

Wan Mahmood Mat Yunus
Department of Physics, Faculty of Science, Universiti Putra Malaysia, Serdang, Selangor, Malaysia

Majid Reayi
School of Chemical Sciences and Food Technology, Faculty of Science and Technology, Universiti Kebangsaan Malaysia, Bangi, Selangor D.E., Malaysia

Afarin Bahrami
Department of Physics, Faculty of Science, Universiti Putra Malaysia, Serdang, Selangor, Malaysia
Department of Physics, Faculty of Science, Islamic Azad University, Eslamshahr Branch, Iran

Acknowledgement

The authors would like to thank the Ministry of Higher Education of Malaysia for financial supporting (as Graduate Research Fellowship). We also gratefully acknowledge the Department of Chemistry and Physics, University Putra Malaysia for their continued support throughout this project.

7. References

[1] Hadziioannou G., Malliaras G.G. (2007) Semiconducting polymers, chemistry, physics and engineering; 1, ed.; WILEY-VCH.
[2] Nicolas M., Faber B., Simonet J. (2001) Electrochemical sensing of F- and Cl- with a boronic ester-functionalized polypyrrole. J. Electroanal. Chem 509: 73-79.
[3] Migahed M D., Fahmy T., Ishra M., Barakat A (2004) Preparation, characterization, and electrical conductivity of polypyrrole composite films. Polym Test 23: 361–365.
[4] Baba A., Lubben J., Tamada K., Knoll W. (2003) Optical properties of ultrathin poly(3,4-ethylenedioxythiophene) films at several doping levels studied by in situ electrochemical surface plasmon resonance spectroscopy. Langmuir 19: 9058-9064.
[5] Damos FS, Luz RCS, Kubota LT (2006) Investigations of ultrathin polypyrrole films: Formation and effects of doping/dedoping processes on its optical properties by electrochemical surface plasmon resonance (ESPR). Electrochim Acta 51: 304–1312.
[6] Barisci JN, Wallace GG, Andrews MK, Partridge AC, Harris PD (2002) Conducting polymer sensors for monitoring aromatic hydrocarbons using an electronic nose. Sens. Actuators B: Chem. 84: 252–257.
[7] Souza JEG, Santos FL, Neto BB, Santos CG, Santos MVB, et al. (2003) Freegrown polypyrrole thin films as aroma sensors. Sens. Actuators B: Chem. 88: 246–259.
[8] Mahmud HNME., Kassim A., Zainal Z., Mat Yunus WM (2007) Fouriertransform infrared study of polypyrrole-poly (vinyl alcohol) conducting polymer composite films: Evidence of film formation and characterization. J. Appl. Polym Sci 100 (5): 4107-4113.

[9]   Costa ACR., Siqueira, AF (1996) Thermal diffusivity of conducting polypyrrole. J.Appl. Phys 80: 5579-5580

[10] Kim MS., Kim HK., Byun SW., Jeong SH., Hong YK., Joo JS., Song KT., Kim .K., Lee CJ., Lee JY (2002) PET fabric/polypyrrole composite with high electrical conductivity for EMI shielding. *Synth. Met* 126: 233–239.

[11] Kim H K., Kim M S., Chun S Y., Park YH., Jeon BS., Lee JY (2003) Characteristics of electrically conducting polymercoated textiles. Mol. Cryst. Liq. Crys 40: 161–169.

[12] Baba A., Advincula R.C., Knoll W (2002) In situ investigations on the electrochemical polymerization and properties of polyaniline thin films by surface plasmon optical techniques. J Phys Chem. B 106: 1581-1587.

[13] Homola J., (2006) Surface Plasmon Resonance Based sensors; ed.; Springer: Berlin Heidelberg.

[14] Peyghambarian N, Koch SW, Mysyrowicz, A (1993) Introduction to semiconductor optics. Cliffs E, ed. Prentice Hall: Newjersey.

[15] Garland JE, Assiongbon KA, Pettit CM, Roy D (2003) Surface Plasmon resonance transients at an electrochemical Interface: Time resolved measurements using a bicell photodiode. Anal Chim Acta 475: 47–58.

[16] Wang, S., Boussaad S., Wong S., Tao, N.J (2000) High-Sensitivity Stark Spectroscopy Obtained by Surface Plasmon Resonance Measurement. Anal. Chem 72: 4003-4008.

[17] Bredas JL., Scott JC., Yakushi K., Street GB (1984) polarons and bipolarons in polypyrrole: evolution of the band structure and optical spectrum upon doping. Phys. Rev. B 30: 1023-1025.

[18] Abdi MM., Mahmud HNME., Abdullah LC., Kassim A, Rhman MZ., Chyi JLY (2012) Optical band gap and conductivity measurement of polypyrrole-chitosan composite thin film. Chinese Journal of Polymer Science 30(1): 93-100

[19] Jorn CC Yu., Edward PC Lai (2004) Polypyrrole film on miniaturized surface Plasmon resonance sensor for ochratoxin A detection. Synthetic Metals 143: 253–258.

[20] Guedon P., Livache T., Martin F,. Lesbre F., Roget A., Bidan G., Levy Y (2000) Characterization and Optimization of a Real-Time, Parallel, Label-Free, Polypyrrole-Based DNA Sensor by Surface Plasmon Resonance Imaging. Anal. Chem. 72: 6003–6009.

[21] Thiel AJ., Frutos AG., Jordan C., Corn R., Smith L (1997) In situ surface Plasmon resonance imaging detection of DNA hybridization to oligonucleotide arrays on gold surface. Anal. Chem.69: 4948-4956.

[22] Zizlsperger M., Knoll W (1998) Multispot Parallel on-Line Monitoring of Interfacial Binding Reactions by Surface Plasmon Microscopy. Prog. Colloid Polym. Sci. 109: 244- 253.

[23] Sathawara NG, Parikh DJ, Agarwal YK (2004) Essential Heavy Metals in Environmental Samples from Western India. Bulletin of Environmental Contamination and Toxicology 73: 756–761.

[24] Pearce JM (2007) Burton's line in lead poisoning. European neurology 57(2): 118–9.

[25] Clifton JC (2007) Mercury exposure and public health. Pediatr Clin North Am 54 (2): 237–69.

[26] Skoog DA, Holler FJ, Nieman TA (1992) Principles of Instrumental Analysis. 5th ed. Philadelphia: Saunders College Publishing. pp 206–229.

[27] Taylor HE (2001) Inductively Coupled Plasma Mass Spectrometry: Practices and Techniques, Academic Press, San Diego.

[28] Tang Y, Zhai YF, Xiang JJ, Wang H, Liu B, et al. (2010) Colloidal gold probe based immunochromatographic assay for the rapid detection of lead ions in water samples. Environmental Pollution 158: 2074–2077.

[29] Bard AJ., Faulkner LR (2000) Electrochemical Methods: Fundamentals and Applications, 2nd ed, John Wiley & Sons, New York. pp. 458–464.

[30] Zhang Y, Xu M, Wang , Toledo F, Zhoua F (2007) Studies of metal ions binding by apo-metallothioneins attached onto preformed self-assembled monolayers using a highly sensitive surface plasmon resonance spectrometer Sensors and Actuators B. 123: 784–792.

[31] Yu JCC, Lai EPC, Sadeghi S (2004) Surface plasmon resonance sensor for Hg(II) detection by binding interactions with polypyrrole and mercaptobenzothiazole. Sens. Actuators B 101: 236–241.

[32] Matejka P, Hruby P, Volka K (2003) Surface plasmon resonance and Raman scattering effects studied for layers deposited on Spreeta sensors, Anal. Bioanal. Chem. 375: 1240–1245.

[33] Park JS, Lee CM, Lee KY (2007) A surface plasmon resonance biosensor for detecting Pseudomonas aeruginosa cells with self-assembled chitosan-alginate multilayers. Talanta 72: 859–862.

[34] Kim S H., Han S K., Kim J H., Lee M B., Koh K N., Kang S W (2000)A self-assembled squarylium dye monolayer for the detection of metal ions by surface plasmon resonances Dyes. Pigm. 44: 55-61.

[35] Sadrolhosseini AR., Moskin MM., Mat Yunus WM, Talib ZA., Abdi MM (2011) Surface Plasmon Resonance Detection of Copper Corrosion in Biodiesel Using Polypyrrole–Chitosan Layer Sensor. Optical review 18 (4): 1–7.

[36] Kittle C (1996) introduction to Solid State Physics. Wiley, New York. p. 390.

[37] Azzam RMA., Bashara N M (1997) Ellipsometry and Polarized Light. Elsevier, Amsterdam. p. 269.

[38] Schasfoort RBM, Tudos AJ (2008) Handbook of surface Plasmon resonance, RSC publishing UK: Cambridge. pp 15–25.

[39] [39] Blonder B. Sensing Application of Surface Plasmon Resonance. Available online: http://www.eduprograms.seas.harvard.edu/reu05_papers/Blonder_Benjamin.pdf

[40] Jaaskelainen AJ., Peiponen KE., Raty JA (2001) On reflectometric measurement of a refractive index of milk. J. Dairy Sci. 84: 38–43.

[41] Pedrorotti LF, Pedrotti LS, Pedrotti LM (2007) Introduction to optics. 3th ed. New York: Addison Wesley. pp 476–482.

[42] Abdi MM, Mahmud HNME, Kassim A, Mat Yunus WM, Talib ZBMohd, et al. (2010) Synthesis and Characterization of a New Conducting Polymer Composite. Polymer Science Ser B 52: Nos. 11–12, 662–669.

[43] Abdi MM, Kassim A, Mahmud HNME, Mat Yunus WM, Talib ZB, et al. (2009) Physical, optical, and electrical properties of a new conducting polymer. J Mater Sci 44: 3682–3686.

[44] Sanjay B, Krishna R, Chang W (1994) Redox polymer films for metal recovery applications US Patent 5368632.

[45] Nomanbhay SM, Palanisamy K (2005) Removal of heavy metal from industrial wastewater using chitosan coated oil palm shell charcoal. Electron j biotechn 8: 44–53.

[46] Zhang X, Bai R (2002) Adsorption behavior of humic acid onto polypyrrolecoated nylon 6,6 granules. J Mater Chem 12: 2733–2739. SPR Sensor Based on PPy-CHI

Chapter 10

Application of Surface Plasmon Polaritons in CMOS Digital Imaging

Qin Chen, Xiaohua Shi, Yong Ma and Jin He

Additional information is available at the end of the chapter

1. Introduction

Recent years there has been a rapid expansion of research into nanophotonics based on surface plasmon polariton (SPP), which is a collective electron oscillation propagating along a metal-dielectric interface together with an electromagnetic wave [1]. What distinguishes SPPs from photons is that they have a much smaller wavelength at the same frequency. Therefore SPPs possess remarkable capabilities of concentrating light in a nanoscale and the resulted significant enhancement of localized field [2]. SPPs can be excited by an incident electromagnetic wave if their wavelength vectors match. This is usually achieved by nanopatterning the metal film. The resonant frequency of SPP is determined by the metal materials, dielectric materials, profiles and dimensions of the patterns, etc. As a result, the tunability of SPP enables its application as a colour filter in the visible range. Actually this was used in the stained glass manufacture hundreds years ago. Since the extraordinary optical transmission through a nanohole array in a thin metal film was reported by Ebbesen [3], the plasmonic photon sorting has been explored for the potential applications in digital imaging and light display [4,5]. In addition, SPP based light manipulating elements like planar metallic lenses, beam splitters, polarizers have been investigated both theoretically and experimentally [6-7].

In solid state digital imaging, complementary metal oxide semiconductor (CMOS) image sensors (CISs) are the leading mass-market technology. CMOS image sensors with smaller pixels are expected to enable digital imaging systems with better resolution and possibly high photosensitive area in the pixel (fill-factor). In present colour CMOS digital imaging systems, dye-doped polymer filters and curved dielectric microlenses are used to disperse light of different wavelengths and manipulate the light beams, respectively . With the down scaling of pixel size to the sub-2 µm range, these optical elements suffer from performance degradation, such as colour cross-talk, due to the large distance between the optical elements and the photodiodes underneath [8]. Therefore most high-resolution digital cameras have to use the backside illuminated CISs that are fabricated with complicated processes [9]. Furthermore different colour polymer filters have to be fabricated successively in several process steps using the back-end-of-line process. Continuous development of new applications for CISs requires that they are able to be manufactured at low cost.

The idea of introducing plasmonics into a CIS was first proposed by Catrysse *et al* [10]. They were able to show the potential of metallic nanodevices in a CIS by detailed numerical simulations, and provided preliminary experimental results [10,11]. There are several advantages: (1) this technique has the potential to produce all the required colour filters in a single metal layer by one lithography step at a low cost; (2) colour cross-talk can be reduced by integrating plasmonic colour filters in Metal 1 layer which is very close to the CMOS photodiodes; (3) the thickness of the plasmonic device is one or two orders of magnitude thinner than that of the colourant one used in a CIS or a LCD; (4) localized field enhancement effect of metallic nanostructures may increase the photodetection sensitivity; (5) the light filtering can be readily obtained by varying the nanostructures that enables tuning of the resonant wavelengths and thus achieves a large spectral design freedom; (6) other optical elements such as lenses and polarizers can be achieved based on SPP in metal layers in CMOS technology. Planar metallic lenses integrated in the metal layers in standard CMOS technology can readily tune the phase of incident light on a pixel level to achieve a great control on the light beam divergence therefore the cross-talk especially for the pixels at the edge of the whole sensor; (7) plasmonic metal like aluminium is a CMOS compatible metal and it is more stable compared to polymer.

In this chapter, we briefly review the research progress on various plasmonic optical elements for the application in digital imaging and describe our work on a plasmonic CIS (pCIS). Section 2, 3 and 4 focus on plasmonic colour filters, PLs and wire-grid polarizers, respectively. Section 5 presents our work on the integration of plasmonic colour filters on CISs. Finally, we conclude this chapter and discuss the outlook of this technique.

2. Plasmonic colour filters

2.1. Background

A colour filter that selectively transmits or reflects input light is an important element in a CIS. Established colour filtering technologies for CISs use dye-doped polymers. Each colour filter for red (R), green (G) and blue (B) must be fabricated successively in several process steps. Because of the encroaching difficulties of cross-talk and the cost of manufacture, it is desirable to find new methods for building colour filters into CISs. Other colour filtering techniques for imaging arrays have been investigated. Guided-mode resonance filters based on subwavelength dielectric gratings were shown to work as a bandpass filter [12]. They can be designed to work in both reflection and transmission modes. Hybrid metal-dielectric gratings were found to offer excellent optical transmission (87%) property in the midinfrared range due to the Fano transmission resonance [13]. One-dimensional (1D) periodic metal-insulator-metal (MIM) waveguide array supports a surface plasmon antisymmetric mode which showed colour filtering effect [14]. However, devices based on a 1D structure have an intrinsic polarization dependency. Photonic crystal colour filter was proposed for the application in a MOS image sensor, where two multilayer stack mirrors separated by a defect layer were used to form a cavity resulting a passband filtering [15]. But the dielectric mirrors have a relatively narrow bandwidth limited by the index contrast in the stack and the multilayer deposition process is complex. Silver mirrors were proposed to replace the dielectric mirrors and a CIS based on this technique was demonstrated where the different colours were achieved by tuning the cavity length [16]. The device requires multi-lithography steps to make R, G, B pixels. A bull's eye structure consisting of concentric grooves with a central hole was proposed as another candidate for colour filters [5]. This method gave very good narrow band wavelength filtering, but the low fill ratio resulted in poor transmission efficiency.

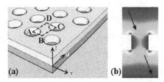

Figure 1. (a) A square-lattice hole array in a metal film. (b) Cross-section field distribution around a hole where light couples to SPPs.

Ebbeson *et al.* investigated the subwavelength holes in silver films as shown in Fig. 1(a) and observed extraordinary optical transmission (EOT) of light through the holes and wavelength filtering due to the excitation of surface plasmon resonance (SPR) [3]. The incident light first couples to the SPR at the top surface of the metal film in the presence of the 2D hole array. The SPR at the top surface then couples to the one at the bottom surface of the metal film. Finally, light reemits from the bottom surface of the patterned metal film as shown in Fig. 1(b). The peak position, λ_{max}, of the transmission spectrum at normal incidence can be approximated by

$$\lambda_{max} = \frac{a}{\sqrt{i^2 + j^2}} \sqrt{\frac{\varepsilon_m \varepsilon_d}{\varepsilon_m + \varepsilon_d}} \quad (square \ lattice) \tag{1}$$

$$\lambda_{max} = \frac{a}{\sqrt{\frac{4}{3}(i^2 + ij + j^2)}} \sqrt{\frac{\varepsilon_m \varepsilon_d}{\varepsilon_m + \varepsilon_d}} \quad (triangular \ lattice) \tag{2}$$

where a is the period of the array, ε_m and ε_d are the dielectric constants of the metal and the dielectric material in contact with the metal respectively, and i and j are the scattering orders of the array [3,4]. As shown in Eq. (1) and (2), the period determines the transmission peak positions of SPRs for a given material configuration. Following this discovery, researchers started to use 2D hole arrays in metal films as transmissive colour filters. A square-lattice hole array surrounded by a square-lattice dimples at the same period in a silver film was reported to show colour filtering function [4]. Without surrounding dimples, a square array of circular holes were designed and fabricated in a thin aluminium film as red and green colour filters [17], where unfortunately the green filter devices exhibited a yellow colour due to the high colour cross-talk.

In this section, we focus on transmitted colour filters consisting of triangular-lattice hole arrays in aluminium films that are compatible with standard CMOS technology. Both numerical simulation and experimental results are presented.

2.2. FDTD simulation

Eq. (1) and (2) theoretically predict the resonant wavelengths of SPRs. But the thickness of the metal film and the coupling between the SPRs at two interfaces has not been considered in these equations. To accurately optimize the design, a finite-difference time-domain (FDTD) algorithm based commercial software, Lumerical FDTD Solutions [18], was used to investigate the colour filters and identify the transmission peaks in the measured spectra. As shown in Eq. (1) and (2), the wavelength interval between the first two SPR peaks in a triangular array is larger than that for a square array of a same period. Therefore we focus on the triangular lattice in the following part. In our work, we focus on a triangular-lattice hole array in an aluminium film on glass. Compared to silver and gold, aluminium is compatible with standard CMOS technology and cheap although it has a relatively higher absorption loss. Aluminium also has good adhesion to many substrates making fabrication easier.

The effects of the structure dimensions were investigated numerically. As shown in Fig. 2(a), the transmission efficiency was strongly affected by the size of the circular hole. As the hole size increases, the magnitude of the main transmission peak increases from 24% to 75%. There is also a red shift and an increase in the full-width at half-maximum (FWHM) from 60 nm to 300 nm. The reflections at the transmission peaks are almost zero in all three cases. Therefore the loss at the transmission peak is larger at a smaller radius, where the absorption within the metal due to the non-zero imaginary component of the permittivity increases [4]. A tradeoff between transmission and FWHM must be considered to optimize the RGB filters for the application in a CIS. The coupling between the SPRs at both sides of the metal film has an important effect on the transmission spectra. As shown in Fig. 2(b), the structure made with a thinner aluminium film has higher transmission but also a much larger bandwidth. The enhanced coupling of SPR in the case of a thin metal film increases the splitting of the two transmission peaks, i.e. the long wavelength side peak has a red shift but the short wavelength side peak has a blue shift. In addition, the transmittance of the very low level at a thickness of 30 nm. The EOT phenomenon is more prominent in the case of a thin metal film where the SPR coupling is strong. However, the FWHM is large for a thick metal film and may cause a spectral cross-talk for the RGB filters. Predicting in Eq. (2), the period has a dominant effect on the transmission peak, i.e. the transmitted colour of the filter. The filters shown in Fig. 2(c) have the peak transmissions around 50% filters in the visible range can be readily obtained by tuning the period of the hole array. Finally, the different hole shapes with a similar area in a same triangular lattice were investigated as shown in Fig. 2(b). The circular hole array has the highest transmittance at the transmission peak and the smallest FWHM. Therefore, we focused on the circular hole array in our experiments. Plasmonic colour filters with spectral responses matching the International Commission on Illumination colour matching functions are useful for effectively communicating color between colour detection and output devices. A fully automated genetic algorithm that incorporated on-demand 3D FDTD simulations were used to determine the structure dimensions [19].

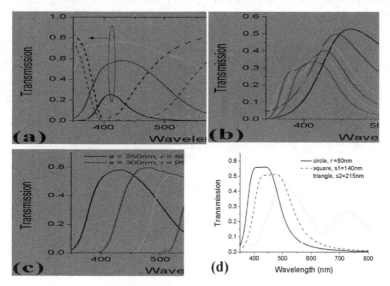

Figure 2. Simulated transmission/reflection spectra for hole arrays in a triangular lattice in an aluminium film on glass with a 200 nm SiO$_2$ cap layer. (a) Different circular hole radius r with a period $a = 250$ nm and a thickness $t = 150$ nm. (b) Different t at $a = 250$ nm and $r = 65$ nm. (c) Different a and r at $t = 150$ nm. (d) Different hole shapes.

2.3. Fabrication of colour filters on glass

Fig. 3(a) is the process flow for fabricating plasmonic colour filters. The first step is to evaporate a 150 nm aluminium film on a clean glass substrate at a rate of 0.3 nm/s by electron beam evaporation. Then ZEP520A electron beam resist was spin-coated on to the sample and exposed. After development, aluminium was etched using SiCl$_4$ in a Plasmalab System 100. Finally, a 200 nm SiO$_2$ layer is deposited on top of the patterned aluminium film after removal of the residual resist to enhance the transmission due to the symmetric SPR coupling (the refractive index of SiO$_2$ is close to that of glass). Fig. 3(b) shows a scanning electron microscope (SEM) image of patterned aluminium film after removing the residual resist. Vertical and smooth sidewalls can be seen from the inset SEM image for which the sample was tilted at 30º [20].

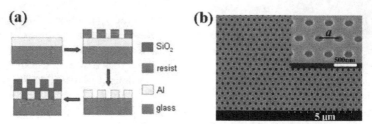

Figure 3. (a) Process flow for fabricating plasmonic colour filters. (b) A SEM image of etched holes in a triangular array with a = 430 nm in an aluminium film on glass. The inset is a SEM image for a sample tilted at 30°.

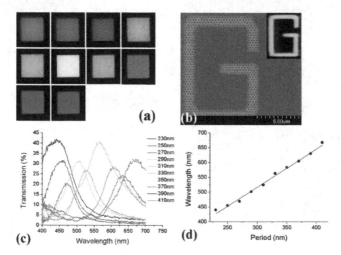

Figure 4. Performance of various plasmonic colour filters fabricated on glass. (a) Images of various plasmonic colour filters taken in microscope transmission mode under a white light illumination. (b) A SEM image of holes composing the letter 'G' with a = 330 nm. The transmitted light image of the structure with a 200 nm SiO_2 cap layer is shown in the inset. (c) Transmissive spectra of the filters shown in (a). (d) Wavelengths at which transmission through each peak versus the period of the plasmonic nanostructures. Reprint from [21].

The colour images of plasmonic filters were examined using an Olympus BX51 microscope fitted with a broadband halogen lamp. Fig. 4(a) shows the well-defined colour squares with a size of 50 μm × 50 μm consisting of holes in a triangular lattice with different periods. A letter 'G' consisting of holes with a = 330 nm was fabricated as shown in Fig. 4(b), where the period number of hole array is as small as three. The clear green letter 'G' appeared under the white light illumination as shown in the inset. Unlike the structures in earlier work [4], there is no dimple around the etched holes in our structure. This means that the pixel size could be as small as 1 μm on a high-resolution image sensor. Spectral measurement was carried out on a microscope spectrophotometer TFProbe MSP300 that can sample the signal from a minimum area of 10 μm × 10 μm. Different colour filters were written on to the same sample to ensured consistent measurement of the different structures. Due to the symmetrical properties of the triangular-lattice hole array and the circular holes, the transmittance shows no difference for illumination at different polarization angles. An unpolarized light beam from a halogen lamp was launched normally on to the back-side of the sample. The transmitted light was collected using a 40× lens with a NA=0.9 and guided to the spectrometer and image camera. The transmission spectra from the sample in the visible range from 400 nm to 700 nm are shown in Fig. 4(c), where transmittances are between 20% and 40%, with full width at half maximum between 70 nm and 110 nm. As predicted in Eq. (2), Fig. 4(d) shows that the wavelength at which transmission through each filter peaks was found to linearly increase with the period of the holes. This means that a complete colour filter set can be reliably made using a single lithographic step.

3. Plasmonic polarizers

3.1. Background

Polarization is a general property of light and contains information about reflecting objects that traditional intensity-based sensors like human being's eyes ignore. However, polarization offers a number of advantages for imaging [22-24]. Polarization filtering has long been used in photography through haze. Difficult computer vision tasks such as image segmentation and object orientation are made tractable with polarization vision techniques. Investigation of the polarized light backscattering enables noninvasive surface and beneath-the-surface imaging of biological systems. Traditionally, the polarization image was obtained by taking an objective at the same place twice at different polarization angle with an external polarizer mounted in front of the camera. Mathematic analysis on these data finally generates a polarization image. Obviously, this method has low efficiency, low accuracy and high cost. Micropolarizer array was proposed to implement the polarization imaging like a Baye array in colour filter array [25,26]. In each 2×2 cell, there are four polarizers at different polarization angles to collect the required polarization information. Similar to a colour image, polarization image can be obtained from a polarization image sensor with micropolarizer arrays by just one shot of the scene. In the past, micropolarizer array was fabricated using polymer which is usually above 10 μm [25]. The large thickness causes large cross-talk between neighboring pixels. Furthermore, polymer polarizer faces a stability issue. Alternatively, metal wire grids are an ideal polarizer, which has been applied for IR polarization imaging [26]. Optimization of metal wire grids for a large extinction ratio (ER), defined as the transmission ratio of TM modes (the polarization of light perpendicular to the metal wire grid) to TE modes (the polarization of light parallel to the metal wire grid), is important to a high-contrast polarization image.

In this section, we numerically investigate the plasmonic polarizers and demonstrate the experimental results of aluminium grating polarizers.

3.2. Simulation

2D FDTD simulation is used to investigate the performance of metal wire grids. Wire grids with different periods were simulated and the results are shown in Fig. 5. In all cases, the effect. With the decreasing of the period, the transmission of TM polarization keeps at a high value but that of TE polarization decreases significantly. As a result, the ER increases with the decrease of the period. The state-of-the-art CMOS technology enables a feature size below 50 nm. However, a thick metal layer increases the aspect ratio of the gratings and therefore increases the difficulty in fabrication. The aspect ratio also affects the ER of the wire grids. As shown in Fig. 6, with a period of 100 nm, the transmission of TE polarization decreases more than one order of magnitude and the ER of Al wire grids increases approximately 15 dB when t increases 50 nm. All the above results are for wire grids with a duty cycle (metal filling ratio) of 50%. In Fig. 7, we show the effect of the duty cycle.

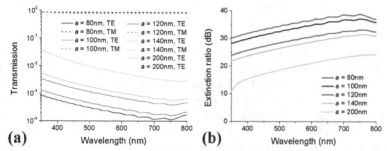

Figure 5. Transmission spectra (a) and ER (b) of Al wire grids on glass with different periods at t = 100 nm and f = 50%.

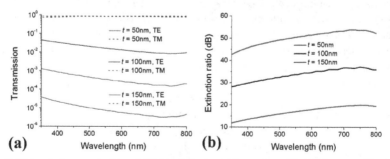

Figure 6. Transmission spectra (a) and ER (b) of Al wire grids on glass with different metal thicknesss at a = 100 nm and f = 50%.

Transmissions of both TE and TM polarizations reduce when the duty cycle increases. Furthermore, ER is higher at a higher duty cycle.

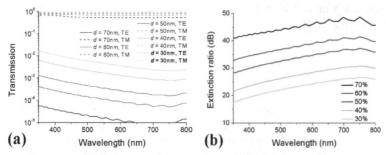

Figure 7. Transmission spectra (a) and ER (b) of Al wire grids on glass with different duty cycle at $a =$ 100 nm and $t = 100$ nm.

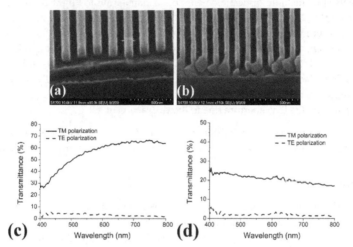

Figure 8. SEM images of fabricated Al wire grids with a period of 200 nm (a) and 100 nm (b). (c) and (d) Measured transmission spectra of polarizers in (a) and (b).

3.3. Experiments

The process flow is similar to that of plasmonic colour filters. The SEM images of devices with periods of 200 nm and 100 nm are shown in Fig. 8(a) and (b), respectively. Spectral measurement was carried out on a same microscope spectrophotometer TFProbe MSP300 as the plasmonic colour filter experiments. The incident light was polarized using a linear polarizer amounted before the sample in the light path. The transmitted light was collected using a 40× lens with a $NA=0.9$ and guided to the spectrometer. As the simulation results, the TM polarization has higher transmission than TE polarization as shown in Fig. 8(c) and (d). The measured ER of the device with a period of 100 nm is approximately 16, enabling a 4-bit polarization imaging. Actually, our polarizers may have better performance because the transmittance of TE polarization was closed to the background noise limited by our detector. In addition, dual grating structures may be considered to obtain a higher ER [27].

4. Plasmonic Lenses (PLs)

4.1. Background

Traditionally light is manipulated using dielectric optical elements such as refractive lens, diffractive gratings, mirrors and prisms. For example, dielectric microlens arrays are used in current CMOS image sensor to increase the light collection efficiency. However, diffraction may put the usefulness of the microlens in question in sub-2μm pixels [28]. Recent progress in nanotechnology and the theory of plasmonics has led to rapidly growing interest in the implementation of metallic optical elements on a nano-scale for light beam manipulation [29-39]. Especially, lots of theoretical predictions [30-35] and experimental demonstration [28,36-39] have been reported on both one-dimensional (1D) and 2D PLs. However, most experimental results showed a large deviation of focal length from the design [28,36-39]. The authors attributed this phenomenon to the finite size of the lenses and the resulting diffraction. The effect of the lens size on the focal length was theoretically investigated in a plano-convex refractive microlens [40]. It turned out that the upper limit of the focal length was determined by the lens aperture due to the diffraction effect.

In this section, we discuss the diffraction effect in PLs and experimentally demonstrate PLs with accurate control of the focal lengths that are important for the application in a CIS.

4.2. Theory

A 1D PL, as shown in Fig. 9(a), is a group of nano-slits in metal that form zones to modulate the phase delay distribution across the device surface. It focuses light only in the x-direction. The distribution of the phase delay at the exit of each slit is designed to provide constructive interference at the focus. Incoming laser light excites SP modes at the slit entrances that propagate through the slits before emitting into light at the exits, forming a convergent focus. In general, the SP modes can be excited in a metal/dielectric/metal waveguide, as shown in the inset of Fig. 9(b), under TM-polarized illumination with the electric field perpendicular to the slits. In a narrow slit, only the fundamental SP mode exists with a complex propagation constant, β, that can be calculated from the equation

$$\tanh(\sqrt{\beta^2 - k_0^2 \varepsilon_d}\, \frac{w}{2}) = \frac{-\varepsilon_d \sqrt{\beta^2 - k_0^2 \varepsilon_m}}{\varepsilon_m \sqrt{\beta^2 - k_0^2 \varepsilon_d}} \tag{3}$$

where ε_d and ε_m are the permittivity of the dielectric inside the slit and the metal, w is the slit width, and k_0 is the wave vector of light in free space [41]. The real and imaginary part of the effective refractive index n_{eff}, defined to be β/k_0, determine the phase velocity and the propagation loss of the SP modes, respectively. The real part of n_{eff} and the corresponding loss are shown in Fig. 9(b) for the example of Al/air/Al structure as a function of the slit widths. We use ε_m = -56.12 + i21.01 for aluminum at a wavelength of 633 nm [19]. A phase delay modulation on the surface of a patterned metal film can be realized by simply adjusting the widths of patterned nano-slits. Light focusing can therefore be obtained using a planar structure. The phase delay βd (d is the thickness of the metal film) of the SP modes inside the slits dominates the phase delay distribution across the device surface. For a PL with a focal length f in air, the various slits and their positions can be readily obtained according to the constructive interference principle

$$\beta(x)d + 2\pi\sqrt{f2+x2}\big/\lambda = \beta(0)d + 2\pi f\big/\lambda + 2(m-1)\pi \qquad (4)$$

where λ is the wavelength of the illumination, β is a function of the position, and m is the zone number. $\beta(0)$ is the propagation constant of the SP mode in the central slit, which is usually the narrowest slit with the largest value of β. Light coming from slits in a same zone, i.e. a same m (including the case of two identical slits at x and $-x$), has a same phase at the focus. But light from slits in different zones has a phase shift of an integer times of 2π at the focus.

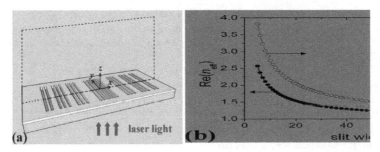

Figure 9. (a) Schematic of a PL. (b) Dependence of the real part of neff and the loss of a SP mode on the slit width. Reprint from [42].

Using Eq. (4), PLs with focal lengths of 0.5 µm, 1 µm, 2 µm, 3 µm, 6 µm and 12 µm at 633 nm were designed in a 200 nm thick aluminium film on a glass substrate. All PLs have an aperture size D of approximately 10 µm. The minimum width of the slit was 10 nm and the minimum gap between two neighboring slits was 50 nm, which was thicker than the skin depth. The central slit, with a width of 10 nm, was at $x = 0$ and the whole structure was symmetrical in x on the the yz plane. For an ideal planar lens with a focal length f at λ, the phase delay caused by the light path difference is

$$\phi = 2\pi\sqrt{f2+x2}\big/\lambda - 2\pi f\big/\lambda \quad (x \geq 0) \qquad (5)$$

ϕ for each lens calculated from Eq. (5) is shown in Fig. 14 as a solid line. Light emitting from a position at a larger x to the focus has a larger phase delay. By counting from the center we divide the phase delay into zones with a range of 2π. As a result, the slits in a same zone have a same m in Eq. (4). Optimization of the positions and widths of nano-slits gives a phase delay in the slits of the SP mode, $(\beta(x)d - \beta(0)d)$, obeying the constructive interference principle as shown in Eq. (4). The symbols in Fig. 10 show the phase delay, ϕ', for each slit in different zones. We can see that the symbols and the lines agree very well, i.e. $\phi=\phi'$. For a fixed aperture size, the PL with the longest (shortest) f has the smallest (largest) ϕ'_{max}. A PL with a longer focal length has more slits in each zone but has fewer zones. We note that the PL with $f = 12$ μm has almost the same number (21) of slits as the PL with $f = 0.5$ μm that has 22 slits, but the former has 2 zones and the later has 7 zones.

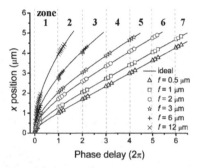

Figure 10. The phase delay ϕ and ϕ' for each slit in different zones (x ≥ 0) for similar aperture PZPLs with $f = 0.5$ μm, 1 μm, 2 μm, 3 μm, 6 μm and 12 μm at 633 nm. Red crosses are related to the structure shown in Fig. 18(b). Reprint from [42].

4.3. FDTD simulation

A 2D FDTD simulation was chosen to be an adequate approximation since the length of the uniform 10 μm slits in the y-direction in our experiment is much larger than the light wavelength. In the simulation, a uniform cell of $\Delta x = \Delta z = 1$ nm was used in the metal slab and a nonuniform cell was used elsewhere. The simulation domain was bounded by perfectly matched layers. A TM-polarized plane wave source at 633 nm was launched with normal incidence to the PL surface from the glass substrate side of the metal film.

4.3.1. PLs with variant focal lengths

The Poynting vector, P_z, distribution of the PL with variant focal lengths is plotted in Fig. 11. As can be seen, P_z for each PL has a light spot at the expected focal point. The constructive interference of the light from each slit at the focus is clearly demonstrated. For the PLs with $f = 0.5$ μm, 1 μm, 2 μm, 3 μm and 6 μm, P_z has its maximum value at the design focus. However, although there is a local maximum in P_z around the design focus for the PL with $f = 12$ μm, the maximum P_z occurs at a distance of 5.4 μm from the PL surface. In [36-39], the shorter focal lengths compared to the designs were explained as a similar phenomenon found in conventional refractive microlens [49]. The effect of diffraction at the lens aperture limits the maximum of the focal length. Using FDTD method, an aperture of 10 μm in the same aluminium film was simulated and showed a peak radiation at a distance of approximately

40 µm. Although the designed focal length of 12 µm is well below this value, the decrease of the focal length still occurs. Furthermore, close examination of the field pattern for the PL with $f = 6$ µm shows an unwanted local maximum at a distance of 2.1 µm from the PL surface that has an amplitude slightly smaller than that at the focus. Although the PL with $f = 12$ µm has almost the same number of slits as the PL with $f = 0.5$ µm in a similar size aperture, the focusing quality of the former is much poorer. The much smaller $NA = 0.35$ for the PL with $f = 12$ µm compared to 0.99 for the PL with $f = 0.5$ µm may be one reason. Furthermore, ϕ'_{max} is 12.5 π for the PL with $f = 12$ µm and 2.5 π for the PL with $f = 0.5$ µm.

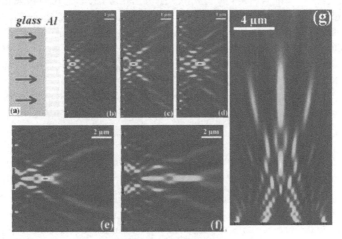

Figure 11. (a) Schematic of the PL. (b)-(g) Simulation results for the Poynting vector Pz distributions of the PZPLs with $f = 0.5$ µm, 1 µm, 2 µm, 3 µm, 6 µm and 12 µm respectively.

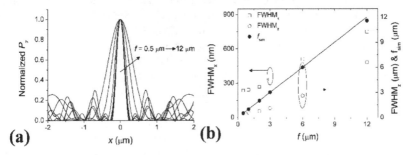

Figure 12. (a) Normalized simulation results for the Poynting vector Pz distributions along the x axis through the foci of PLs with $f = 0.5$ µm, 1 µm, 2 µm, 3 µm, 6 µm and 12 µm respectively. The optical axis is at x = 0. (b) The simulation results of the focal length, FWHMx and FWHMz for PLs with the design $f = 0.5$ µm, 1 µm, 2 µm, 3 µm, 6 µm and 12 µm respectively. fdesign is plotted as a solid line. Reprint from [42].

In Fig. 12(a) the normalized P_z distribution through the centre of the focus spot along the x axis is shown for each PL in Fig. 11. As can be seen, P_z is tightly confined in the x-direction although there are some side lobes. The simulation results of the focal length and FWHM of the beam extension in both x- and z- directions are shown in Fig. 12(b). We can see that the focal length obtained using the FDTD method is in an excellent agreement with the design. Moreover, the FWHM of the beam extension on the z-axis (FWHM$_z$) increases with the increasing f, starting from 0.49 µm for $f = 0.5$ µm to 6.6 µm for $f = 12$ µm. The FWHM of the beam in the x-axis at the focal point (FWHM$_x$) also increases from 236 nm (0.37λ) for $f = 0.5$ µm to 750 nm (1.18λ) for $f = 12$ µm. As a further test of the PL quality, we compared FWHM$_x$ with what we would expect for a diffraction limited beam. The Rayleigh limit for the resolving power of a lens is given by $d_s = \lambda f/D$ for a line, where d_s is the line-width, D is the lens aperture and λ is the wavelength. For $f = 6$ µm, the PL has FWHM$_x$ = 510 nm while a conventional lens with the same aperture has $d_s = 440$ nm. The PL therefore achieves focus resolution close to the diffraction limit. One advantage of the planar PL is the ability to achieve a large aperture for a small focal length and therefore a high resolution. We would like to mention that the above formula for d_s can not be approximated as $\lambda/2NA$ because NA is not even close to $D/2f$ in the case of near-field application. The performance of the proposed plasmonic Fresnel zone plate (FZP), working beyond the diffraction limit in [32], was therefore overestimated.

4.3.2. The effect of the number of zones on the PL performance

Varying the number of zones in the structures, PLs with $f = 0.5$ µm and $f = 6$ µm were simulated. As shown in Fig. 13(a), the simulated focal length has a large deviation from the design for a PL with a small zone number. In the case of a one-zone PL with $f = 6$ µm, NA is 0.23 and ϕ'_{max} is 0.5 π. Increasing to three zones, the focal lengths become stable and agree well with the design for both PLs with a focal length approximately 0.8λ and 10λ, respectively. In the case of three zones in the PL with $f = 6$ µm, NA is 0.59 and ϕ'_{max} is 4.4 π. The increasing NA and ϕ'_{max} provide an accurate focal length as predicted in the design. As shown in Fig. 13(b), the lens aperture ($f = 6$ µm) was found to almost linearly increases with the increasing zone number but FWHM$_x$ reduces. FWHM$_x$ decreases from 510 nm for 3 zones to 258 nm for 16 zones. The resolution approximately doubles at a cost of a 3.3 times increase of the aperture. Although $d_s = \lambda/2NA$ is not valid in the near-field case, the FWHM$_x$ is found to be nearly proportional to $1/NA$, where the trend predicts a resolution of approximately 210 nm at $NA = 1$. So the focal length of a PL agrees with the design very well if the zone number is larger than 3. However, the extension of the focus in the focal plane, i.e. FWHM$_x$, is sensitive to the number of zones. Other authors also found the dependence of the focal lengths of a FZP on the zone numbers [34]. A straightforward understanding is that more zones integrated in the PL aperture, i.e. a larger NA, give a focus with a higher contrast because of more constructive light interference from each slit dominates the interference between the diffraction wave from the zones. We should notice that a 3-zone PL with $f = 6$ µm has a same $NA = 0.59$ as a one-zone PL with $f = 0.5$ µm. As can be seen from Fig. 13(a), the accuracy of the focal length of the former is much better than the later. So the number of zones or the total phase shift range of light involving in the interference at the focus is the real factor determining the device performance.

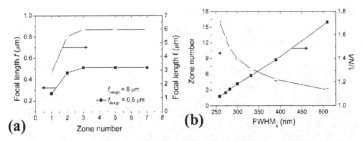

Figure 13. (a) Simulation results for the focal lengths in the PL s with variant zone numbers. The two PLs were designed to have a f of 0.5 μm and 6 μm respectively. (b) Calculated FWHMx and NA for the PL (f = 6 μm) with variant zone numbers. Reprint from [42].

Simulation results for the Poynting vector P_z distributions of the PL with f = 6 μm and in total 7 zones are shown in Fig. 14(a). Compared to the 3-zone PL shown in Fig. 11(f), the aperture size increases to 15.8 μm from 8.6μm. Accordingly, NA increases from 0.59 to 0.80 and ϕ'_{max} increases from 4.4 π to 12.3 π. The 7-zone PL has a more confined focus both in the focal plane and the light propagating direction. The amplitude ratio to the high-order foci increase as well because the additional zones contribute substantially to the main focus but negligible amount to the high-order foci due to the increasing incline. To achieve a good focusing performance, increasing the lens aperture is a straightforward way. But the footprint increases at the same time. In order to get complete constructive interference at the focus, the amplitude of the diffracted light waves from each zone should be the same. The SP modes in slits with different widths have different loss as shown in Fig. 9(b). Although the phase match at the focus in the design is nearly perfect, the amplitudes of light emitting from each zone are different. Amplitude compensation for the narrow slit is able to balance the constructive interference at the focus and therefore improve the contrast of the focus without increase of the device size. A simple method is proposed to demonstrate this effect as shown in Fig. 14(b), where some additional slits shown as red crosses in Fig. 12 were introduced to the original structure. Two more slits with a width of 10 nm were added neighbouring to each 10 nm slit in the original PL with a gap of 50 nm to enhance the transmitted light contribution at the focus. In this case, both NA and the number of zones are unchanged but very obvious improvement of the contrast around the focus is seen. In addition, the accuracy of the focal length, FWHMx and FWHMz are almost the same as the original device.

Although a well-defined focus has been demonstrated above, the transmission efficiency of a PL is still low and the side lobes are obvious. Integrating more slits in a limited lens aperture is a straightforward way to solve this problem. Etching narrower slits in a thicker metal film is one way to integrate more slits due to the increased phase delay across the lens surface but the fabrication is very difficult at a large aspect ratio. The current design has very limited modulation of the phase delay due to the quickly decreasing dependence of the mode propagation constant on the slit width (>30 nm) as shown in Fig. 9(b). Nano-slits narrower than 30 nm are very difficult to fabricate in a thick metal film (>200 nm). The dependence of Re(n_{eff}) on the refractive index of material inside the slits apparently implies a method of phase delay modulation by selectively filled the slits with different materials

[43]. Combining both the slit width and the refractive index tuning, more slits, i.e. a wider phase delay range, can be obtained. The selective refractive index tuning is feasible by applying the masked planarization and the etch-back process.

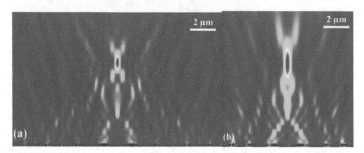

Figure 14. Simulation results for the Poynting vector Pz distributions of the modified structures of the PL shown in Fig. 11(f). (a) A PL with an extended aperture including seven zones compared to three zones in the reference PL. (b) A PL with additional slits shown as red crosses in Fig. 10 compared to the reference PL. Reprint from [42].

4.3.3. Experiments on PLs

The PLs in this work were designed to operate at 633 nm. To simplify the fabrication process, the minimum slit width was 50 nm and the minimum gap was 100 nm in the experiment. Thus the contrast around the focus will degenerate due to the less slits but the accuracy of the focal length keeps in a proper design. Electron beam lithography and dry etch were used to fabricate the nano-slits in a 200 nm thick aluminium film on glass. Fig. 15 shows SEM images of a typical PL structure that was patterned, where the focal length $f = 6$ μm. The nanoslits were 10 μm in length and the PL width, or aperture, was 10.84 μm. The device has in total 25 slits. From the central slit to the last one on the right hand side, the first four slits have widths of 50 nm, 54 nm, 68 nm, 125 nm and all the others are 100 nm. An enlarged image of the region inside the dash line rectangle of Fig. 15(b) is shown in Fig. 15(c). The roughness on the sidewalls of the gratings is predominantly due to the large grain size of the evaporated aluminium. The error of the slit widths of the fabricated slits is within 10%.

The far-field focusing pattern produced by the lenses was measured using a WITec alpha300S confocal scanning optical microscope (CSOM). A pure confocal mode was used for the experiments because the probe for near-field scanning optical microscopy may have caused a perturbation of the local fields. Sample illumination was with a collimated laser beam operating at 633 nm. The laser source was polarized in the TM mode with its electric field perpendicular to the slits. The light that was transmitted through the sample was collected using a 100×, $NA = 0.9$ objective. A multi-mode fiber with a core diameter of 25 μm was used to couple the transmitted light into a photomultiplier tube that had a sample integration time is 0.5 ms. The core of the fiber acted as the CSOM pinhole. The sample was scanned in the x and y directions using a piezoelectric scan table, and the microscope working distance was scanned to obtain the z-axis data. The step size in any direction was 200 nm.

Fig. 16(a) and (b) show the focusing light pattern in the xz plane measured by the CSOM for lenses designed to have $f = 3$ μm and $f = 6$ μm respectively. The diffracted light distribution clearly shows focusing for both lenses. The constructive interference at the focal point can be clearly seen. The positions of the foci and the side lobes agree extremely well with the simulated electric field intensity distribution shown in Fig. 16(c) and (d). The simulation results showed that the field intensity at the focus of the lens with $f = 6$ μm was 1.9 times that of the incident light. The overall transmission through the lens at 633 nm is 27% that is much larger than the slit filling ratio of 10%. The significant enhancement of the transmission is mainly due to the excitation of SPs [3]. The relatively stronger intensity of the side lobes observed in the experiment was caused by the tolerance limits of fabrication that most strongly affected the narrower slits towards the lens centre. It can be seen that the transmitted light from the central slits is weaker. Theoretical work has also shown that multilayer metal/dielectric films may give a much stronger light intensity at the focus [45].

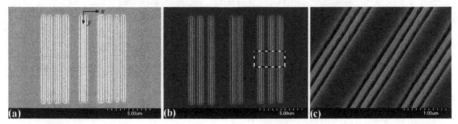

Figure 15. Scanning electron micrographs of PLs in an aluminium film. (a) and (b) are lenses designed to have $f = 3$ μm and $f = 6$ μm, respectively. (c) An enlarged image of the region inside the dash rectangle in the tilted PL in (b).

The normalized light intensity distributions through the centre of the focus spot along the x and z axis, is shown for both lenses in Fig. 17. The experimental focal lengths for the lenses (Fig. 17(a) and (b)) were 3.1 μm and 6.1 μm, which compare favorably with the simulated values of 3 μm and 6 μm, respectively. The deviation from the intended focal length is less than 3.5% for both lenses. The FWHM of the beam extension in the z-direction for the lens with a focal length of 3.1 μm is 1.3 μm, and the light intensity at the focus centre is approximately seven times that of the nearest side lobe. The extension of the focus of the lens with the focal length of 6.1 μm is larger as shown in Fig. 17(b). In the x-direction, a line plot across the focal plane of each lens shows that the light intensity drops quickly with the distance from the optical axis (at $x = 0$) as shown in both Fig. 17(c) and (d). The FWHM linewidth for the $f = 3.1$ μm lens is 470 nm, and for the $f = 6.1$ μm lens it is 490 nm. The resolution of the measured data is potentially limited by the CSOM scan step size of 200 nm and the resolving power of the objective, but as can be clearly seen from Fig. 17, the measured profiles compare very well with the simulations. As a further test of lens quality, we have compared the line-width at the focal point with what we would expect for a Gaussian limited beam. The Rayleigh limit for the resolving power of a lens is given by $d_s = \lambda f/D$, where d_s is a half of the line width, λ is the wavelength, f is the focal length and D is the lens aperture. For the lens with $f = 3.1$ μm, $d_s = 200$ nm, whereas for the lens with $f = 6.1$ μm, $d_s = 356$ nm. The results therefore suggest that the $f = 6.1$ μm lens works close to its theoretical limit. Subdiffraction focusing elements based on near-field effects were reported to be able to focus electromagnetic waves to a spot of size less than $\lambda/10$ [46-48]. However, the focal length is limited to dimensions much less than the wavelength, which is about 100 times less than our devices. Alternatively, we focused on the far-field lenses.

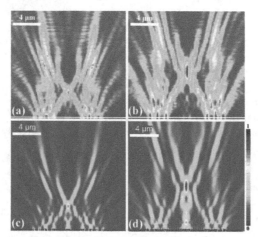

Figure 16. (a) Focusing light pattern in the xz plane obtained by the CSOM for a lens designed to have $f = 3$ μm and (b) for a lens designed to have $f = 6$ μm. The horizontal white line in (a) and (b) shows the position of the sample surface. (c) Simulation results for the $f = 3$ μm lens and (d) simulation results for the $f = 6$ μm lens. Reprint from [44].

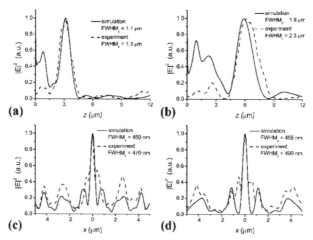

Figure 17. (a) Normalized simulation and experimental results for the light intensity distributions of the lenses. The optical axis is at $x = 0$ and the lens plane is at $z = 0$. (a) and (b) show the distribution along the z direction through the foci of the lenses with $f = 3.1$ μm and $f = 6.1$ μm respectively. (c) and (d) show the distribution along the x-direction through the foci of the lenses with $f = 3.1$ μm and $f = 6.1$ μm respectively. Reprint from [44].

To investigate the diffraction effect discussed in Section 4.3, a PL consisting of the central seven slits in the one shown in Fig. 15(b) was fabricated and characterized. As shown in Fig. 18(a), the PL with a single zone shows a light spot far away from the design focus, with an error of 90%. The pattern was also well predicted by the FDTD simulation result as shown in Fig. 18(b). We found that NAs are 0.67 and 0.1 and ϕ'_{max} s are 8π and 0.08π for the PLs in Fig. 16(b) and Fig. 18(a), respectively. For a light focusing optical element with the focal length and the aperture size comparable to the wavelength, the number of the zones or ϕ'_{max} play an important role. These experimental results confirm our conclusion that the focal length of a PL is sensitive to the number of zones. More zones give a higher accuracy of the focal length.

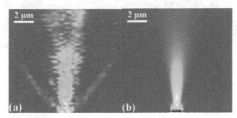

Figure 18. (a) Focusing light pattern in the xz plane obtained by the CSOM for a PL consisting of the central seven slits in the one shown in Fig. 15(b). (b) Simulation results for the Poynting vector P_z. Reprint from [44].

5. Plasmonic CMOS image sensors

The above plasmonic optical elements were finally integrated on to CISs. One CIS used in this work has a single-pixel photodiode manufactured with a United Microelectronics Corporation (UMC) 0.18 μm process. The fabrication procedure for pCIS must be modified from that used to make plasmonic filters on glass. To increase the transmission of the colour filters integrated on the CIS, we deposited a layer of SiO₂ on top of the SiNx surface passivation layer of the CIS before depositing a 150 nm film of aluminium by evaporation. This SiO₂ layer also protected the bond pads of the chip during processing. Before spin-coating with ZEP520A electron beam resist, a thin layer of SiO₂ was added on to the aluminium film to improve adhesion. The sample was exposed using a Vistec VB6 UHR EWF electron beam lithography tool. After development in o-xylene, the sample was etched using CHF₃ and Ar in a Plasmalab 80 plus and then etched using SiCl₄ in a Plasmalab System 100. After deposition of a 200 nm thick SiO₂ cap layer, a further mask and etch step was required to reopen windows over the bond-pads of the integrated circuits. Throughout processing the chip was bonded to a Si carrier to aid handling.

As shown in Fig. 19(a)-(c), reflection microscope images of the processed CIS showed different colours of the pixels. The three primary colour filters integrated on the CIS in Fig. 19(a)-(c) were designed to transmit blue (sample S1), green (sample S2) and red (sample S3) light through to the photodiodes. In accordance with this, the reflection spectra showed a complementary minimum in the reflection coefficient for each colour filter. In Fig. 19(c) we can also see a colour variation across the whole photodiode area. This is caused by non-uniformity in the fabrication of nanohole array. The non-uniformity arose because the CIS was bonded to a carrier using photoresist and there was a non-negligible tilt error during electron beam lithography. As we can see in Fig. 19(c), there are four small slightly different coloured sections in the large pixel area. The boundaries between different sections are the boundaries of the electron beam writing fields between which there is the largest non-uniformity. These non-uniformities would be eliminated in wafer-scale manufacturing.

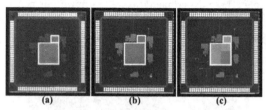

Figure 19. (a)-(c) are microscope images of CIS with integrated blue (S1), green (S2) and red (S3) plasmonic colour filters. Reprint from [49].

Simulated transmission spectra of all three samples are shown in Fig. 20(a). A complete layer stack was modelled using FDTD method to replicate the CIS structure. The stack parameters were derived from the design data of UMC 0.18 μm process. The periods and radiuses are 250nm/80nm, 340nm/90nm and 420nm/110nm for S1, S2 and S3, respectively. Since we could not measure the transmission spectra on the CIS directly, the reflection spectrum was measured instead. The result for sample S2 is shown in Fig. 20(a). As we can see, the experiment and simulation results match very well. The minimum in the reflection coefficient observed in the experimental results is slightly wider than that in simulation. This is mainly caused by the non-uniformity of the nanostructures across the 1 mm^2 photodiode area, as discussed. Photocurrent measurements for the CMOS photodiodes were conducted using a tungsten bulb, a monochromator and an Agilent/HP 4155B. The experimental wavelength resolution of 5 nm was determined by the grating and slit of the monochromator. To test the electrical variations of the photocurrent measurement and themechanical variations of the monochromator, 15 repetitive scans were conducted for one sample; the results showed negligible change. The red dash line in Fig. 20 (b) shows that the photocurrent spectrum of an unprocessed reference CIS has significant fluctuations; most notably two main dips labelled C (520 nm) and D (695 nm). A simple simulation for the whole dielectrics stack using the transfer matrix method showed that the experimental dips approximately match the dips A and B in the simulated spectrum in Fig. 20(b). The dips can be attributed to FP resonances in the CIS dielectric stack. This result is not unexpected since the CMOS process we used has not been optimized, as is usual for commercial CIS. Because we cannot directly measure the transmission spectrum, we have determined the relative transmission using the ratio of the photocurrent for sample S2 to that of an unprocessed reference CIS. The relative transmission spectrum of S2 in Fig. 20(b) has an obvious transmission band for the green colour with an average transmission of approximately 30% and a full-width at half-maximum of 130 nm. There is a sharp transmission peak near 700 nm that is caused by the shift of the labelled dip D in the photocurrent of S2 due to the change in the stack dielectric structure that arises from our processing as compared to the reference CIS. Optimisation of the CIS dielectric stack would remove this unwanted dip. The relative transmission of CIS S1, S2 and S3 is shown in Fig. 20(c). We can see the transmission bands for blue, green and red respectively. These bands are wider than the simulation results shown in Fig. 20(a). We attribute the poor performance to the non-uniformity of the nanostructures across the whole photodiode area and variation in the fabrication tolerance for the hole sizes required. Note that the unwanted transmission peaks labelled F and G are also a consequence of FP resonances in the unoptimised the unwanted transmission peaks labelled F and G are also a consequence of FP resonances in the unoptimised layer stack. The colour cross talk was evaluated in the same way as a conventional CIS in [50]. Our plasmonic CMOS photodetectors have higher cross talk due to the wider passbands of the fabricated colour filters. But these can be reduced by optimizing the plasmonic filters and the fabrication process. Note that the integration of plasmonic colour filters in a CIS would reduce the colour cross talk between the neighbouring pixels, which is not included in the calculation because the devices discussed here are single-pixel photodetectors.

A CIS with a 100×100-pixel photodiode array was investigated after the initial work on a single-pixel photodetector. As shown in Fig. 21(a), the pixel size is 10 µm × 10 µm and each pixel contains a 4.5 µm × 9 µm photodiode with an inter-photodiode gap of 0.7 µm between some pixels. The topography of the pixels has a 1.1 µm vertical step between the photodiode and the circuit regions within each pixel due to the top metal layer in the AMS 0.35 µm process as shown in Fig. 21(b). Because the CMOS chips use aluminium, it is difficult to register accurately to the pixels using our 100 kV EBL system that has a back-scatter detector. We therefore deposit approximately positioned gold electron beam markers around the CIS using a standard lift-off process. A registration EBL step is then used to write a dummy pattern on to the pixel array. With this pattern we determine the positional error between our gold markers and the pixel array using a high resolution SEM Hitachi S4700 SEM. This data is then used in a third EBL step to write the final pattern on a 150 nm aluminium film evaporated on the photodiode pixel array. Of course, only one lithography step would be needed if this process was implemented in the manufacturing flow of the foundry. A final mask and etch step was required to reopen windows over the bond-pads of the integrated circuits. A microscope image of the processed pixels is shown in Fig. 21(c), where a series of varying plasmonic components were repeated across the photodiode pixel array. In this reflection image it is possible to see the various colours of the pixels due to the SPR generated by differently patterned nanostructures. It is also possible to see black rectangles that correspond to reference pixels, from which the aluminium film above the photodiodes has been totally removed. A SEM image of the pixels with plasmonic filters is shown in Fig. 21(d), where we can see a good alignment of the plasmonic structure to each pixel. The inset in Fig. 21(d) shows a hole array with a period of 230 nm. This filter transmits a dark blue colour.

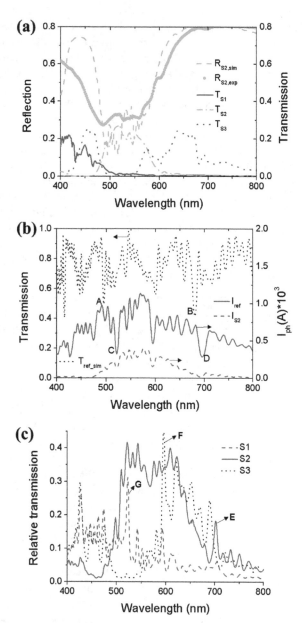

Figure 20. (a) Simulated transmission spectra of CIS S1, S2 and S2. Simulated and measured reflection spectra of S2 are shown as well. (b) Measured photocurrent of the reference sensor and S2. The simulated transmission spectrum of the reference sensor is shown for comparison. (c) Relative transmission of sample S1, S2 and S3. Reprint from [49].

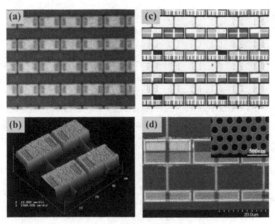

Figure 21. (a) A microscope image of the photodiode pixel array on an unprocessed chip. (b) An AFM image of the topography of pixels before processing. (c) A microscope image of the photodiode pixel array with plasmonic colour filters fabricated on top. (d) A SEM image of the plasmonic colour filters on top of a CIS. The inset is an enlarged image of a filter with the period of 230 nm. Reprint from [21].

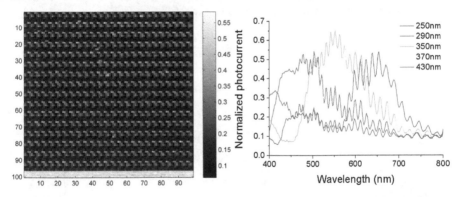

Figure 22. (a) Photocurrent distribution across the 100×100-pixel photodiode array. (b) Normalized photocurrents of photodiodes with various plasmonic colour filters. Reprint from [21].

To test the pixel array, a tungsten bulb and a monochromator were used to illuminate the array with wavelengths in the range between 400 nm and 800 nm. The slits used within the monochromator were chosen to limit the range of wavelengths during each experiment to 5 nm, whilst the centre wavelength was changed in 2 nm increments. At each centre wavelength the response of each pixel, was measured using a data-acquisition system, specifically a NI USB-6218. In order to correct for fixed pattern noise the response of each pixel was determined from the change in the output voltage after an integration time of 125 ms. In addition, to compensate for the effects of the currents that flow within each pixel even in the absence of light, the response of each pixel after an integration time of 125 ms in the absence of light was subtracted from the response at each centre wavelength. The response distribution across the pixel array when it is illuminated by light with a wavelength of 550 nm is shown in Fig. 22(a) to demonstrate that, with the exception of a very few defective pixels, there is a good uniformity to the plasmonic components. The spectral responses we present have been determined by averaging the responses of all the pixels with the same plasmonic filter, across the whole pixel array. Since it is impossible to directly measure the transmission spectrum of the filters, the performance of the filters has been assessed by normalizing the average response of each pixel design using the average response of the unpatterned reference pixels. Results for pixels with five different plasmonic filters (Fig. 22(b)), show that the filters act as bandpass filters with different centre wavelengths. The normalized responses of these filters are above 50% of the signal measured from an unprocessed pixel and the FWHMs are between 110 nm and 150 nm. One obvious feature of these results that was not seen in the results in Fig. 4(c) is the oscillations in all the responses. The fact that these oscillations were previously observed in the single-pixel plasmonic CMOS photodetector and that they are also observed in the responses of pixels before any back end of line processing has occurred leads to the conclusion that they are attributable to FP resonances in the CIS dielectric stack. The resonances occur because the layers in the dielectric stack have different permittivities, hence there is a reflection at each interface. This result is not unexpected since the CMOS process we used has not been optimised to reduce this effect, as is usual for commercial CIS.

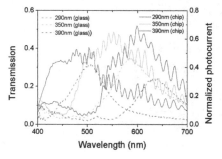

Figure 23. Normalized photocurrents of three sets of photodiodes with peak responses close to 445 nm, 555 nm and 600 nm. Transmission spectra of the same colour filters on glass are shown for comparison. Reprint from [21].

Average results of three groups of pixels with the peak response wavelengths close to Red (600 nm), Green (555 nm), Blue (445 nm), according to the 1931 International Commission on Illumination 2° standard observer colour matching functions [51], are shown in Fig. 23. The transmission spectra of the same filters on glass are shown for comparison. All photodiode pixels show a blue-shift of their peak responses compared to the peak transmission wavelengths of the plasmonic colour filters on glass. Two factors contribute to this phenomenon; one is the smaller hole sizes of the filters integrated on chip than those on glass, the other is the high-index substrate loading effect in the case of CMOS chips. This effect can be eliminated by calibration of the process for colour matching. In addition the transmission bands of the filters on the CIS seem to be wider than the equivalent filters when they are made on glass. The phenomena that could be contributed to this broadening include the possibility that a wider range of angles of incidence of light occur when the pixels are tested and cross-talk caused by the large vertical separation, 8 μm, between the filters and the photodiodes underneath. These effects mean that the results obtained are not necessarily representative of those that will be obtained in an optimized manufacturing process, especially if the plasmonic colour filters are integrated in lower metal layers in a standard CMOS process. In addition, it is anticipated that the broadening will be reduced by using narrow band filters such as low loss silver filters. The optoelectronic efficiency can be further improved if these plasmonic nanostructures are fabricated close enough to the photodiodes, where the localized field can be greatly enhanced due to the SPR effects [52]. It is anticipated that in future state-of-the-art CMOS technology (ITRS roadmap 2010) the half-pitch in Metal 1 will scale down to 32 nm in 2012, enabling mass manufacture of suitable SPR structures.

6. Conclusion

We demonstrated a detailed study of plasmonic optical elements for the application in CISs and the first plasmonic CIS with plasmonic colour filters replacing conventional polymer colour filter array. The plasmonic optical elements such as colour filters, polarizers and lenses showed promising performance where the complete control on wavelength filtering, polarization and the phase distribution were achieved by carefully optimizing the metallic nanostructures. The complete compatibility with the CMOS technology of these metallic optical devices facilitates the plasmonic CIS integration. Replacing the conventional optical elements, plasmonic devices offer various advantages such as less cross-talk, low cost and multifunction, etc. It would be an important step forward to apply nanophotonics in the CMOS imaging. It could be a new way to bring plasmonics research from the lab to the foundry.

Author details

Qin Chen
Corresponding Author
Suzhou Institute of Nano-Tech and Nano-Bionics, Chinese Academy of Sciences, China
University of Glasgow, United Kingdom
SOC Key Laboratory, Peking University Shenzhen Research Institute, China

Xiaohua Shi
Suzhou Institute of Nano-Tech and Nano-Bionics, Chinese Academy of Sciences, China

Yong Ma
University of Glasgow, United Kingdom

Jin He
Peking University Shenzhen SOC Key Laboratory, PKU-HKUST Shenzhen-Hongkong Institution, Shenzhen, P.R. China

7. References

[1]   Raether H. Surface Plasmons. Berlin: Springer; 1988.

[2]   Gramotnev DK, Bozhevolnyi SI. Plasmonics beyond the diffraction limit. Nature Photonics 2010;4 83-91.

[3]   Ghaemi HF, Thio T, Grupp DE, Ebbesen TW, Lezec HJ. Surface plasmons enhance optical transmission through subwavelength holes. Physical Review B 1998;58 6779-6782.

[4]   Genet C, Ebbesen TW. Light in tiny holes. Nature 2007;445 39-46.

[5]   Laux E, Genet C, Skauli T, Ebbesen TW. Plasmonic photon sorters for spectral and polarimetric imaging. Nature Photonics 2008;2 161-164.

[6]   Xu T, Du C, Wang C, Luo X. Subwavelength imaging by metallic slab lens with nanoslits. Applied Physics Letters 2007;91 201501.

[7]   Verslegers L, Catrysse PB, Yu Z, Fan S. Planar metallic nanoscale slit lenses for angle compensation. Applied Physics Letters 2009;95 071112.

[8]   Koo C, Kim H, Paik K, Park D, Lee K, Park Y, Moon C, Lee S, Hwang S, Lee D, Kong J. Improvement of crosstalk on 5M CMOS image sensor with 1.7x1.7µm² pixels. In Proc. Of SPIE 2007;6471 647115.

[9]   Prima J, Roy F, Leininger H, Cowache C, Vaillant J, Pinzelli L, Benoit D, Moussy N, Giffard B. Improved colour separation for a backside illuminated image sensor with 1.4 µm pixel pitch. Proc. International Image Sensor Workshop, Bergen, Norway; 2009.

[10] Catrysse PB, Wandell BA. Integrated colour pixels in 0.18um complementary metal oxide semiconductor technology. Journal of the Optical Society of America A 2003;20 2293-2306.

[11] Catrysse PB. Monolithic integration of electronics and sub-wavelength metal optics in deep submicron CMOS technology. Materials Research Society Symposium Proceedings 2005;869 D1.5.1-D1.5.12.

[12] Kanamori Y, Shimono M, Hane K. Fabrication of transmission color filters using silicon subwavelength gratings on quartz substrates. IEEE Photonic Technology Letters 2006;18 2126-2128.

[13] Collin S, Vincent G, Haidar R, Bardou N, Rommeluere S, Pelouard J. Nearly perfect Fano transmission resonances through nanoslits drilled in a metallic membrane. Physical Review Letters 2010;104 027401.

[14] Xu T, Wu Y, Luo X, Guo LJ. Plasmonic nanoresonators for high-resolution colour filtering and spectral imaging. Nature Communications 2010;1 59.

[15] Inaba Y, Kasano M, Tanaka K, Yamaguchi T. Degradation-free MOS image sensor with photonic crystal colour filter. IEEE Electron Device Letters 2006;27 457-459.

[16] Frey L, Parrein P, Raby J, Pellé C, Hérault D, Marty M, Michailos J. Color filters including infrared cut-off integrated on CMOS image sensor. Optics Express 2011;19 13073-13080.

[17] Lee HS, Yoon YT, Lee SS, Kim SH, Lee KD. Colour filter based on a subwavelength patterned metal grating. Optics Express 2007;15 15457-15463.

[18] Lumerical FDTD Solution, http://www.lumerical.com/

[19] Walls K, Chen Q, Collins S, Cumming DRS, Drysdale TD. Automated design, fabrication and characterization of colour matching plasmonic filters. IEEE Photonics Technology Letters 2012;24 602-604.

[20] Chen Q, Cumming DRS. High transmission and low colour cross-talk plasmonic colour filters using triangular-lattice hole arrays in aluminium films. Optics Express 2010;18 14056-14060.

[21] Chen Q, Das D, Chitnis D, Walls K, Drysdale TD, Collins S, Cumming DRS. A CMOS Image Sensor Integrated with Plasmonic Colour Filters. Plasmonics, DOI: 10.1007/s11468-012-9360-6.

[22] Andreou AG, Kalayjian ZK. Polarization imaging: principles and integrated polarimeters. IEEE Sensors Journal 2002;2 566-576.

[23] Schechner YY, Narasimhan SG, Nayar SK. Polarization-based vision through haze. Applied Optics 2003;42 511-525.

[24] Tokuda T, Sato S, Yamada H, Ohta J. Polarization analyzing CMOS sensor for microchamber/microfluidic system based on image sensor technology. IEEE International Symposium on Circuit and Systems, May 18-21, 2008, Seattle, USA.

[25] Zhao X, Boussaid F, Bermak A, Chigrinov VG. Thin Photo-Patterned Micropolarizer Array for CMOS Image Sensors. IEEE Photonics Technology Letters 2009;21 805-807.

[26] Nordin GP, Meier JT, Deguzman PC, Jones MW. Micropolarizer array for infrared imaging polarimetry. Journal of the Optical Society of America A 1999;16 1168-1174.

[27] Ekinci Y, Solak HH, David C, Sigg H. Bilayer Al wire-grids as broadband and high-performance polarizers. Optics Express 2006;14 2323-2334.

[28] Huo Y, Fesenmaier CC, Catrysse PB. Microlens performance limits in sub-2μm pixel CMOS image sensors. Optics Express 2010;18 5861-5872.

[29] Lezec HJ, Degiron A, Devaux E, Linke RA, Martin-Moreno L, Garcia-Vidal FJ, Ebbesen TW. Beaming Light from a Subwavelength Aperture. Science 2002;297 820-822.

[30] Sun Z, Kim HK, Refractive transmission of light and beam shaping with metallic nano-optic lenses Applied Physics Letters 2004;85 642-644.

[31] Shi H, Wang C, Du C, Luo X, Dong X, Gao H. Beam manipulating by metallic nano-slits with variant widths. Optics Express 2005;13 6815-6820.

[32] Fu Y, Zhou W, Lim L, Du CL, Luo XG. Plasmonic microzone plate: Superfocusing at visible regime. Applied Physics Letters 2007;91 061124.

[33] Chen Y, Zhou C, Luo XG, Du C. Structured lens formed by a 2D square hole array in a metallic film. Optics Letters 2008;33 753-755.

[34] Mote RG, Yu SF, Ng BK, Zhou W, Lau SP. Near-field focusing properties of zone plates in visible regime - New insights. Optics Express 2008;16 9554-9564.

[35] Verslegers L, Catrysse PB, Yu Z, Shin W, Ruan Z, Fan S. Phase front design with metallic pillar arrays. Optics Letters 2010;35 844-846.

[36] Verslegers L, Catrysse PB, Yu Z, White JS, Barnard ES, Brongersma ML, Fan S. Planar Lenses Based on Nanoscale Slit Arrays in a Metallic Film. Nano Letters 2009;9 235-238.

[37] Lin L, Goh XM, McGuinness LP, Roberts A. Plasmonic Lenses Formed by Two-Dimensional Nanometric Cross-Shaped Aperture Arrays for Fresnel-Region Focusing. Nano Letters 2010;10 1936-1940.

[38] Fu Y, Liu Y, Zhou X, Xu Z, Fang F. Experimental investigation of superfocusing of plasmonic lens with chirped circular nanoslits. Optics Express 2010;18 3438-3443.

[39] Goh XM, Lin L, Roberts A. Planar focusing elements using spatially varying near-resonant aperture arrays. Optics Express 2010;18 11683-11688.

[40] Ruffieux P, Scharf T, Herzig HP, Völkel R, Weoble KJ. On the chromatic aberration of microlenses. Optics Express 2006;14 4687-4694.

[41] Gordon R, Brolo AG. Increased cut-off wavelength for a subwavelength hole in a real metal. Optics Express 2005;13 1933-1938.

[42] Chen Q. Effect of the Number of Zones in a One-Dimensional Plasmonic Zone Plate Lens: Simulation and Experiment. Plasmonics 2011;6 75-82.

[43] Chen Q. A novel plasmonic zone plate lens based on nano-slits with refractive index modulation. Plasmonics 2011; 6 381-385.

[44] Chen Q, Cumming DRS. Visible light focusing demonstrated by plasmonic lenses based on nano-slits in an aluminum film. Optics Express 2010;18 14788-14793.

[45] Kim HC, Ko H, Cheng M. High efficient optical focusing of a zone plate composed of metal/dielectric multilayer. Optics Express 2009;17 3078-3083.

[46] Grbic A, Jiang L, Merlin R. Near-Field Plates: Subdiffraction Focusing with Patterned Surfaces. Science 2008;320 511-513.

[47] Eleftheriades GV, Wong AMH. Holography-Inspired Screens for Sub-Wavelength Focusing in the Near Field. IEEE Microwave and Wireless Components Letters 2008;18 236-238.

[48] Gordon R. Proposal for Superfocusing at Visible Wavelengths Using Radiationless Interference of a Plasmonic Array. Physical Review Letters 2009;102 207402.

[49] Chen Q, Chitnis D, Walls K, Drysdale TD, Collins S, Cumming DRS. CMOS Photo Detectors Integrated with Plasmonic Colour Filters. IEEE Photonic Technology Letters 2012;24 197-199.

[50] Agranov G, Berezin V, Tsai RH. Crosstalk and microlens study in a color CMOS image sensor. IEEE Transactions on Electron Devices 2003;50 4-11.

[51] CIE Free Documents for Download: CIE 1931 Standard Colorimetric Observer Data. http://www.cie.co.at/main/freepubs.html.

[52] Schaadt DM, Feng B, Yu ET. Enhanced semiconductor optical absorption via surface plasmon excitation in metal naonparticles. Applied Physics Letters 2005;86 063106.

Chapter 11

Plasmonic Rectenna for Efficient Conversion of Light into Electricity

Fuyi Chen, Jian Liu and Negash Alemu

Additional information is available at the end of the chapter

1. Introduction

Solar energy has been widely studied as a green energy source. Present photovoltaic techniques are mainly semiconductor solar cells. The silicon solar cell has a good power conversion efficiency of 24%, and the semiconductor solar cell based on GaAs multijunction has the highest conversion efficiency of 32% at present [1]. Though these semiconductor solar cells have relatively high conversion efficiency, their high cost and some poisonous byproduct produced during the manufacturing process limit their commercial use. As an alternative for large scale application, the dye-sensitized solar cell (DSSC) has obtained conversion efficiency of 12% at maximum in these days [2], only a few percentages has been increased since the Grazel's research, in which the conversion efficiency of DSSC is improved using nanoporous photoanodes [3].

In 1998, Stuart et al. [4] reported the first application of surface plasmons in the photovoltaic technology. It was found that the photocurrent of the thin-film Silicon-on-insulator (SOI) device was 18 times improved at the wavelength of 800 nm by metallic nanoparticles. So far, the plasmonic effect of the metal nanostructure is one of the main directions [5, 6] to improve the conversion efficiency for DSSC. The United States of America (USA), the European Union and many other countries undertake some related projects, such as the SOLAMON project, and some typical research achievements are obtained in the plasmonic DSSC field, which are described as follows according to the role of metal surface plasmons:

Scattering effect of localized surface plasmon resonance (LSPR). Brown et al. [7] reported that adding Au-Si nanoparticles with the shell-core structure with size of 20 nm into the typical solid electrolyte DSSC led to an improvement of the conversion efficiency; Hagglund et al. [8] studied the influence of Au nanoplate arrays on the photoelectric conduction of the dye-sensitized TiO_2 thin-film.

Concentration effect of light using localized surface plasmon resonance (LSPR). Du et al. [9] theoretically demonstrated that the Au nanodimer improved the ultraviolet absorption of the TiO₂ nanoparticles; Guilatt et al. [10] studied the light absorption property of the Au nanoshell in the Si thin-film.

Trapping effect of surface plasmon polariton (SPP). Fu et al. [11] reported the conversion efficiency of the typical DSSC with the Pt fishnet arrays as the counter electrode, Durr et al. [12] studied the effect of the Ag thin-film on the short circuit current in DSSC with a planar waveguide structure.

It is clear that metallic nanostructures possess the excellent light-harvesting capability in the visible spectrum, and they can collect almost all the photon energy which is concentrated on the nanoscale plasmonic hotspots from an incident light wavelength region of several hundred nanometers. Therefore, the electromagnetic fields in these regions are enhanced exponentially.

However, the total energy conversion efficiency of DSSC with the plasmonic optical effect is not increased simultaneously as we expect. As shown in the experiments, the power conversion efficiencies were only 1.95 % [7] and 5.77 % [11] with the plasmonic effect considered. We show in next section that the parasitic absorption of small metal nanoparticles is the main reason, and the power conversion efficiency enhanced by metallic nanoparticles is much less than their enhanced efficiency of light collection because the metallic nanoparticles absorb one part of the incident light during surface plasmon resonance (SPR). The parasitic absorption had been mentioned in previous thin film devices [13, 14] which demonstrated that the metallic absorption had an adverse effect on the solar cell performance.

2. Parasitic absorption of metal nanoparticles in the plasmonic DSSC

Our numerical calculations were performed using the finite difference time domain technique (FDTD) to solve the three-dimensional vector Maxwell equations in the optical structure of the plasmonic DSSC, as showed in Figure 1. The distance of the two transparent conductive oxides (TCO) electrodes was 1600 nm, the dye-sensitized layer was 500 nm in thickness. The FCC-arranged Ag spherical nanoparticle array was on top of the dye-sensitized layer with a 2 nm space. The plasmonic DSSC was illuminated with plane wave from the top TCO cathode with the electric field in the Ag spherical nanoparticle array. A perfectly matched layer and a conformal mesh region self-adapted to the structure were used.

The transmission parameters were calculated as the ratio of the power transmitted through the structure to the power incident, and the reflection parameters were calculated as the ratio of the power reflected from the structure to the power incident. The particle swarm optimization (PSO) was used to evaluate different design geometries to locate an optimum solution. PSO, based on the movement and intelligence of swarms, was a robust stochastic evolutionary computation technique inspired in the behavior of bee flocks [15]. It has been shown in certain instances to outperform other methods of optimization like genetic algorithms (GA) in the electromagnetic community [16].

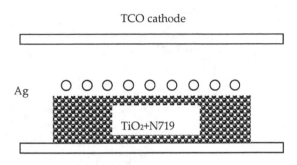

TCO cathode

Ag

TiO$_2$+N719

TCO anode

Figure 1. Schematic of the plasmonic DSSC structure. Dye-sensitized layer is 500 nm in thickness, Ag nanoparticle is 210 nm in diameter, the spacing between sensitized layer and Ag nanoparticles is 2 nm, the distance of the two transparent conductive oxides (TCO) electrodes is 1600 nm.

Figure 2 shows the monochromatic absorption enhancement and the integral quantum efficiency (IQE) enhancement as a function of Ag nanoparticles diameter and period of the Ag nanoparticles arrays in the wavelength range of 400~1100 nm. The optimal optical structure designed based on PSO for the present study is the Ag nanoparticles array of 210 nm in diameter and 500 nm in period, which has an improved integral quantum efficiency (IQE) enhancement of 11.3%. The monochromatic absorption enhancement of the Ag nanoparticles of diameter = 210 nm and period = 500 nm is improved by 20-30 % in the wavelength range 630 nm to 1100 nm, it has a maximum enhancement of absorption at the wavelength of 796 nm, and it is far higher than that of the Ag nanoparticle arrays of diameter = 70 nm and period = 250 nm which does not improve the conversion efficiency significantly.

Figure 3 shows the optical properties, the electric near-field distribution and the visible light absorption distribution at wavelength of 796 nm in the optimized plasmonic DSSC. It can be seen from Figure 3 (left) that even under optimization condition the metal absorption takes up 20-30 % of total absorption in the visible bands of 400-700 nm, which is consistent with the monochromatic absorption enhancement spectrum. This may be because of the electric field is mainly distributed at the outside of the Ag nanoparticles, and the metal nanoparticle absorption intensity is higher than the dyes, which can be seen from Figure 3 (right). It is clear from Figure 3 (middle) that the silver particle increases absorption of the dye by the enhanced forward scattering due to the surface plasmon resonance.

Although the conversion efficiency of photoactiviated thin film can be improved by about 15-20% theoretically by optimized metal nanoparticle arrays, the absorption of metal nanoparticles (210 nm in diameter) take up about 20-30% of total absorption of the DSSC in visible light, i.e., lights absorbed by DSSC does not completely convert into photocurrent. A part of the energy is lost in the form of Joule heat (ohmic loss), which led to the low external quantum efficiency of photoelectric conversion.

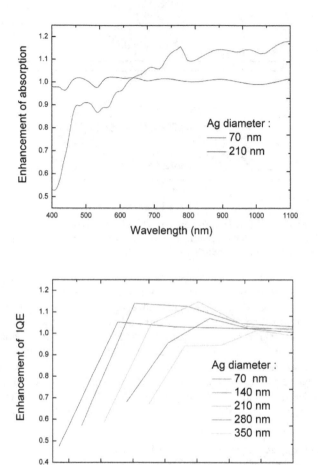

Figure 2. (a) Monochromatic light absorption enhancement of the Ag nanoparticle arrays of 70 nm and 210 nm in diameter and (b) the integral quantum efficiency enhancement of Ag nanoparticle arrays with different period (70-800 nm) and diameter (70-210 nm).

In typical DSSC device, the dye sensitized TiO_2 is 10-80 nm in diameter [17], in order to build the electrostatic localized surface plasmon resonances of metallic nanoparticles inside the dye sensitized TiO_2, the smaller Ag nanoparticles were buried to improve the absorption of dye sensitized TiO_2 nanoparticles, the intrinsic heat loss of small metal nanoparticles (such as 70 nm) at visible light bands is far greater than the electromagnetic energy scattered into the optical activity medium by these metal nanoparticles. The problem has to be considered in the plasmonic DSSC device, which has been highlighted in a recent review [18].

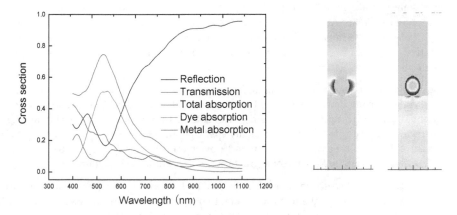

Figure 3. Optical properties of the plasmonic DSSC (left) under optimized conditions (Ag diameter of 210 nm, period 500 nm), the electric near-field distribution (middle) and the visible light absorption distribution (right) in DSSC at wavelength of 796 nm where the enhancement of monochromatic absorption is maximum.

To solve the problem of metal parasitic absorption in DSSC caused by surface plasmon optical effect, Guilatt et al. [10] theoretically studied the method of using metal shell-core nanostructure to substitute pure metal nanoparticles. But the potential development of this method was limited by the metallic skin absorption property. Designing and fabricating the metal-insulator-metal (MIM) sandwich structure optical rectenna as photon acceptor and electron donor may be a research direction for the further development of the plasmonic photovoltaic techniques. In this chapter, we further develop the idea of optical rectification (OR) that can convert visible/near-infrared (VIS/NIR) wavelengths to direct current. Section 3 includes the antenna properties of the noble metal (Ag) and transition metal (Ti), and metal-insulator-metal (MIM) nanostructure and Section 4 includes the optical rectification properties of Ag-TiO$_2$-Ti and AgCu-TiO$_2$/Ti metal-insulator-metal (MIM) diode devices. We exploit these contents to address the effect of plasmonic properties on the electronic transport behaviors in noble-transition metal heterodimers in order to put forth a new theory on the plasmonic optical rectification.

3. Antenna properties of the noble, transition metal nanowires, metal-insulator-metal (MIM) nanostructures

3.1. Review of optical rectification (OR) effect

Metallic nanostructures provide a means to transduce free-propagating electromagnetic waves into localized surface plasmon resonance modes [19] and can function as optical antennas [20]. In the traditional radiofrequency and microwave regime, antennas (such as antennas in mobile phones of our everyday life) are usually used to convert electromagnetic radiations into electric currents, however, most of the optical antennas studied so far operate on a "light-in and light-out" basis [21], the utilizing of metallic optical antennas to convert the optical radiations into photocurrents is one of the most imperative tasks for the state-of-the-art plasmonic technology. Conversion of visible/near-infrared (VIS/NIR) wavelengths to effective direct currents (DC) is a second-order nonlinear optical phenomena, e.g., optical rectification (OR) effect.

The search for OR effect in metallic optical antennas was started in 1964, when Brown of the Raytheon Company demonstrated the first flight of a microwave-powered helicopter using rectenna [22], a high up to 90% efficiency was obtained for a rectenna operating at single microwave frequency in 1976 [23]. Generally speaking, the OR effect could achieve up to 80 percent efficiency of energy conversion and power transmission in microwave range 300 MHz to 300 GHz, and typical antenna structure size was around few millimeters.

The endeavor for conversion of light radiation to DC power was started in 1968 [24], following the example of the microwave point-contact diode rectification structure, much research has been performed to extend the concept and application of rectenna into ultrahigh (visible) frequencies (1000THz), and progress has been made in the fabrication and characterization of the metal-insulator-metal (MIM) diodes for use in rectification device. It has been demonstrated that optical antennas can couple electromagnetic radiation in the visible in the same way as radio antennas do at their corresponding wavelength, three kinds of the metallic nanostructures with OR effect in the visible/near-infrared wavelength range had been located: whisker diodes [24,25], Schottky diodes [26,27] and metal-insulator-metal tunneling diodes [28,29]. Meanwhile, for the application in the solar cell technology, Bailey originally proposed that broadband rectifying antennas can be used for direct conversion of solar energy to DC power [30], the first patent on solar rectification was issued to Marks [31].

It is clear that metallic optical antennas possess as good capability of receiving electromagnetic energy in the visible/near-infrared spectrum as the radiofrequency antennas do, and that they can collect almost all the photon energy in a wavelength range of several hundreds of nanometers. However, in comparison with the high conversion efficiency of the radio frequency antennas, the theory and application of OR effect of visible and infrared radiation is still in its infancy, a major technological difficulties that impede the performance of OR device at high frequency is that in the infrared and visible wavelength, metals are no longer perfect conductor.

The main difference between low frequency and ultrahigh frequencies electromagnetic waves when it comes to interaction with the free electrons in a finite piece of metal is the fact that electrons have an effective mass which causes them to react with increasing phase lag to an oscillating electromagnetic field as the frequency increases. For increasing frequency of the excitation, they exhibit an increasing oscillation amplitude as well as an increasing phase lag. As soon as the phase lag approaches 90° the amplitude of the charge oscillation goes through a maximum and is only limited by the internal (ohmic and radiation) damping of the system.

This resonance corresponds to the surface plasmon resonance, for certain metals (such as gold, silver and copper) which consist of the anode and cathode in the metal-insulator-metal (MIM) structure optical rectennas, the surface plasmon resonance happens to appear in or close to the visible spectral range. Plasmon resonances do not appear in perfect conductors (metals at low enough frequencies) since in those materials by definition no phase lag exists between excitation and charge response. The presence of plasmon resonance is therefore characteristic for optical frequencies and give rise to drawbacks of antenna systems in this frequency range, such as enhance Ohmic losses compared to the radiofrequency regime.

3.2. Modeling of the nanowire antenna

Nanowires which are made up of either noble or transition metals exhibit distinctive electromagnetic properties that make them appropriate candidate for nanoantenna at higher frequencies. This nanoantenna is put to practical application for efficient conversion of visible-near infrared spectrum energy into direct current electricity since its operation is on the basis of the wave nature of optical radiation.

In this section, long vertical nanowires were theoretically investigated as a model system to understand the behavior of dipole optical antennas. Figure 4(a) shows the geometry of the nanowire structure. The nanowire consists of the perfectly conductor (PEC), silver (Ag) or titanium (Ti) with the diameter of 100 nm and the length of 2000 nm. The plane wave source is located to the left of the structures in vacuum. The results of field enhancement and vertical components of the Poynting vector of antenna radiation simulations are depicted in Figures 5-11.

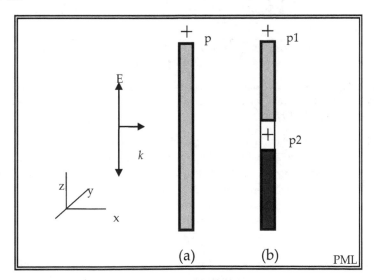

Figure 4. Schematic of the nanowire (a) and its heterodimer (b) as the calculation model for the antenna properties using the finite difference time domain (FDTD) method. The polarization direction (E) is along the long symmetric axis of the nanowires. The position of p and p1 are 5 nm from the nanowire apex, and p2 is at the middle of the dimer. The mesh is truncated using PML absorbing boundary conditions.

3.3. Perfectly conducting nanowire

According to classical antenna theory, metal antennas resonate at a wavelength λ when their length equals approximately $(2n+1)\lambda/2$, where n=0, 1, 2, etc. At resonance the charge distribution inside the antenna intensifies the incident electric field (E-field) locally at its end. In Figure 5 and 6, we plot the calculated values for the field enhancement $|E|/|E_0|$, normalized to the incident electric field, at the apex of the nanorod, for a 2000 nm perfectly conducting nanowires.

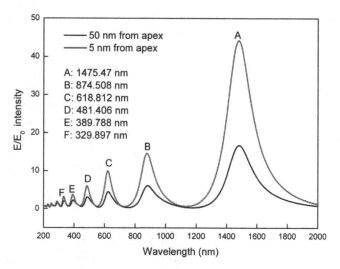

Figure 5. Multiple resonance spectrum around a perfectly conducting nanowire. The resonance is shown as a value of $|E|/|E_0|$ at a point of 5 or 50 nm above the nanowire as a function of wavelength of incident light (200 nm to 2000 nm).

Figure 5 and 6 show the simulated local field intensity enhancement spectra for perfectly conducting nanowire, its intensity distribution and Poynting vector corresponding to the higher harmonic mode resonances. Multiple resonances have seen in the field enhancement spectrum (Figure 5) at the given wavelength range. Each resonance corresponds to specific harmonic modes in the near field.

3.4. Ag nanowire

Any specific structure to consider as an optical antenna is supposed to be able to localize and enhance the propagating electromagnetic wave within a certain bandwidth. The results for silver nanowire with the parameters mentioned above are presented in Figure 7, 8 and 11. The results show that silver nanowire has enhancement spectra with very distinct peaks. The peak at shorter wavelength (350.477nm) attributed to surface plasmon resonance and for the wavelengths in the remaining band strong multiple resonances are observed.

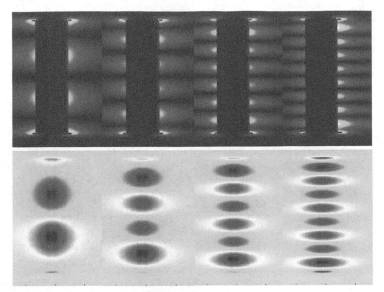

Figure 6. Simulated results for electric field profiles (top four panels) and vertical components of the Poynting vector of antenna radiations (bottom four panels) around a perfectly conducting nanowire, corresponding to the antenna mode ($3\lambda/2$, $5\lambda/2$, $7\lambda/2$ and $9\lambda/2$ resonances). The electric field profile running through the center of the nanowires and the Poynting vector calculated at 55 nm above the nanowire axis.

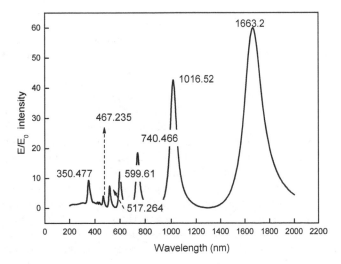

Figure 7. Multiple resonance spectrum around a silver nanowire. The resonance is shown as a value of $|E|/|E_0|$ at a point of 5 nm above the nanowire as a function of wavelength of incident light (200 nm to 2000 nm).

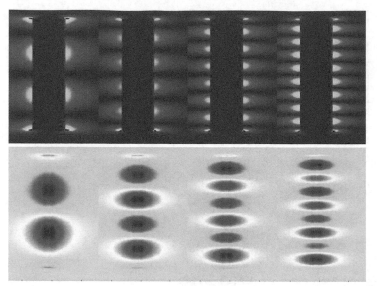

Figure 8. Simulated results for electric field profiles (top four panels) and vertical components of the Poynting vector of antenna radiations (bottom four panels) around a silver nanowire, corresponding to the antenna mode ($3\lambda/2$, $5\lambda/2$, $7\lambda/2$ and $9\lambda/2$ resonances). The electric field profile running through the center of the nanowires and the Poynting vector calculated at 55 nm above the nanowire axis.

3.5. Ti nanowire

Figure 9 and 10 show the calculated result of amplitude enhancement, electric field profile and poynting vector as a function of wavelength for Ti nanowire. In the spectral range of calculation a feature of the Ti nanowires is the broaden peaks and weak field distribution as compared to perfectly conducting and Ag nanowire structures.

When comparing the field enhancement vs. wavelength for the silver and titanium nanowires to the perfect conducting nanowire, we can understand the optical antennas based on the background of both well-developed radiowave antenna engineering and the plasmonic behavior. According to the classical antenna theory that a fundamental mode occurs when an antenna length is half of the wavelength ($\lambda/2$), the fundamental antenna mode ($\lambda/2$ resonance) for a 2000 nm long perfect conducting nanowire is around 4000 nm, the observed modes (A-D) shown in Figures 5 and 6 should be assigned as higher harmonic modes ($3\lambda/2$, $5\lambda/2$, $7\lambda/2$, and $9\lambda/2$ resonances). Comparing the multiple resonances spectrum between the perfect conducting and silver nanowires (Figures 7 and 8), it can be pointed out that a silver nanowire is a superb candidate for broadband optical antenna, which support four harmonic resonance modes in spectral range 517 to 1663 nm and one surface plasmon resonance (SPR) at 350.477 nm, as shown in Figure 11. The silver nanowire antenna is a one-dimensional SPR cavity, support the plasmon mode by reflecting the SPR currents at both ends of the nanowires.

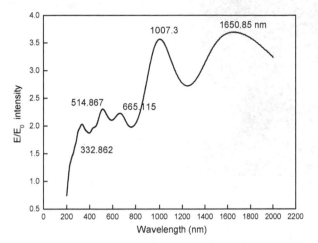

Figure 9. Multiple resonance spectrum around a titanium nanowire. The resonance is shown as a value of |E|/|E₀| at a point of 5 nm above the nanowire as a function of wavelength of incident light (200 nm to 2000 nm).

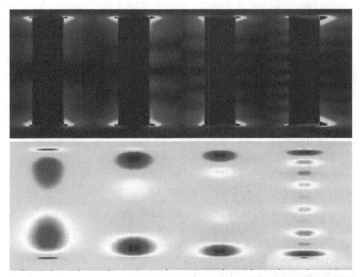

Figure 10. Simulated results for electric field profiles (top four panels) and vertical components of the Poynting vector of antenna radiations (bottom four panels) around a 2000 nm long titanium nanowire, corresponding to the antenna mode ($3\lambda/2$, $5\lambda/2$, $7\lambda/2$ and $9\lambda/2$ resonances). The electric field profile running through the center of the nanowires and the Poynting vector calculated at 55 nm above the nanowire axis.

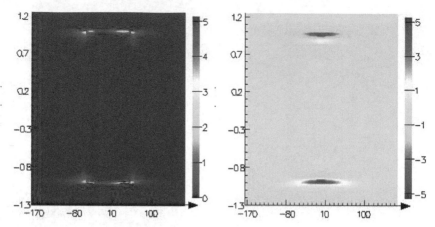

Figure 11. The electric field profiles (left) and vertical components of the Poynting vector of antenna radiations (right) around a silver nanowire at 350.477 nm.

For the titanium nanowire (Figures 9 and 10), their $|E|/|E_0|$ profiles and Poynting vector components are much weaker than the perfect conducting and silver nanowires, the possible research is that titanium has interband electronic transition at visible range [32]. The noble metals silver has higher energy of interband transitions 3.8ev (326nm) with respect to its surface plasmon resonance energy 3.38 eV (350nm) which reduces the damping and enhances the appearance of the plasmon mode, however, the transition metal titanium has lower energy of interband transition 2.0ev (610nm) which is liable for strong damping.

3.6. Ag-TiO$_2$-Ag MIM nanostructures

The end to end coupling of two nanowires with a small distance between them can create highly localized and strongly enhanced optical fields in the gap and due to this phenomenon such structures are well competent for optical antenna. The coupling between two nanowire antennas robustly increases with decreasing the gap distance. To study this effect in the spectral response and its intensity enhancement we therefore calculated the symmetric and asymmetric metal-insulator-metal (MIM) nanostructures

MIM nanostructures were theoretically investigated as model systems for split-dipole optical antennas. Figure 4(b) shows the geometry of the heteronanowire structure. The heteronanowire is a split-dipole nanoantennas consisting of a silver nanowire (diameter, 100 nm) and identical titanium (Ti) nanowire with the length of 950 nm, aligned on the long axis along the dimer symmetric axes. The gap in the dimer is bridged by the titania (TiO$_2$) nanowire with the same diameter and 100 nm in length. The position p1 are 5 nm from the heterodimer apex, and the point p2 is at the middle of the titania nanowire. The nanoantenna-insulator-metal substrate structure imitated the nanorectenna, where an ac voltage could generate in the insulator and stimulate a tunnel current in an asymmetric

MIM tunneling diode. To observe the antenna electric-field profiles, full-field finite-difference-time-domain (FDTD) electromagnetic simulations were carried out for Ag-TiO$_2$-Ag and an Ag-TiO$_2$-Ti asymmetric MIM nanostructure under plane wave excitation.

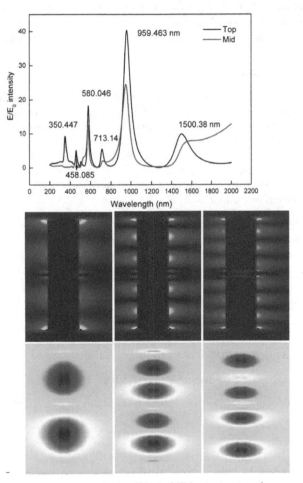

Figure 12. Resonance spectrum around a Ag-TiO$_2$-Ag MIM nanostructure shown as a value of |E|/|E$_0$| at a point of 5 nm on top and in the middle of the homodimer as a function of wavelength (upper panel), the electric field profiles (middle panel) and vertical components of the Poynting vector of antenna radiations (bottom panel) corresponding to the antenna mode ($\lambda/2$, $3\lambda/2$, and $5\lambda/2$ resonances)

Figure 12 shows the local field enhancement spectra along with field profile and poynting vector component for the Ag-TiO$_2$-Ag symmetric. The Ag-TiO$_2$-Ag nanostructure is resonant at wavelength of 1500.38 nm, 959.463 nm and 713.14 nm and etc, corresponding to the first three order of the antenna resonances, and the near-field profiles at these $\lambda/2$, $3\lambda/2$ and $5\lambda/2$

resonances antenna mode are viewed from the electric field profiles and vertical components of the Poynting vector of antenna radiations. The near-field profiles clearly show that the electric field of the 2000 nm long Ag-TiO$_2$-Ag MIM antenna has a standing wave pattern, which is quite similar to that of single Ag nanowire antenna with 2000 nm length, their antenna radiations along the nanowires have same number of lobes and the lobe spacing of the fundamental mode is a little enlarged in the MIM antenna due to the gap between nanowires.

3.7. Ag-TiO$_2$-Ti MIM nanostructures

For asymmetric Ag-TiO$_2$-Ti MIM nanostructure, we also calculated the electric field, filed amplitude enhancement and vertical component of the Poynting vector antenna radiation corresponding to resonance modes, and the results are presented in Figure 13. It can be seen that the Ag-TiO$_2$-Ti nanostructure is resonant at wavelength of 1401.05 nm, 914.077 nm and 683.527 nm and etc, corresponding to the $\lambda/2$, $3\lambda/2$ and $5\lambda/2$ resonances antenna mode and the near-field profiles at first three order of the antenna resonances shows that the main contribution of the dimer antenna intensity field enhancement and field distribution are from silver in the asymmetric dimer.

For the Ag-TiO$_2$-Ti MIM antenna, the electric field is dominated by the Ag nanowires, the antenna radiations along the Ag nanowire of the MIM structure have same lobe patterns to the single Ag nanowire antenna with 950 nm lengths, and a fundamental antenna mode $\lambda/2$ can be identified unambiguously at the wavelength of 1401.25 nm. The electric field enhancement is bigger at the second harmonic mode ($3\lambda/2$) and reaches up to 22 and 7.5 in the middle of 100 nm thick TiO$_2$ insulators for the Ag-TiO$_2$-Ag and Ag-TiO$_2$-Ti MIM structures.

4. Optical rectification properties of metal-insulator-metal (MIM) devices

4.1. Review of metal-insulator-metal (MIM) diodes

The characterization and design of the optical rectification device focus mainly on the following areas:

1. The antenna;
2. The rectifier;
3. System integration.

The rectifier can transform an ac voltage to a dc voltage by means of a non-linear device, such as diode. To convert electromagnetic energy efficiently, a diode should be coplanar and couple to an optical antenna in order to take full advantage of the enhanced electric-field at the top of the metal electrodes or in the center gap of the antenna. Schottky diodes are routinely used in high frequency rectifiers because of their fast response time. These diodes have an operating frequency upper limit of approximately 3 THz, which is far below visible light frequencies. Antenna-couple metal-insulator-metal (MIM) diodes have been the subject of increasing interest due to their small size, CMOS compatibility, and ability to offer full functionality without cooling and applied bias [33-36]. The diode of choice for the optical frequency rectifier is a metal-insulator-metal (MIM) diode, it is previously considered that these diodes are the fastest available diodes for detection in the optical region.

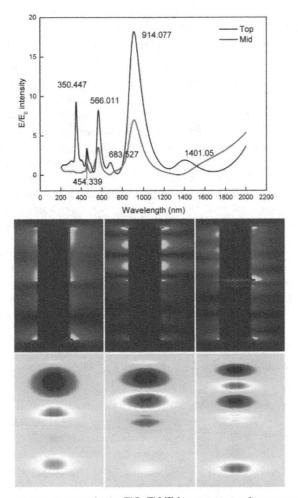

Figure 13. Resonance spectrum around a Ag-TiO$_2$-Ti MIM nanostructure shown as a value of $|E|/|E_0|$ at a point of 5 nm on top and in the middle of the heterodimer as a function of wavelength (upper panel), the electric field profiles (middle panel) and vertical components of the Poynting vector of antenna radiations (bottom panel) corresponding to the antenna mode ($\lambda/2$, $3\lambda/2$, and $5\lambda/2$ resonances)

However, plasmon absorption in metal-TiO$_2$ Schottky diode structures have recently been applied to photovoltaic devices [37, 38] and photocatalysts [39, 40] in the UV-visible and wavelength range, and an enhanced light harvesting property and a visible-light-induced charge separation are obtained for the benefit of the metal nanoparticles. Furthermore, different explanations have been presented about the role of metal nanoparticles in the observed improvement in light conversion efficiency. These include (i) metal nanoparticles increased absorption due to surface plasmons and light trapping effects [41], (ii) metal nanoparticles functioned as electron donor promoting electron transfer from metal to semiconductor [37, 38, 42] and (iii) metal nanoparticles served as electron trapping media that can minimize the surface charge recombination in semiconductor [43 , 44].

Meanwhile, two main mechanisms have been mentioned for the electron transfer between the metal nanoparticle and the semiconductor during the energy conversion process. First, Tatsuma et al. [37, 38, 45] and other workers [42, 46] proposed that the photoexcited electrons in the metal nanoparticles transferred from the metal particle to the TiO$_2$ conduction band since the photoresponse of these metal-TiO$_2$ diode structures was consistent with the absorption spectra of Au or Ag nanoparticles. Second, Kamat et al [43, 44] and Li et al. [47] have suggested that the noble metal nanoparticles act as electron sinks or traps in the metal-TiO$_2$ diode structures to accumulate the photogenerated electrons, which could minimize charge recombination in the semiconductor films. Obviously, a better understanding of these effects is crucial in exploiting the beneficial aspects of metal nanoparticles in photovoltaics.

In this section, we modeled and fabricated a MIM diode located at the mid-point of an Ag-TiO$_2$-Ti MIM antenna. The photodeposition was used to synthesis Ag-Cu nanoparticles in the TiO$_2$ nanotube grown from Ti foils to obtain the AgCu-TiO$_2$/Ti MIM diode structures, and their photoelectronic properties were measured under simulated sunlight and visible light. The Schottky barrier analyses indicated that the electron transport direction was the electrons transferred from metal nanoparticles to TiO$_2$.

4.2. Theoretical modeling of Ag-TiO$_2$-Ti MIM diodes

As shown in Figure 14, a MIM diode consists of three sections: a polished metal on a substrate as the base (anode), a natural oxide layer as the insulator (barrier), and another metal layer (cathode). When the barrier layer is extremely thin (10 to 50 Å), some quantum confinement effects may result in the tunnelling of electrons through the insulator layer. In this device, metals are treated as perfect electrical conductors, and are assumed to be in thermal equilibrium. Typically, the electrons in the metal are driven out of equilibrium by the absorbed power. The absorbed photons energize the electrons in the metal, allowing them to overcome the Schottky barrier and be collected in the cathode, eventually leaving through the back contact as measured current.

However, in these recent metal-TiO$_2$ Schottky diode structures [37-47], it would appear that the barrier layer was actually quite a bit thicker than 10 nm (probably in excess of 1 um) and further details was unable to be found in these papers. Semi-classical models did not account for non-equilibrium energy distributions of carriers, or do so through a localize lattice temperature. This problem in general is difficult for a traditional drift-diffusion model to capture the plasmonic effect in the illuminated metal layer. The logical approach to modeling the MIM rectifier would be to start with a customary potential-energy diagram.

Figure 14 shows the energy-band diagram of an Ag-TiO₂-Ti MIM diode. For smooth electron transportation through the interface between the semiconductor TiO₂ and the metal electrodes, an intervening ohmic contact is necessary. Since TiO₂ is an n-type semiconductor and its conduction band level is close to 4.2 eV, the electron affinity of TiO₂ is little greater than 4.2 eV, the work function of Ti is 4.33 eV, and the Ti-TiO₂ interface therefore becomes Schottky contact. Ohmic contact can be formed using Ag with the work function of 4.0 eV.

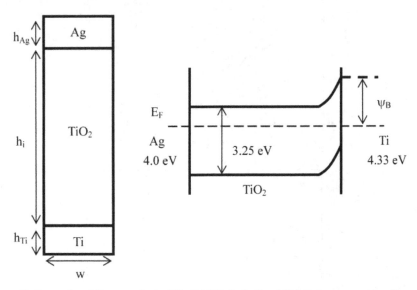

Figure 14. Energy-band diagram of a Ag-TiO₂-Ti MIM diode. Two MIM diodes are considered here, one is conventional diode with the diameter w= 2 μm, the TiO₂ layer thickness h_i = 60 μm, and the metal electrode thickness h_{Ag} = h_{Ti} = 2 μm, the other is MIM nanodiode with the diameter w= 100 nm, the TiO₂ layer thickness h_i = 100 nm, and metal electrode thickness h_{Ag} = h_{Ti} = 100 nm.

In order to understand the electron transport properties in the Ag-TiO₂-Ti MIM nanodiode, we performed the two-dimensional finite element (FEM) calculation based on the drift and diffusion equation for electrons and holes. In Figure 15, the calculated I-V curves are depicted for a conventional diode (a) and a MIM nanodiode (b) under various n-doping concentrations. It is show that the I-V curves show clear nonlinear and asymmetric current characteristic. The nonlinearity of current is basic mechanism for the rectification of an incident wave to a DC output.

As shown in Figure 15(a), for a conventional Ag-TiO₂-Ti MIM diode, the high doping leads to the high current density at zero bias. When doping is added, it will shift the Fermi level in the TiO₂, which will then define the Schottky barrier height (SBH) at the Ti-TiO₂ interface, the greater SBH results in larger charge transfer across metal-semiconductors interface, creating a large potential drop across depletion width and allowing a more efficient collection of electrons and holes.

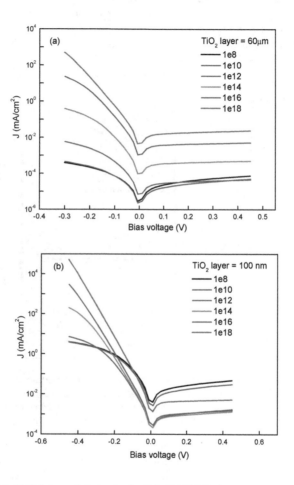

Figure 15. Simulated I-V characteristic for the Ag-TiO₂-Ti MIM diode. (a) is conventional diode with the diameter w= 2 μm, the TiO_2 layer thickness h_i = 60 μm, and the metal electrode thickness h_{Ag} = h_{Ti} = 2 μm; (b) is MIM nanodiode with the diameter w= 100 nm, the TiO_2 layer thickness h_i = 100 nm, and the metal electrode thickness h_{Ag} = h_{Ti} = 100 nm.

4.3. Experimental work of AgCu-TiO₂/Ti MIM nanostructures

The AgCu-TiO₂/Ti MIM nanoantenna heterostructures were fabricated using an electrochemical process and several measurements were carried out to determine if the MIM diode was producing direct current through plasmonic rectenna action. The TiO₂ nanotube layer was grown by anodizing 300-600 nm of titanium [48] on which the bimetallic Ag-Cu nanoparticle film were deposited using a photodeposition method. We chose the bimetallic Ag-Cu nanoparticles as the plasmonic medium where Cu alloying was used to prevent the natural oxidation of the silver nanoparticles and keep its good plasmonic property [49-51]. The surface morphologies of samples were characterized by scanning electron microscopy (JSM, 6390A) with energy dispersive X-ray spectroscopy (SEM-EDS). The photocurrent density - voltage curves (J-V) were measured at a potential sweep rate of 10 mV/s, with the Pt net as counter electrode, and a saturated calomel electrode (SCE) as reference electrode. Electrochemical Impedance Spectroscopy (EIS) was used to evaluate the properties of the different electrodes under AC polarization. The frequency range was 0.1 Hz to 100 kHz for amplitude of 5 mV in a DC potential of –0.2 V$_{SCE}$.

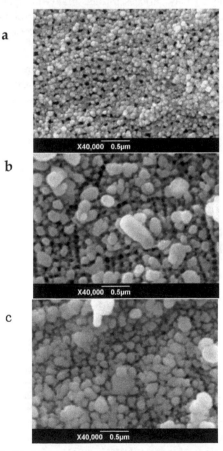

Figure 16. SEM images of Cu (a), Ag (b) and Ag-Cu (c) nanoparticle layer deposited on TiO₂/Ti substrate.

Figure 16 (a, b and c) shows the surface morphology of the Cu, Ag and Ag-Cu nanoparticles deposited on TiO₂ nanotube, respectively. The Cu nanoparticles were evenly and densely distributed on the surface of TiO₂ nanotube layer, the measured nanoparticle size ranged 46 nm to 120 nm. The Ag nanoparticles are bigger than Cu nanoparticles and are not in uniform size with the big ones over 500 nm and the small ones below 100 nm. The possible reason is that the standard electrode potential of Ag^+/Ag^0 (0.78 eV) is higher than that of Cu^{2+}/Cu^0 (0.34 eV), so Ag^+ is reduced more rapidly and grows faster than Cu^{2+}. For the surface morphology of the Ag-Cu nanoparticles on TiO₂ nanotube layers, it is obvious that the Ag-Cu nanoparticles were denser than Ag nanoparticles. The energy dispersive X-ray spectroscopy (EDS) show that the atom ratios of Ag to Cu was 1.68:3.72 and the Cu content was less than that in the electrolyte because the reduced Cu can be further oxidized by Ag^+. Actually, this galvanic reaction has a dominated effect on the formation of Ag-Cu nanoparticles. Once the Cu nanoparticles formed firstly, the Ag^+ will be reduced at the surface of Cu nanoparticles and form the core-shell structure, an alloyed Ag-Cu nanoparticles formed after the reductive reaction and the atomic mutual diffusion at the interface of Cu and Ag atoms.

Figure 17 shows the measured J-V curves under simulated sunlight and visible light for the AgCu-TiO₂/Ti MIM structure, which had a short current density of -1.201 mA/cm² under simulated sunlight, and decreased to -0.734 mA/cm² under visible light, indicating that Ag-Cu nanoparticles were very photosensitive to the UV light.

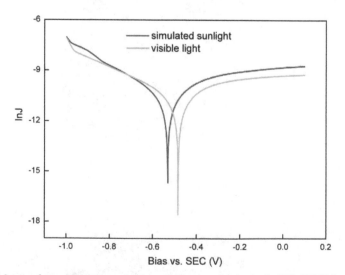

Figure 17. Measured current density and voltage (J-V) curves for the AgCu-TiO₂/Ti MIM structure under visible light and under simulated sunlight. The applied bias is versus the saturated calomel electrode (SCE).

The work function difference between the metal and the n-TiO₂ results in electrons transferred from TiO₂ to the metal nanoparticles yielding a Schottky junction. We can determine the electron transport direction by comparing the SBH changes under different irradiation conditions. The SBH values were calculated in lnJ-V diagram shown in Figure 17 by the following Equation:

$$J_s = A^* T^2 \exp(-q\phi_{SBH} / k_B T) \qquad (1)$$

where ϕ_{SBH} is SBH at the zero bias, A^* is the Richardson constant, k_B is the Boltzmann constant, and J_s is the zero bias saturation current density. The calculated SBH values were 1.021eV and 1.006 eV for the simulated sunlight and under visible light irradiations, *i.e.*, the ϕ_{SBH} values calculated under the simulated sunlight are higher than that measured under visible light and less electrons were produced and transferred to TiO₂ under visible light. This result clearly indicates that the metal nanoparticles can be photoexcited as electron donors, and that the electrons transferred from metal nanoparticles to TiO₂.

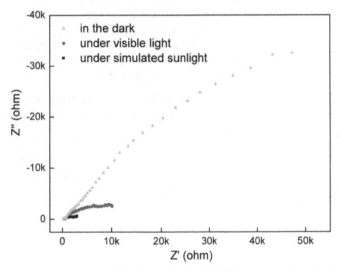

Figure 18. Nyquist diagram measured for the AgCu-TiO₂/Ti MIM structure under simulated sunlight, visible light, and in the dark.

Figure 18 exhibits that impedance measured under simulated sunlight and visible light were sharply decreased in comparison with that measured in the dark at the low frequency region (< 100Hz), indicating a decrease of charge transfer resistance. This can be justified by the bending of the impedance arc in the low frequency region (the second arc) in the Nyquist diagram, because the bending of the arc in this region indicated a process of charge transfer while the linear relationship between the imaginary and real component of the impedance mean a diffusion process controlled step. Among the three photoelectrodes, the low frequency region arc of the AgCu-TiO₂/Ti MIM structure was largely bended under both light irradiations, which indicated that faster charge transfer was obtained in the AgCu-TiO$_2$/Ti diode.

5. Summary

The plasmonic effect of the metal nanostructure has been explored to improve the conversion efficiency for DSSCs. We found that metal nanoparticles in the plasmonic DSSC can lead to serious parasitic absorption even under the optimized conditions, the absorption of metal nanoparticles takes up about 20–30% of the total absorption of the DSSC at visible wavelengths. To solve the problem of the metal parasitic absorption in the DSSC caused by the surface plasmon optical effect, designing and fabricating metal-insulator-metal sandwich structure as photon acceptor and electron donor may be a research direction for the further development of the plasmonic photovoltaic techniques.

The metal-insulator-metal (MIM) nanostructures were theoretically investigated as model systems for optical antennas. The field enhancements vs. wavelength for the silver and titanium nanowires with 2000 nm length have shown the higher harmonic modes ($3\lambda/2$, $5\lambda/2$, $7\lambda/2$, and $9\lambda/2$ resonances). A silver nanowire is a superb candidate for broadband optical antenna, which support four harmonic resonance modes in spectral range 517 to 1663 nm and one surface plasmon resonance (SPR) at 350.477 nm. The Ag nanowires dominated the electric field for the Ag-TiO$_2$-Ti MIM antenna.

The current density and voltage curves of Ag-TiO$_2$-Ti (MIM) nanostructure have been modeled at the micrometer and nanometer scale. We are carrying out experiments on the optical rectification properties of AgCu-TiO$_2$/Ti metal-insulator-metal (MIM) diode devices. Cu, Ag, and Ag-Cu nanoparticle films exhibits different morphological characters. The measured J-V curves under simulated sunlight and visible light for the AgCu-TiO$_2$/Ti MIM structure exhibit a short current density of -1.201 mA/cm^2 under simulated sunlight, and decreased to -0.734 mA/cm^2 under visible light, indicating that the electrons transferred from metal nanoparticles to TiO$_2$ layer.

Author details

Fuyi Chen[*], Jian Liu and Negash Alemu
State Key Laboratory of Solidification Processing,
Northwestern Polytechnical University, Xian, China

Acknowledgement

This study was supported by the National Natural Science Foundation of China (Grant Nos. 50971100 and 50671082), the Research Fund of State Key Laboratory of Solidification Processing in China (Grant No. 30-TP-2009), and the NPU Foundation for Fundamental Research (Grant No. NPU-FFR-ZC200931).

---

[*] Corresponding Author

6. References

[1] Takamoto T, Kaneiwa M, Imaizumi M, et al. InGaP/GaAs-based Multijunction Solar Cells. Progress in Photovoltaics 2005; 13(6) 495-511

[2] Nazeeruddin M K, DeAngelis F, Fantacci S, Selloni A, Viscardi G, Liska P, Ito S, Takeru B, Gratzel B M. Combined Experimental and DFT-TDDFT Computational Study of Photoelectrochemical   Cell Ruthenium Sensitizers. Journal of American Chemistry Society 2005; 127(48)16835-16847

[3] O'Regan B, Gratzel M, A Low-cost, High-efficiency Solar Cell Based on Dye-sensitized Colloidal TiO_2 Films. Nature 1991; 353(24)737-739

[4] Stuart H R, Hall D G. Island Size Effects in Nanoparticle-enhanced Photodetectors. Applied Physics Letters 1998; 73, 3815,

[5] Atwater H A, Polman A. Plasmonics for Improved Photovoltaic Device. Nature Materials 2010; 9,205-213

[6] Ferry V E, Munday J N, Atwater H A. Design Consideration for Plasmonic Photovoltaics. Advanced Materials 2010; 22, 4794-4808

[7] Brown M D, Suteewong T, Kumar R S S, Innocenzo V D, Petrozza A, Lee M M, Wiesner U, Snaith H J. Plasmonic Dye-sensitized Solar Cells Using Core-shell Metal-insulator Nanoparticles. Nano Letters 2011; 11(2)438-445

[8] Hagglund C, Zach M, and Kasemo B. Enhanced Charge Carrier Generation in Dye sensitized Solar Cells by Nanoparticle Plasmons. Applied Physics Letters 2008; 92, 013113

[9] Du S, Li Z. Enhanced Light Absorption of TiO_2 in the Near-ultraviolet Band by Au Nanoparticles. Optics Letters 2010; 35(20)3402-3404

[10] Guilatt O, Apter B, Efron U, Light Absorption Enhancement in Thin Silicon Film by Embedded Metallic Nanoshells. Optics Letters 2010; 35(8)1139-1141

[11] Fu D, Zhang X, Barber R L and Bach U. Dye-sensitized Back-contact Solar Cells. Advanced Materials 2010; 22, 4270–4274

[12] Durr M, Menges B, Knoll W, Yasuda A, and Nelles G. Direct Measurement of Increased Light Intensity in Optical Waveguides Coupled to a Surface Plasmon Spectroscopy setup. Applied Physics Letters 2007; 91, 021113

[13] Atwater H A, Polman A. Plasmonics for Improved Photovoltaic Device. Nature Materials 2010; 9,205-213

[14] Yuan L, Chen F Y, Zheng C F, Liu J, Alemu N. Parasitic Absorption Effect of Metal Nanoparticles in the Dye-sensitized Solar Cells. Physica Status Solidi A 2012; 209,1376-1379.

[15] Robinson J and Rahmat-Samii Y. Particle Swarm Optimization in Electromagnetics. IEEE Transaction on Antennas and Propagation 2004; 52, 397.

[16] Robinson J, Sinton S and Rahmat-Samii Y. Particle Swarm, Genetic Algorithm, and their Hybrids: Optimization of a Profiled Corrugated Horn Antenna. In: Proc. IEEE Int. Symp. Antennas Propagation, San Antonio, TX, 1(314-317), 2002

[17] O'Regan B, Gratzel M. A Low-cost, High-efficiency Solar Cell Based on Dye-sensitized Colloidal TiO_2 Films. Nature 1991; 353,737

[18] Green M A and Pillai S. Harnessing Plasmonics for Solar Cells. Nature Photonics 2012; 6, 130

[19] Schuck P J, Fromm D P, Sundaramurthy A, Kino G S, Moerner W E. Improving the Mismatch between Light and Nanoscale Objects with Gold Bowtie Nanoantennas. Physics Review Letters 2005; 94, 017402

[20] Muhlschlege P, Eisler H, Martin O, Hecht B, Pohl D. Resonant Optical Antennas. Science 2005; 308, 1607-1609

[21] Novotny L, van Hulst N. Antennas for Light. Nature Photonics 2011; 5, 83-90,

[22] Brown W C. The history of Power Transmission by Radio Waves. IEEE Transactions on Microwave Theory and Techniques 1984; 32(9)1230-1242

[23] Brown W C. Optimization of the Efficiency and other Properties of the rectenna element. IN: 1976 IEEE-MTT-S International Microwave Symposium, 142-144, 1976

[24] Hocker L O, Sokoloff D R, Daneu V, Szoke A and Javan A. Frequency Mixing in the Infrared and Far-infrared using a Metal-to-metal Point Contact Diode. Applied Physics Letters 1968; 12, 401-402

[25] Matarrese L M and Evenson K M. Improved Coupling to Infrared Whisker Diodes by Use of Antenna Theory. Applied Physics Letters 1970; 17, 8-10

[26] Fetterman H R, Clifton B J, Tannenwald P E and Parker C D. Submillimeter detection and mixing using Schottky diodes. Applied Physics Letters 1974; 24, 70

[27] Fetterman H R, Tannenwald P E, Clifton B J, Parker C D, Fitzgerald W D and Erickson N R. Far-ir Heterodyne Radiometric Measurements with Quasioptical Schottky Diode Mixers. Applied Physics Letters 1978; 33, 151

[28] Small J G, Elchinger G M, Javan A, Sanchez A, Bachner F J and Smythe D L, An Electron Tunnelling at Infrared Frequencies: Thin-film M-O-M Diode Structure with Broad-band Characteristics. Applied Physics Letters 1974; 24,275-279

[29] Gustafson T K, Schmidt R V, and Perucca J R. Optical Detection in Thin-film metal - oxide - metal diodes. Applied Physics Letters 1974; 24, 620

[30] Bailey R L. A Proposed New for a Solar-energy Converter. Journal of Engineering for Power 1972; 94, 73-77

[31] Marks A M. Device for Light Power to Electric Power. U.S. Patent 4 445 050, 1984

[32] Chen F Y, Alemu N, Johnston R L. Collective Plasmon Modes in a Compositionally Asymmetric Nanoparticle Dimer. AIP Advances 2011; 1,302134

[33] Fumeaux C, Herrmann W, Rothuizen H, De Natale P and Kneubühl F K. Mixing of 30 THz Laser Radiation with Nanometer Thin Film Ni-NiO-Ni Diodes and Integrated Bow-tie antennas. Applied Physics B Laser and Optics 1996; 63(2)135–140.

[34] Codreanu I, González F and Boreman G. Detection Mechanisms in Microstrip Dipole Antenna-coupled Infrared Detectors. Infrared Physics and Technology 2003; 44(3)155–163.

[35] Hobbs P C, Laibowitz R B and Libsch F R. Ni-NiO-Ni Tunnel Junctions for Terahertz and Infrared Detection. Applied Optics 2005; 44(32)6813–6822

[36] Chen F Y, Yuan L, Johnston R. L. Low-loss Optical Magnetic Metamaterials on Ag-Au Bimetallic Fishnets. Journal of Magnetism and Magnetic Materials 2012; 324, 2625-2630.

[37] Tian Y, Tatsuma. T. Plasmon-induced Photoelectrochemistry at Metal Nanoparticles Supported on Nanoporous TiO_2. Chemistry Communication 2004; 16, 1810-1811.

[38] Tian Y, Tatsuma T. Mechanisms and Applications of Plasmon-Induced Charge Separation at TiO_2 Films Loaded with Gold Nanoparticles. Journal of American Chemistry Society 2005; 127, 7632-7637.

[39] Awazu K, Fujimaki M, Rockstuhl C, Tominaga J, Murakami H, Ohki Y, Yoshida N, Watanabe T A. Plasmonic Photocatalyst Consisting of Silver Nanoparticles Embedded in Titanium Dioxide. Journal of American Chemistry Society 2008; 130, 1676-1680

[40] Irie H, Kamiya K, Shibanuma T, Miura S, Tryk D A, Yokoyama T, Hashimoto K. Visible Light-Sensitive Cu(II)-Grafted TiO_2 Photocatalysts: Activities and X-ray Absorption Fine Structure Analyses. Journal of Physical and Chemistry C 2009; 113,10761–10766.

[41] Stuart H R and Hall D G. Island Size Effects in Nanoparticle-enhanced Photo Detectors. Applied Physics Letters 1998; 73, 3815-3817.

[42] Mubeen S, Hernandez-Sosa G, Moses D, Lee J, Moskovits M. Plasmonic Photosensitization of a Wide Band Gap Semiconductor: Converting Plasmons to Charge Carriers. Nano Letters 2011; 11, 5548–5552.

[43] Chandrasekharan N, Kamat P V. Improving the Photoelectrochemical Performance of Nanostructured TiO_2 Films by Adsorption of Gold Nanoparticles. Journal of Physical and Chemistry B 2000; 104, 10851–10857.

[44] Takai A, Kamat P V. Capture, Store, and Discharge. Shuttling Photogenerated Electrons across TiO_2 Silver Interface. ACS Nano 2011; 5, 7369-7376.

[45] Sakai N, Fujiwara Y, Takahashi Y, Tatsuma T. Plasmon-Resonance-Based Generation of Cathodic Photocurrent at Electrodeposited Gold Nanoparticles Coated with TiO_2 Films. ChemPhysChem 2009; 10, 766 – 769.

[46] Furube A, Du L, Hara K, Katoh R, Tachiya M. Ultrafast Plasmon-Induced Electron Transfer from Gold Nanodots into TiO_2 Nanoparticles. Journal of American Chemistry Society 2007; 129, 14852-14853.

[47] Liu L, Wang G, Li Y, Li Y, Zhang J. Z. CdSe Quantum Dot-Sensitized Au/TiO2 Hybrid Mesoporous Films and Their Enhanced Photoelectrochemical Performance. Nano Research 2011; 4, 249–258.

[48] Chen F Y, Liu J. A Plasmonic Rectenna and its Fabrication Method. Chinese Patent, 201210002179.5, 2012.

[49] Chen F Y, Zheng C F, A Silver-copper Nanoalloy and its Electrical Synthesis Method. Chinese Patent, 201110310987.3, 2011

[50] Chen F Y, Johnston R L. Charge Transfer Driven Surface Segregation of 13-atom Au-Ag Nanoalloy and its Relevance to Structural, Optical and Electronic properties. Acta Materialia 2008; 56, 2374-2380

[51] Chen F Y, Johnston R L. Energetic, Electronic and Thermal effects on Structural Properties of Ag-Au Nanoalloys. ACS Nano 2008; 2, 165-175

Merging Plasmonics and Silicon Photonics Towards Greener and Faster "Network-on-Chip" Solutions for Data Centers and High-Performance Computing Systems

Sotirios Papaioannou, Konstantinos Vyrsokinos, Dimitrios Kalavrouziotis,
Giannis Giannoulis, Dimitrios Apostolopoulos, Hercules Avramopoulos,
Filimon Zacharatos, Karim Hassan, Jean-Claude Weeber, Laurent Markey,
Alain Dereux, Ashwani Kumar, Sergey I. Bozhevolnyi, Alpaslan Suna,
Oriol Gili de Villasante, Tolga Tekin, Michael Waldow, Odysseas Tsilipakos,
Alexandros Pitilakis, Emmanouil E. Kriezis and Nikos Pleros

Additional information is available at the end of the chapter

1. Introduction

In recent years it has become evident that the increasing need for huge bandwidth and throughput capabilities in Data Center and High Performance Computing (HPC) environments can no longer be met by bandwidth-limited electrical interconnects. Besides their limited capacity, electrical wiring technologies impose great energy and size limitations originating from the requirements to handle the vast amount of information that needs to be exchanged across all hierarchical communication levels within Data Centers and HPCs, i.e. rack-to-rack, backplane, chip-to-chip and on-chip interconnections. This reality has inevitably led to a clearly shaped roadmap for bringing optics into the spotlight and replacing electrical with optical interconnects, thereby overtaking the bottleneck in traffic exchange imposed by electrical wires [1]. Hence, the optical technology should penetrate into intra-rack and board-to-board transmission links, considering that the optical fiber as a large-bandwidth transmission medium [2] is used only in commercial systems for rack-to-rack communication [3].

Nevertheless, data communication and power consumption are still daunting issues in Data Centers and HPCs. According to recent predictions made in [4], the barrier of 10PFlops computing performance should have been overcome in 2012 by a supercomputer that consumes 5MW of power [5]. In addition, [4] predicted that exascale supercomputing machines would consume 20MW having a power efficiency of 1mW/Gb/s [5]. Nonetheless, power consumption in such environments has been proven to be even higher than expected: Today's top-ranked supercomputer, the "K computer", has already reached the 10PFlops performance benchmark but at the expense of excessive consumed power that is more than twice [6] the value that was predicted in 2008. All the above imply that the use of optics at inter-rack communication level is not enough for delivering the necessary performance enhancements. Therefore, the optical technology should now be exploited at shrinked networking environments: The penetration of low-energy photonic solutions at board-to-board, chip-to-chip and eventually intra-chip interconnects would yield remarkable savings in energy consumption [7]. The current mainstream photonic route with high integration and low-cost perspectives relies on the Silicon-on-Insulator (SOI) photonics platform, whose growing maturity is soon expected to release Tb/s-scale data transmission and switching capabilities in datacom and computercom units ensuring low latency, low power consumption and chip-scale integration credentials [8].

Even so, photonic devices cannot reach the compactness of their electronic counterparts: the dimensions of traditional optical structures are limited by the fundamental law of diffraction, preventing the way towards high density integration for interfacing with electronics at the nanoscale. This gap in size between photonic and electronic components is called to be bridged by a promising disruptive technology named plasmonics [9]-[11]. The emerging discipline of plasmonics has started to gain ground as the "beyond photonics" chip-scale platform that can enter the interconnect area [12]-[14], holding a great promise for additional reductions in circuit size and increase in energy efficiency. Plasmonics relies on the excitation of surface plasmon polaritons (SPPs) that are electromagnetic waves coupled to oscillations of free electrons in a metal and propagate along a metal-dielectric interface at near the speed of light. These "hybrid" surface waves have transverse magnetic (TM) polarization in nature and their exhibited electromagnetic field intensity reaches its maximum value at the metal surface whereas it decays exponentially while moving away from the metal-dielectric interface [15]. In this way strong intrinsic confinement is feasible even at sub-wavelength scale [16], breaking the size barriers of diffraction-limited optics and enabling the development of compact integrated nanophotonic circuits [17]. Plasmonic technology does not only succeed in providing light manipulation at sub-wavelength dimensions but at the same time allows for the injection of electrical pulses via the metallic layer, offering thereby a seamless energy-efficient platform for merging light beams with electrical control signals towards "active" operations [18]-[25].

Among the various plasmonic waveguide structures proposed so far (i.e. band-gap structures [26], metallic nanowires [27], V-groove waveguides [28]), the low-energy credentials of plasmonics has been mainly highlighted in the case of dielectric-loaded SPP (DLSPP) waveguides, where a dielectric (e.g. polymer) ridge is deposited on top of a smooth metallic film. As a result, strong sub-wavelength (DLSPP) mode confinement is achieved at the metal-dielectric interface. Nonetheless, this performance comes at the expense of excessively increased propagation losses compared to conventional dielectric waveguides due to radiation absorption in the metallic stripe, yielding quite short propagation lengths lying in the order of few tens of micrometers [29],[30]. Apart from the strong mode confinement, the main advantage of the DLSPP waveguide technology stems from the nature of the chosen dielectric material which, depending on its thermo-optic (TO) [25],[31]-[35] or electro-optic [24],[36] properties, allows for the exploitation of the corresponding effects, enabling in this way the deployment of highly functional active plasmonic elements. Regarding the TO effect, the underlying metallic layer, which is in direct contact with the polymer and therefore with the largest part of the DLSPP mode field, serves as an energy-efficient electrode that yields immediate change of the effective index of the propagating mode at the presence of electric current, leading to fast TO responses. These advantages render the DLSPP waveguides ideal for TO manipulation of the plasmonic waves in various functional circuitry implementations like modulation [32],[33], ON/OFF gating [25] and switching [34],[35]. The progress made so far on active DLSPP-based circuits renders the TO effect as the most mature mechanism for bringing low-energy active plasmonics in true data traffic environments [25],[31].

To this end, the roadmap towards practical employment of active plasmonic circuits in the development of "greener" and faster "Network-on-Chip" (NoC) solutions for Data Centers and HPCs seems to be the following: the synergy between plasmonic, electronic and photonic components for the realization of NoC deployments with minimized power consumption, size and enhanced throughput capabilities, simply by taking advantage of the virtues of each technology. In this perspective, silicon photonics can be used for low-loss optical transmission in passive interconnection components, plasmonics can provide small footprint and low power consumption in active switching modules and electronics can be employed for intelligent decision-making mechanisms. However, considering that plasmonic technology is a premature technology with only a few years of development compared to silicon photonics and with limited functionality so far, the way towards the implementation of such hybrid NoC environments requires priorly: a) the interconnection between silicon and plasmonic waveguide structures in a SOI platform in order to avoid the employment of high-loss plasmonic waveguides for passive circuitry [30],[37], b) the proof of the credentials of plasmonics in wavelength division multiplexing (WDM) applications so as to support the complete portfolio of optics [38], c) the demonstration of functional active plasmonic circuitries in realistic data traffic environments [25],[31].

This chapter aims to cover all these issues starting from the 4×4 silicon-plasmonic router architecture, where its main building blocks and their principle of operation are described briefly. The chapter continues with the geometrical specifications of the silicon and plasmonic waveguides of the hybrid routing platform as well as with the silicon components and sub-systems, namely the photodiodes (PDs) and the multiplexers (MUXs), respectively. The silicon-plasmonic asymmetric Mach-Zehnder interferometer (A-MZI) is then presented and analyzed thoroughly since it is the constituent component of the 4×4 switching matrix of the router. Subsequently, the WDM credentials of plasmonics are experimentally confirmed by demonstrating WDM data transmission over a DLSPP waveguide as well as WDM data switching operation via a hybrid Si-DLSPP A-MZI, along with its switching performance metrics. Various optimization procedures are also listed. The chapter concludes with an estimation of the router's total power consumption and losses.

2. Tb/s-scale hybrid Router-on-Chip platform

Data Center and HPC communication systems require the development of NoC environments with compact size, high-throughput and enhanced power saving capabilities for efficient traffic management. In this perspective, the benefits of silicon photonics and plasmonics can be combined towards realizing a 4×4 "Router-on-Chip" platform [13] with Tb/s-scale data capacity and ultra-low energy requirements for chip-to-chip and on-chip optical interconnects. This hybrid routing platform employs a silicon motherboard hosting all various optical and electrical circuitry, where the "heart" of this router chip is a hybrid 4×4 switching matrix comprising 2×2 TO Si-DLSPP switches, and exploits the best out of all technology worlds: the low optical loss of Si-waveguide motherboard for its passive interconnection parts, the high energy efficiency of plasmonics in the router's "active" parts, and the intelligence provided by electronics for its decision-making mechanisms. The following sections describe in detail the Tb/s-scale Si-DLSPP router architecture and its basic building blocks, the experimental evidences of the WDM data transmission and switching capabilities of the DLSPP waveguides, various optimization approaches and finally the performance of the 4×4 router with estimations in power consumption and optical insertion losses.

2.1. 4×4 silicon-plasmonic router architecture

The 4×4 silicon-plasmonic router architecture aims at achieving 1.12Tb/s aggregate throughput with small footprint and low power consumption. Figure 1 illustrates a schematic layout of this hybrid router architecture. The router consists of SOI MUXs, Si PDs, a hybrid Si-DLSPP 4×4 switching matrix and an integrated circuit (IC) microcontroller. Size minimization and power savings are promoted by the employment of four SOI MUXs composed of cascaded Si ring resonators (RRs) placed in parallel at the frontend of the router and four identical thermo-optically addressed Si-DLSPP A-MZIs interconnected in a Beneš topology. The plasmonic components are used only where switching functionality is required in order to achieve both low optical losses and low power consumption. The Si technology is also responsible for coupling light in and out of the platform, the interconnection of all subsystems and the optoelectronic (OE) signal conversion of the incoming packets' headers. Four 7×1 MUX circuits based on SOI RRs are used to multiplex 28 time-overlapped wavelengths, spaced by 100GHz and modulated with 40Gb/s non-return-to-zero (NRZ) packet traffic, into four optical data streams. Consequently, each one of these multi-wavelength data sequences carries an aggregate traffic of 280Gb/s that will enter one of the four inputs of the plasmonic switching device and will then follow the same route through the entire platform. Apart from the 28 input ports that are utilized for the insertion of the data carrying packets, the router also incorporates 4 additional input ports where they are inserted four discrete low-rate wavelength beams containing the header information of the four incoming data streams. The header for each stream is a two pulse combination within $1\mu s$ time interval and specifies the desired output port, while the duration of the payload of the data packets can be variable. These header pulses are optoelectronically converted via low-rate all-silicon PDs into respective electrical signals that subsequently feed the IC control unit. Here the low speed electronics operating at the data packets rate carry out the intelligent decision-making process that is essential for the routing functionality: The IC microcontroller performs header processing operations and generates differential control electronic signals for driving appropriately the switching matrix. In this way, the incoming optical data streams are routed to the desired output ports as a result of fruitful cooperation between the silicon photonic, plasmonic and electronic technologies.

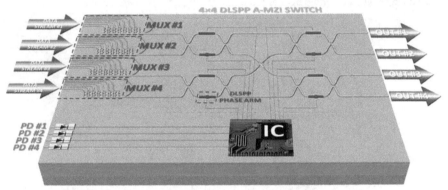

Figure 1. Schematic layout of the 4×4 silicon-plasmonic router architecture.

2.2. Silicon and DLSPP waveguides

The different technologies and building blocks involved in the hybrid Tb/s-scale router require the employment of the appropriate waveguide technology in order to merge all of them successfully on the same motherboard. The existence of a slab layer in rib waveguides compared to the strip counterparts facilitates the fabrication process for the hetero-integration of all processing components, namely the PDs, the SOI MUXs and the DLSPP waveguide structures, rendering this waveguide technology suitable for the low-loss Si-based communication links needed in the SOI routing platform. Taking into consideration the intrinsic TM nature of plasmonic waves, the geometrical dimensions of the Si rib waveguides need to be carefully chosen for compatibility with plasmonic waveguides. In this perspective, rib waveguides featuring a cross section of 400nm width by 340nm height with a 50nm-thick remaining slab are the Si waveguide technology of choice for the routing platform in order to support low-loss TM light propagation [37].

Nevertheless, the coexistence of silicon photonics and plasmonics on the same platform imposes proper formation of the SOI motherboard for efficient hetero-integration of the

DLSPP waveguides [31]: By etching the SOI motherboard down to 200nm (Figure 2(a)), the formed recess in the 2μm-thick buried oxide (BOX) of the SOI substrate serves as the hosting region of a DLSPP waveguide. Towards selecting the dimensions of the DLSPP waveguides, it should be taken into account that the DLSPP waveguide characteristics, i.e., the mode field confinement, effective index and propagation length, are strongly influenced by the width and height of the employed dielectric ridge [29] as a result of the SPP field confinement that occurs when this ridge is placed on top of a metal film. Considering methyl-methacrylate (PMMA) ridges placed on gold stripes to be operated at telecom wavelengths (~1550nm), the optimum ridge dimensions, ensuring tight mode confinement (<1μm) and relatively long propagation (~50μm) of DLSPP modes, are about 500nm and 600nm for the ridge width and height, respectively [29]. Moreover, the underlying gold stripes are about 3μm wide and 60nm thick. A microscope image taken from a Si-DLSPP waveguide is shown in Figure 2(b) where the recessed area of the BOX and the underlying gold film are evident.

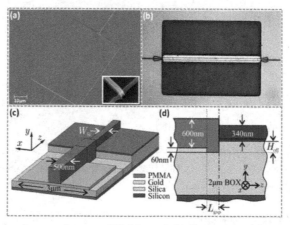

Figure 2. (a) Scanning electron microscope (SEM) image in bird's eye view of the silicon cavity after etching, including as inset the Si waveguide tip after the cavity patterning process, (b) microscope image of a straight DLSPP waveguide coupled to Si waveguides. (c) Bird's eye and (d) side schematic views of a Si-to-DLSPP waveguide transition.

Apart from the geometrical specifications of these two different waveguide technologies, it is necessary to implement an efficient transition from one type of waveguide to the other. In our case, the interface between the silicon and plasmonic waveguides is implemented by a butt-coupling approach, as depicted in bird's eye and side views in Figure 2(c)-(d), respectively. The design of this approach relies on numerical modeling conducted by means of full vectorial three-dimensional finite element method (3D-FEM) simulations. Towards finalizing the design specifications, various parameters should be examined, i.e., the Si waveguide's width, the vertical offset between the two types of waveguides, the existence of a longitudinal gap in the metallic stripe as usually exists in fabricated structures, so as to optimize the matching of the optical modes during transition [37]. The validity of the simulation outcomes is achieved by their comparison with cut-back measurements for a variety of fabricated all Si and hybrid Si-DLSPP waveguide samples [31],[37]. As a result, a coupling loss per interface variance with a mean value of -2.5dB is feasible for 175nm-wide Si waveguides with 200nm vertical offset and DLSPP waveguides with 0.5μm metal gap. The same cut-back evaluation procedure leads to silicon and plasmonic propagation losses of approximately 4.5dB/cm and 0.1dB/μm, respectively, being in good agreement with typical results reported in the literature [30].

2.3. Silicon photodiodes

Since the silicon-plasmonic routing platform comprises also microelectronics, it is necessary to develop integrated C-band (1530-1565 nm) photodiodes that enable the OE conversion of the optical header pulses into electrical signals. However, a complementary metal-oxide-semiconductor (CMOS) -compatible SOI-based integration of a material that absorbs inside this wavelength window is really challenging. One detection scheme within the C-band on a SOI platform is the hybridization of integrated Si structures with other materials (e.g. Ge [39], InGaAs [40]) for direct (linear) absorption. This hybrid approach can lead to high-performance but complex devices, requiring complicated and cost-intensive fabrication procedures and modifications compared to the standard CMOS process. For this purpose, monolithical and fully CMOS-compatible detector concepts have been proposed for the development of comparably simple devices with less fabrication costs. In this perspective, two photon absorption detection can be exploited inside a non-resonant or resonant Si device [41], revealing however non-linear effect and requiring high light intensity. Alternatively, the generation of direct (linear) absorption inside the telecom window can be achieved by the introduction of midgap energy states via Si+-ion implantation [42]. Despite the lower (direct) absorption compared to the hybrid detectors, this type of monolithic Si detector approach, the defect state Si+-ion implanted detector, offers the best compromise for simple integration and low-rate detection.

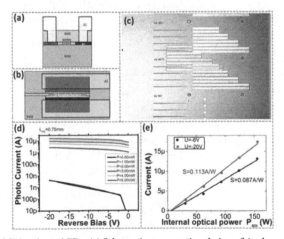

Figure 3. Integrated Si-implanted PDs: (a) Schematic cross sectional view, (b) schematic top view, (c) test chip, (d) I-V curves at various input optical power levels, (e) sensitivity at different reverse bias.

Taking into account that the 4×4 silicon-plasmonic router architecture employs low-rate optical header pulses, the defect state all-silicon implanted PDs are suitable for the procedure of the OE conversion, pursuing a trade-off between minimum signal distortion during header detection and integration complexity. Schematic illustrations of this type of PD interconnected with a Si waveguide are shown in cross sectional and top views in Figure 3(a)-(b), respectively, whereas the top view of a test chip including fabricated all-Si implanted PDs with different absorption lengths is depicted in Figure 3(c). Figure 3(d)-(e) illustrate the performance of a 0.75mm-long Si$^+$-ion implanted PD, exhibiting linear absorption and sensitivity values of 0.09A/W and 0.11A/W for -6V and -20V bias voltage, respectively. By increasing the device length, the sensitivity is getting higher. However, even for a three times longer device, the dark current is well below 100nA.

2.4. Silicon multiplexers

In order to support Tb/s-scale bandwidth, the hybrid router takes advantage of SOI multiplexing devices. Till now various multiplexing/demultiplexing configurations based on SOI technology have been proposed in the scientific community, such as echelle diffraction gratings (EDGs) and arrayed waveguide gratings (AWGs) [43], first or higher order RRs [44]-[49] and A-MZIs [50]. In many cases, due to fabrication limits and discrepancies, thermal tuning is performed to attain ideal spectral positioning of the MUX resonances. This controlling technique is adopted for tunability of the second-order SOI RRs employed in the silicon-plasmonic router, aiming at optimized multiplexing performance. Considering that the router is planned to handle 40Gb/s line-rates, the 3-dB passband bandwidth for each data channel should be at least 40GHz and the adjacent channel crosstalk should be kept lower than -15dB. These issues can be met by the 2nd order RRs employed in the router, since such structures enable increased flexibility in the formation of transmission peaks' spectral shape and bandwidth within the same channel spacing constraints due to the additional coupling stage employed between the two RRs [51].

A fabricated 2nd order RR featuring the Si waveguide specifications of Section 2.2 is illustrated in Figure 4(a). Indicative experimental spectra obtained at the Drop port of 2nd order ring structures are presented with blue lines in Figure 4(b)-(c). In particular, Figure 4(b) refers to a fabricated 2nd order RR with 9μm radius, 190nm bus-to-ring gaps and 380nm ring-to-ring gap, exhibiting 42.5GHz 3-dB bandwidth and 9.2nm free spectral range (FSR). In the case of a ring structure with 12μm radius, 170nm bus-to-ring gaps and 300nm ring-to-ring gap, the corresponding 3-dB bandwidth and FSR values are 48.5GHz and 6.9nm (Figure 4(c)). Replicas of these measured transfer functions are also illustrated in Figure 4(b)-(c) with red dashed line by using the circuit-level modeling tool described extensively in [13].

The aforementioned 2nd order RRs are employed as the building blocks for the design of the 7:1 SOI MUX devices required in the router. Figure 5 depicts a more generic 8:1 multiplexing layout. The MUX design comprises eight cascaded thermo-optically tunable 2nd order RRs clustered in two groups with two different radii (R1 and R2). Based on this clustering approach, where the ring structures of each group have the same radius, it is feasible to comply with the limits dictated by the fabrication resolution and the maximum possible thermo-optically induced wavelength tuning of the rings. The 100GHz spacing between the MUX stages as well as the wavelength resonance matching for each pair of RRs forming the MUX's stages are achieved by considering micro-heaters on top of each ring. However, the Si-DLSPP router requires also four different SOI MUXs operating efficiently at different spectral regions within the C telecom band. This requirement can be addressed by choosing appropriate ring radii combinations and values for the bus-to-ring (*gap1*) and ring-to-ring (*gap2*) gaps and for the lengths (*L1* to *L7*) of the inter-ring waveguide sections during the design procedure.

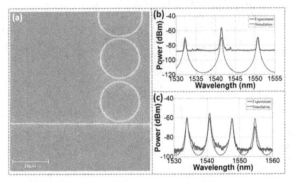

Figure 4. (a) SEM image of a fabricated 2nd order RR. Experimental spectra (blue solid line) at the Drop port for RRs with (b) 9μm and (c) 12μm radii. Red dashed lines illustrate the corresponding simulated spectral responses.

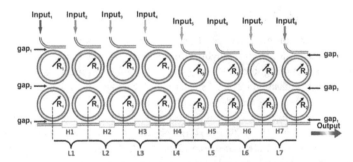

Figure 5. Generic layout of a SOI MUX based on 8 cascaded 2nd order thermo-optically tunable RRs that are grouped in two different radii clusters (*R1* and *R2*). The sections with *H* illustrate the heaters controlling the straight inter-ring waveguide lengths.

Following this rationale, the desired spectral responses of the four MUXs are generated within the 1530-1565 nm wavelength window, revealing 100GHz channel spacing, at least -20dB optical crosstalk and 40GHz 3-dB bandwidth, as illustrated in Figure 6. The first three designs rely on rings with 12μm and 11.7μm radii for the first and second cluster, respectively, whereas the corresponding ring radii values for the fourth design are 9μm and

9.2μm. Moreover, the power coupling coefficients between ring-to-ring and bus-to-ring waveguides for all MUX designs are in the range of 0.33-0.007, corresponding to gap dimensions between 170-380 nm. It should be mentioned that the low crosstalk values in adjacent channels are the result of uneven straight waveguide sections between the successive MUX stages that are long enough (~100μm) also to prevent thermal crosstalk. Since the slightest variation in their length seems to greatly affect the optical crosstalk, micro-heaters are considered to be placed on the middle of every straight waveguide section so as to restore length deviations.

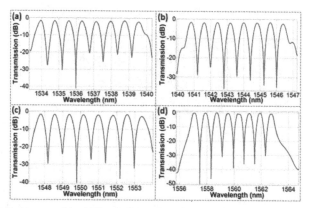

Figure 6. Spectral responses of the (a) first (1533-1540 nm band), (b) second (1540-1547 nm band), (c) third (1547-1554 nm band) and (d) fourth (1557-1563 nm band) 8:1 MUX designs.

2.5. Silicon-plasmonic A-MZIs

Targeting the best compromise between losses, footprint and power consumption, the A-MZI modules of the router rely on a hybrid implementation, where the coupling stages are based on low-loss Si waveguides and only the active phase branches exploit the DLSPP structures. The silicon-plasmonic A-MZI structures play a key role in the accomplishment of routing operation. These are the elements of the router that are capable of performing optical switching by taking advantage of the TO effect at the plasmonic parts. The principle of operation regarding the TO switching is the following: In the 500×600 nm^2 PMMA ridge, the DLSPP mode fills practically the whole ridge, while the largest fraction of the mode is concentrated at the metal-dielectric interface. The strong confinement at the metal-dielectric interface renders the mode very sensitive to any temperature variation induced through current injection in the metal; in this way the underlying gold film acts as a heating electrode for the dielectric ridge, modifying in a very efficient way the DLSPP mode effective index via changing the PMMA refractive index [32]-[34],[52] and requiring only a small amount of consumed power. At the same time, the inherent instantaneous heating of the metal is immediately transferred to the propagating SPP mode, again due to its strong confinement at the metal-dielectric interface, leading to small response times of the TO effect in the DLSPP waveguide. Such low driving powers (~1mW) and response times (~1μs) in

TO modulation and switching have already been shown in theoretical estimations [32] and calculations [52] that make use of these ridge dimensions in DLSPP waveguide structures.

Considering operation at telecom wavelengths (~1550nm) and an increase in temperature of about 61K in PMMA loadings (where the thermo-optic coefficient TOC is $-1.05\cdot10^{-4}K^{-1}$), a full π phase shift can be achieved for a waveguide length of 120μm, according to the formula $L_\pi=\lambda/(2\cdot\Delta T\cdot TOC)$. This length value is almost twice the maximum propagation length of the DLSPP mode and results in high propagation losses [32]. However, this length can be reduced by a factor of 2 by adopting the asymmetric MZI configuration shown in Figure 7(a). In this configuration, a permanent $\pi/2$ phase shift is induced in one of the two branches by slightly widening the PMMA ridge over a short distance, resulting in this way to a higher effective refractive index value in the wider ridge section and to a natural biasing of the MZI at its quadrature point of operation. To this end, the necessary thermo-optically imposed phase shift is reduced down to $\pi/2$. Following this rationale, the length of the phase arms in the employed hybrid A-MZIs has been set to 60μm ($L1$), while the $\pi/2$ phase asymmetry was achieved by widening the lower 600-nm-thick DLSPP waveguide from 500nm ($W1$) to 700nm ($W2$) for a length of 6μm ($L2$), as a result of an increase of the DLSPP mode effective index by ~0.06 [29],[52]. These dimensions ensure the existence of a single TM mode inside the polymer at telecom wavelengths, both for the nominal and the widened DLSPP waveguides, as depicted in Figure 7(b)-(c).

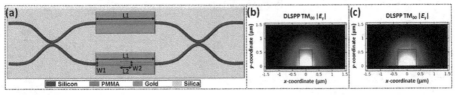

Figure 7. (a) Schematic layout of a Si-DLSPP A-MZI. The lower plasmonic branch is widened in order to introduce a default asymmetry, (b) fundamental quasi-TM mode of the 500×600 nm² PMMA-loaded SPP waveguide, (c) fundamental quasi-TM mode of the 700×600 nm² PMMA-loaded SPP waveguide.

Taking advantage of the default $\pi/2$ phase asymmetry introduced at the lower plasmonic branch, this asymmetric interferometric configuration is capable of performing switching with only a $\pi/2$ phase shift compared to all-plasmonic symmetric MZI switches [52]-[54]. The injection of electric current to a MZI arm yields a temperature rise that alters the effective index of the mode propagating on the heated arm, leading to a phase shift via this TO effect [32],[34],[52]. By electrically controlling the upper A-MZI arm, a negative phase shift is experienced by the propagating DLSPP waveguide mode as a result of the PMMA's TOC. When the induced phase shift equals $\pi/2$, the phase difference between the modes travelling through the two MZI branches equals π and therefore the whole mode power is exported to the BAR output of the device. On the contrary, when the same current level applies only to the lower MZI plasmonic branch, the default MZI phase asymmetry is cancelled out due to the $-\pi/2$ thermo-optically induced phase shift and, thus, the whole mode power emerges at the CROSS port of the MZI. To this end, the ON switching operation is achieved by electrically driving the upper MZI plasmonic arm, whereas the lower branch has to be driven by the same amount of electric current for the OFF switching operation. Nevertheless, due to the default $\pi/2$ phase asymmetry of the MZI, high-performance switching can be reached even if one of the two plasmonic branches is thermo-optically addressed, since the MZI is initially biased at the quadrature point that lies in the linear domain of its output transfer function. Consequently, this asymmetric formation constitutes a simple and passive mechanism for reducing the required energy level and the active plasmonic arm length for a given maximum service temperature.

2.6. WDM data transmission through a DLSPP waveguide

The evaluation of the WDM data transmission capabilities of plasmonics is initially carried out via a 60μm-long PMMA-loaded straight waveguide that is included in a Si-plasmonic chip [31],[38]. This chip comprises Si rib waveguides, DLSPP waveguides and Si-to-DLSPP interfaces that follow the specifications presented in Section 2.2 as well as Si TM grating couplers so as to enable optical communication with the outside world. At 1545nm TM-polarized light the cut-back measurements reveal 0.1dB/μm and 4.6dB/cm plasmonic and silicon propagation losses as well as 2.5dB and 12dB coupling losses for each Si-to-DLSPP and TM grating coupling interface, respectively.

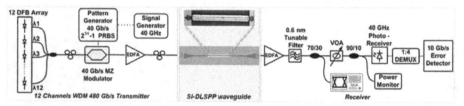

Figure 8. Experimental setup consisting of the WDM transmitter, the Si-DLSPP waveguide and the receiver.

The experimental setup used to transmit a 480Gb/s WDM optical stream through the DLSPP waveguide is presented in Figure 8. The transmitter involves twelve continuous-wave (CW) optical signals stemming from distributed feedback (DFB) lasers and spaced by 200GHz within the 1542-1560 nm spectral range. The multiplexing of these individual light beams into a single optical fiber is realized by using AWG and 3-dB couplers. The multiplexed signal is then encoded by a 2^{31}-1 pseudo-random bit sequence (PRBS) at 40Gb/s NRZ line rate in a Ti:LiNbO$_3$ Mach-Zehnder modulator (MZM). Subsequently, the 480Gb/s WDM signal is amplified by a high-power erbium-doped fiber amplifier (EDFA), providing 24dBm output power towards tackling the high losses induced by the hybrid chip. After ensuring TM polarization conditions for compliance with the inherent TM nature of plasmonics, via a polarization controller, the incoming optical data signal is inserted into the DLSPP waveguide. The multi-wavelength data stream that transmitted through the whole Si-DLSPP waveguide is amplified at the chip's output by a low-noise EDFA and demultiplexed into individual data channels by an optical bandpass tunable filter (OBPF). Each optical data channel is then converted into an electrical signal by a 40GHz 3-dB bandwidth photoreceiver that is connected to a 1:4 electrical demultiplexer, where each electrical data signal is received by a 10Gb/s error detector for bit error rate (BER) measurements.

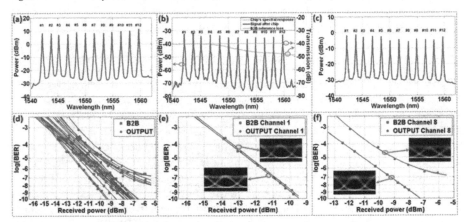

Figure 9. Spectra of the 12×40Gb/s WDM signal (a) at chip's input, (b) at chip's output, in comparison with chip's spectral response and B2B flat losses, (c) after post-chip amplification, (d) BER curves for all 40Gb/s B2B and transmitted channels, (e) BER curves for B2B and transmitted channel 1, (f) BER curves for B2B and transmitted channel 8.

The 12-channel data signal is examined, in terms of its spectrum, in three crucial places along the transmission line: just before the Si-DLSPP chip's input, after the chip's output and after the receiver's EDFA, as depicted in Figure 9(a)-(c). The TM grating couplers of the chip reveal a non-flattened spectral response (red dashed line in Figure 9(b)), dictating thereby the chip's spectral response and resulting in wavelength-dependent fiber-to-fiber transmission losses which range between 40dB and 48dB within the wavelength window of interest. As a consequence, the spectrum of the WDM signal constantly alters, even after the final amplification stage due to the absence of gain flattened filters (GFFs) at the EDFAs. To this end, a wavelength-dependent performance is observed in the receiver with respect to the optical noise-to-signal ratio (OSNR). The performance of the 480Gb/s WDM transmission through the 60μm-long DLSPP waveguide is evaluated via BER measurements that are shown in Figure 9(d)-(f). In particular, Figure 9(d) illustrates an overview of the BER curves obtained for all 40Gb/s discrete channels. Six out of the twelve channels (channels #1-5, 12) perform error-free, exhibiting a power penalty in a range between 0.2dB and 1dB for a 10^{-9} BER value against back-to-back (B2B) measurements. However, the remaining six channels (channels #6-11) reveal an error-floor at $\sim10^{-7}$ BER. Figure 9(e) corresponds to the best performing channel #1 that yields 0.2dB power penalty, whilst Figure 9(f) refers to the worst performing channel #8 that reveals an error-floor at 10^{-7} BER value. The B2B measurements are generated by replacing the hybrid chip by a variable attenuator that induces constant, wavelength-independent losses equal to the losses experienced by channels 1-4 that reside in the 1542-1548 nm spectral region (flat reference loss with green line in Figure 9(b)). The different performances between the channels transmitted via the hybrid chip originate from the unequal OSNR distribution due to the wavelength-dependent spectral responses of the grating couplers and the EDFAs. It should be noted that the error-floor in channels #1-5, 12 can be eliminated by selecting their spectral position to reside within the almost flat low-loss chip's response and by the employment of EDFAs with GFFs.

2.7. WDM data traffic switching with a hybrid Si-plasmonic A-MZI

After verifying the WDM data transmission capabilities of plasmonics through the DLSPP waveguide, the next step towards enriching the WDM portfolio of plasmonics is the demonstration of data switching [25]. In this perspective, the TO electrically controlled PMMA-based A-MZI that analyzed in Section 2.5 is employed towards performing switching of a 4×10Gb/s WDM traffic according to the rationale described in the same section. This A-MZI, along with other structures, is included in a Si-plasmonic chip [25] that is equipped with TM grating couplers, similar to the ones presented in the previous section, in order to couple light in and out of the hybrid chip. At 1542nm TM-polarized light the cut-back measurements reveal 0.1dB/μm and 4.4dB/cm plasmonic and silicon propagation losses as well as 2.5dB and 13dB coupling losses for each Si-to-DLSPP and TM grating coupling interface, respectively.

Before proceeding to the WDM approach, the Si-DLSPP A-MZI is tested in static and single-channel conditions regarding its electrical requirements and switching performance. Figure 10(a) shows the static TO transfer functions for the CROSS and BAR output ports of the A-MZI. The input signal is a 1542nm CW light beam with 6dBm optical power and the control medium is a direct current (DC) that drives the upper MZI arm and takes values up to 40mA. As it is obvious from this figure, the extinction ratio (ER) values for the CROSS and BAR ports are 14dB and 0.9dB when the DC reaches its maximum value. This unbalanced behavior between the two output ports stems from the 95:5, instead of 50:50, Si input/output couplers of the MZI as the result of unfortunate design error. Moreover, by comparing the obtained TO transfer functions with the theoretical transfer functions of a symmetric MZI, the default biasing point of the A-MZI is estimated at ~70°. Following the same procedure, the phase induced by the 40mA DC is ~-90° that corresponds to an increase in temperature by ~60K. These phase conditions in both MZI arms imply a total phase difference of ~160° between the two signal components that travel through the two branches. In terms of power requirements, the hybrid A-MZI consumes ~13.1mW when it is driven by 40mA DC, considering that the resistance of the 60μm-long DLSPP phase arm is found to be 8.2Ω.

After the initial static and single-channel switching characterization, the Si-DLSPP A-MZI is evaluated in terms of switching under dynamic and WDM data conditions. This evaluation is realized by using the experimental setup of Figure 10(b). Four CW optical signals emitted by DFB sources at 1545.1nm, 1546.7nm, 1547.7nm and 1549.1nm wavelengths are combined in pairs by 3-dB couplers in a way that channels 1 and 3 constitute the first pair and channels 2 and 4 form the other one. Subsequently, each channel pair is modulated by a $2^{31}-1$ PRBS at 10Gb/s NRZ line rate in a corresponding Ti:LiNbO$_3$ MZM. The final 4×10Gb/s WDM signal is formed by using a 3-dB coupler to multiplex the two pairs of data sequences and is amplified by a high-power EDFA that provides 31dBm power at the input of the electrically controlled hybrid A-MZI. A polarization controller is used again towards establishing the required TM polarization conditions of the incoming WDM signal. Exploiting the TO effect, the switching state of the A-MZI is controlled dynamically by a pulse generator operating at 20KHz. After its transmission through the A-MZI, the multi-wavelength data signal is amplified by a two stage EDFA that comprises a 5nm midstage OBPF for out-of-band amplified spontaneous emission (ASE) noise rejection. The amplified WDM stream is then demultiplexed into its constituent channels by a 0.8nm OBPF. Each optical data channel is launched into a photoreceiver with 10GHz 3-dB bandwidth for OE conversion. Subsequently, the electrical signal exiting the photoreceiver is then received by a 10Gb/s error-detector for evaluation via BER measurements.

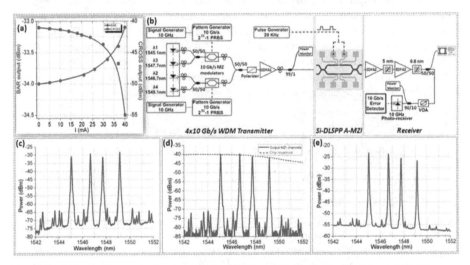

Figure 10. (a) Static TO transfer functions for the CROSS and BAR output ports of the A-MZI, (b) Experimental setup and the 4-channel spectrum at (c) MZI input, (d) directly at the MZI output before entering EDFA2, (e) after the receiver's pre-amplification stage. The spectral response of the chip including the A-MZI and the TM grating couplers is shown with the red dashed line in (d).

As depicted in Figure 10(c)-(e), the spectrum of the 4-channel data signal is observed at three places along the transmission line, that is just before the MZI's input, after the MZI's output and after the pre-amplifier (EDFA2) in the receiver. The high-power signal propagating in the fiber link between EDFA1 and the A-MZI causes the generation of four-wave mixing (FWM) terms, as it is easily noticeable in these figures. Nevertheless, by selecting unequal channel spacing ranging between 125GHz and 200GHz in the source signals, these terms are out-of-band compared to the data channels, giving the opportunity to be removed by subsequent filters. The energy transfer to the FWM components as well as the absence of GFFs after the transmitter's EDFA form the unequal power profile of the WDM signal at the

MZI's input, as shown in Figure 10(c). Due to the spectral response of the chip's TM grating couplers, the WDM spectral power profile at the MZI's output is now different (Figure 10(d)). The same figure also depicts with red dashed line the spectral response of the chip including the A-MZI and the TM grating couplers, revealing a 3dB loss variation within the 1545-1549 nm spectral band. After the receiver's pre-amplification stage, different power levels and also different OSNR values between the four channels are observed, as a result of the gain profile of EDFA2 (Figure 10(e)).

The dynamic control of the Si-DLSPP A-MZI is achieved by driving its upper arm with electrical rectangular pulses of 15μs duration, 20KHz repetition rate and 40mA peak value. The obtained data traces and eye diagrams for channel 1 (λ1) and channel 2 (λ2) signals at the CROSS and BAR output ports of the A-MZI are illustrated in Figure 11(a)-(h), where the electrical control signal is also depicted with red dashed line. According to these figures, inverted mode operation with an ER close to 14dB is attained at the CROSS port. In contrast, only 0.9dB ER performance is recorder at the BAR port as a result of the 95:5 Si couplers of the A-MZI. In terms of response time, the Si-DLSPP A-MZI exhibits fast rise and fall times lying in the 3-5 μs range, as it has been also demonstrated in a 90μm-long PMMA-based A-MZI [25]. It should be noted that similar results for both CROSS and BAR output ports are also obtained for data channels 3 and 4 at λ3 and λ4 wavelengths, respectively.

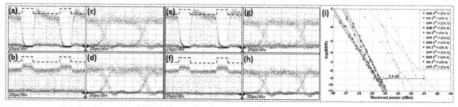

Figure 11. Modulation with 15μs electrical rectangular pulses at 20KHz repetition rate for 10Gb/s: (a) data trace at the CROSS port (channel 1), (b) data trace at the BAR port (channel 1), (c) eye diagram at the CROSS port (channel 1), (d) eye diagram at the BAR port (channel 1), (e) data trace at the CROSS port (channel 2), (f) data trace at the BAR port (channel 2), (g) eye diagram at the CROSS port (channel 2), (h) eye diagram at the BAR port (channel 2), (i) BER curves for ON/OFF operation of the A-MZI.

The performance of the 4×10Gb/s WDM switching via the 60μm-long PMMA-based A-MZI in ON and OFF switching states is evaluated via BER measurements that are depicted in Figure 11(i). In this experiment, the B2B measurements are obtained by using a straight Si waveguide, which is also included in the hybrid chip, as a reference point. In this way, only the signal degradation through the A-MZI is taken into consideration, excluding the lossy input/output TM grating couplers of the chip. Regarding the ON state, a DC value of 40mA is applied to the upper MZI branch, recording at the same time the BER measurements for the BAR output port. In case of no driving current, namely the OFF state, the BER curve is obtained at the CROSS output port of the MZI. As it is shown in Figure 11(i), the ON state reveals a negative power penalty close to the statistical error of 0.2dB for all four channels compared to the B2B curve. However, when operating at OFF state, the four channels exhibit power penalties in a range between 1.7dB and 3.6dB at a 10^{-9} BER value. The enhanced power penalty values compared to the corresponding ON state performance originate mainly from the 8dB lower power level received at MZI's CROSS output (-26dBm) against its BAR port during ON state (-18dBm). Additionally, the power penalty shows a clear tendency to increase with the channel wavelength, due to the wavelength-dependent gain and OSNR profile experienced by the four channels during amplification in EDFA2. Finally, the different slopes observed between the BER graphs of channels 1, 3 and the BER curves of channels 2 and 4 are caused due to the different modulators per channel pair at the 4×10Gb/s transmitter end.

2.8. Switching performance metrics

A clear view about the switching credentials and advantages of plasmonics over the SOI and the polymer-based TO waveguide technology platforms can be obtained by comparing performance metrics regarding the power consumption, switching time and active phase arm length. Table 1 provides an overview of the respective performance metrics reported for different SOI and polymer-based TO MZI switches. The power consumption × switching time metric is also included in this table, since it has been commonly used for the comparison of TO switching elements in applications where both low energy and fast operation are a prerequisite for high performance [55]. This table verifies the advantages of plasmonic technology: The Si-DLSPP A-MZI switch attains the shorter active region and the smallest power-time product over every type of undoped thermo-optically addressed MZI switch. These findings come as a natural consequence of the strong confinement of the DLSPP mode at the metal-dielectric interface, the inherent instantaneous heating of the underlying metallic layer and the asymmetric configuration of the MZI. On the contrary, in SOI TO structures, where the electrode is located at certain distance from the Si waveguide, it is very difficult to achieve simultaneously ultra-low power consumption and response time values. The required power can be greatly reduced by using sophisticated waveguide

Active waveguide technology [reference]	Phase arm length (active region in μm)	Power consumption (P in mW)	Switching time (τ in μs)	Power-time product (P×τ in mW·μs)
SOI [58]	200	20	2.8	56
SOI [56]	1000	0.49	144	70.56
SOI [57]	100	0.54	141	76.14
SOI [61]	700	50	3.5	175
SOI [62]	6300	6.5	14	91
Polymer [55]	300	1.85	700	1295
Polymer [59]	100	4	200	800
PMMA-loaded SPP [Current work]	60	13.1	3.8	49.78

Table 1. Comparison with other SOI and polymer-based TO MZI switches.

engineering (e.g. suspended waveguides), thereby leading to sub-mW switching but at the expense of increased switching times and poor mechanical stability [56],[57]. A very good compromise between power consumption and response time is offered by using integrated NiSi waveguide heaters in single driving SOI-based TO switches [58]. Compared to the latter switches, lower power requirements but higher switching times are observed in polymer-based switches [55],[59]. It should be mentioned that TO switches that use the waveguide itself as the resistive heater and exploit differential driving schemes [60] have been excluded from this study to keep the same conditions for comparison.

2.9. Optimization procedures

The performance of the Si-DLSPP A-MZI switch regarding its BAR and CROSS output ports is determined by the splitting ratio of its Si coupling stages. To this end, the incorporation of 50:50 couplers, instead of the considered 95:5, can greatly improve the switching performance, yielding more than 20dB ER for both output ports. Another aspect that can be optimized is the switching time of this hybrid interferometric element. As it is already mentioned, the adoption of differential driving schemes for actively controlling both of the DLSPP MZI branches has the potential for lower time responses, even in the order of sub-μs [60]. Moreover, the power consumption in switching can be also reduced. One way is to form a kind of "suspended" plasmonic waveguides by etching the BOX layer under the gold film in order to achieve better confinement of the heat in the DLSPP waveguides and therefore enhanced power efficiency. This technique could be also exploited in the SOI MUX devices for energy-efficient thermal tuning of the RRs [63]. Another approach for bringing down the consumed power is the utilization of polymer materials with higher TOC value instead of PMMA loadings. For example, Cycloaliphatic acrylate (Cyclomer) exhibits almost three times higher TOC than PMMA and has been already successfully applied as the polymer loading in DLSPP switching structures [34],[38]. Though yet not optimized for low-energy switching operation, this route is expected to significantly decrease the required energy levels without compromising the switching performance. Alternatively, this scheme can be used to reduce the footprint and as such the losses of the A-MZI device. Considering the same amount of electric current to that used in the aforementioned WDM switching experiment, the length of the active plasmonic regions can be decreased from 60μm to 20μm, resulting in 4dB lower plasmonic propagation losses. Besides, the Cyclomer loadings can boost the TO tuning performance of plasmonic resonant devices compared to respective PMMA-based structures [31],[33] when similar temperature changes occur [38].

Apart from the hybrid A-MZI switch, optimization acts should be carried out for the TM grating couplers that are responsible for coupling light in and out of the chips. The grating couplers employed in the WDM data transmission and switching experiments exhibit 24dB overall insertion losses at best that are too high for practical datacom applications where the employment of amplifiers is prohibited. Towards tackling this issue, new TM grating couplers are considered for a far better efficiency. These coupling structures, filled with Spin-on-Glass (SOG) of 800nm height, rely on a fully etched approach with 0.8 filling factor,

0.71μm grating period, 0.13μm groove width and 10 degrees incident angle of the light. Figure 12(a) presents a layout of the new grating coupler's schematic cross section and Figure 12(b) depicts the SEM image of the fabricated one. These optimized TM grating couplers reveal 3.25dB minimum coupling loss at 1557nm and about 32nm 3-dB bandwidth. The blue line in Figure 12(c) presents also the results obtained from simulation that indicate very good agreement between theory and experiment. These results ensure that the new TM grating coupler is a major achievement since it decreases the router's overall optical losses by 17.5dB.

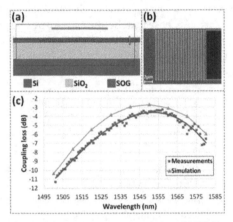

Figure 12. TM grating coupler: (a) Schematic cross sectional view, (b) SEM image, (c) Spectral responses of the simulation and experimental results.

2.10. Router performance in terms of optical insertion losses and power consumption

As a NoC solution for Data Centers and HPC systems, the hybrid Router-on-Chip platform should consume low power towards "greener" computing environments and induce low losses towards intrachip point-to-point connection without using on-chip amplifiers.

Within this framework, an estimation of the total power consumption of the entire 4×4 silicon-plasmonic router is provided here based on the characterization and experimental results obtained from the router's components and subsystems. Considering 60μm-long PMMA-phase arms in the branches of the hybrid A-MZIs that require 40mA driving current, the hybrid router is expected to consume at maximum ~2W. This power consumption stems from 1mW in the four PDs, 335mW in the four frontend SOI MUX devices due to the thermal tuning of their RRs and ~1.67W in the IC that is capable of providing at the same time two differential outputs with 48mA current. Therefore, the electronic IC is the router's component that consumes the largest part of power whereas the optical components reveal a power consumption of only ~400mW, considering the 13mW power requirement of each hybrid A-MZI. In comparison with current multicore central processing units (CPUs) for servers and HPCs that consume 150W [64] and 55W [65] respectively, the hybrid router requires only 2W, rendering this technology ideal for such shrinked networking environments in terms of energy savings. Keeping in mind that the Si-DLSPP router has an aggregate throughput capability of 1.12Tb/s, the power efficiency of this hybrid routing platform is about 1.8mW/Gb/s that is a very low value compared to other hybrid OE routers [66].

The optimization procedures of the previous section could lead to significant reductions in the total power consumption of the silicon-plasmonic router. The usage of Cyclomer, instead of PMMA, loadings in the phase arms of the A-MZIs can bring the IC's power consumption down to ~600mW. Consequently, the total power consumption is limited to ~1W. Keeping the power consumption in the original level of 2W, the same switching performance in the 4×4 DLSPP switching matrix is expected by employing 20μm-long Cyclomer-based A-MZIs but with the profit of 8dB lower losses in the router. Towards minimizing the aggregate power consumption, the approach of suspended arms for the fabrication of the RRs composing the SOI MUXs can lead to increased tuning efficiency that results in ~30mW for the whole multiplexing circuitry and therefore in ~630mW for the entire routing platform. As a result, an outstanding performance with lower than 0.6mW/Gb/s power efficiency can be achieved.

The total insertion losses of the silicon-plasmonic router can be also estimated. The aggregate losses across an input-to-output route are expected to be ~31dB that are analyzed in 6.5dB for the input and output grating couplers, 2dB for the SOI MUX device, 0.5dB for a 90:10 monitor coupler, ~8dB for the silicon waveguides, 10dB for the four Si-to-DLSPP interfaces and 4dB for the two 20μm-long Cyclomer-loaded SPP branches at the A-MZIs. However, the losses concerning the Si waveguide interconnections can be quite lower after finalizing the router prototype, since the value of 8dB relies on first layouts, where the building blocks and their interconnections are not placed in an optimal -in terms of spatial density- way. Nevertheless, even the 31dB losses of the original design can be manageable with high-power vertical-cavity surface-emitting lasers (VCSELs) [67] and high sensitivity Si-PDs [68].

3. Conclusion

Power consumption and size appear as the main set of barriers in next-generation Data Center and HPC environments. Within this framework, the penetration of optics into shrinked networking environments for chip-scale interconnects is now more vital than ever. SOI technology has already demonstrated its low-cost and high-integration credentials towards supporting fundamental operations required for optical interconnect applications. At the same time, the emerging discipline of plasmonics appears as a promising candidate for further size minimization and power savings. To this end, the idea of merging plasmonics and silicon photonics into the same platform seems to be a great solution for the implementation of faster and "greener" NoC environments. Deriving the strengths of each technology, a high-throughput, energy efficient and compact hybrid "Router-on-chip" is feasible. In this perspective, we present a 4×4 silicon-plasmonic router architecture for chip-scale applications with 1.12Tb/s aggregate bandwidth. The hybrid router relies on a SOI motherboard that incorporates Si waveguides for low-loss interconnection of SOI MUXs, PDs and the remaining subsystems, low-energy plasmonic waveguides in active A-MZI switches and an IC control unit for intelligent decision-making operations. Towards the demonstration of the router's functionality, we proceed to experimental evidences that turn the promises of plasmonics into real system-level application benefits: The transmission of a 12×40Gb/s WDM stream through a Si-DLSPP waveguide and the switching of a 4×10Gb/s WDM signal via a Si-DLSPP A-MZI. With the lowest reported power consumption × time response product among undoped SOI and polymer-based TO MZI switches, plasmonic technology becomes suitable for on-chip optical interconnects. According to experimental results, the hybrid router is estimated to consume ~2W with only ~400mW corresponding to the optical parts, yielding a power efficiency of 1.8mW/Gb/s. Moreover, the router is expected to induce about 31dB optical losses. With these power and optical loss characteristics, the Si-DLSPP routing platform seems to be appropriate for NoC environments where fast path establishment and route reconfiguration is necessary for efficient traffic management in Data Centers and HPCs.

Author details

Sotirios Papaioannou
Corresponding Author
Informatics & Telematics Institute, Center for Research & Technology Hellas, Thessaloniki, Greece
Department of Informatics, Aristotle University of Thessaloniki, Thessaloniki, Greece

Konstantinos Vyrsokinos
Informatics & Telematics Institute, Center for Research & Technology Hellas, Thessaloniki, Greece

Dimitrios Kalavrouziotis, Giannis Giannoulis,
Dimitrios Apostolopoulos and Hercules Avramopoulos
School of Electrical & Computer Engineering, National Technical University of Athens, Athens, Greece

Filimon Zacharatos, Karim Hassan, Jean-Claude Weeber,
Laurent Markey and Alain Dereux
Institut Carnot de Bourgogne, University of Burgundy, Dijon, France

Ashwani Kumar and Sergey I. Bozhevolnyi
Institute of Sensors, Signals,and Electrotechnics, University of Southern Denmark, Odense M, Denmark

Alpaslan Suna, Oriol Gili de Villasante and Tolga Tekin
Fraunhofer IZM, Berlin, Germany

Michael Waldow
AMO Gesellschaft für Angewandte Mikro- und Optoelektronik GmbH, Aachen, Germany

Odysseas Tsilipakos, Alexandros Pitilakis and Emmanouil E. Kriezis
Department of Electrical & Computer Engineering, Aristotle University of Thessaloniki, Thessaloniki, Greece

Nikos Pleros
Informatics & Telematics Institute, Center for Research & Technology Hellas, Thessaloniki, Greece
Department of Informatics, Aristotle University of Thessaloniki, Thessaloniki, Greece

Acknowledgement

This work was partially supported by the European FP7 ICT-PLATON (ICT- STREP no. 249135) project.

4. References

[1]   Miller D. A. B. Device Requirements for Optical Interconnects to Silicon Chips. Proceedings of the IEEE 2009; 97(7) 1166-1185.

[2]   Qian D. et al. 101.7-Tb/s (370×294-Gb/s) PDM-128QAM-OFDM Transmission over 3×55-km SSMF using Pilot-based Phase Noise Mitigation. Proceedings of OFC/NFOEC 2011, Paper PDPB5, 6-10 March 2011, Los Angeles, CA, USA.

[3]   Taira Y. et al. High Channel-Count Optical Interconnection for Servers. Proceedings of ECTC 2010; 5490959, 1-4 June 2010, Las Vegas, Nevada, USA.

[4]   Kash J. A. et al. Optical Interconnects in Exascale Supercomputers. Proceedings of IEEE Photonics Society 2010, 483-484, 7-11 November 2010, Denver, CO, USA.

[5]   http://www.igigroup.net/download/FFS09/0930_Benner_IBM.pdf.

[6]   http://www.datacenterknowledge.com/archives/2011/06/20/new-top-500-champ-the-k-supercomputer/.

[7]   Taubenblatt M. A. Optical Interconnects for High-Performance Computing. IEEE/OSA Journal of Lightwave Technology 2012; 30(4) 448-458.

[8]   Paniccia M. Integrating Silicon Photonics. Nature Photonics 2010; 4(8) 498-499.

[9]   Barnes W. L., Dereux A., Ebbesen T. W. Surface plasmon subwavelength optics. Nature 2003; 424(6950) 824-830.

[10] Atwater H. A. The promise of plasmonics. Scientific American 2007; 296(4) 38-45.

[11] Brongersma M. L., Shalaev V. M. The case for plasmonics. Science 2010; 328(5977) 440-441.

[12] Zia R., Schuller J. A., Chandran A., Brongersma M. L. Plasmonics: the next chip-scale technology. Materials Today 2006; 9(7-8) 20-27.

[13] Papaioannou S. et al. A 320Gb/s-throughput capable 2×2 silicon-plasmonic router architecture for optical interconnects. IEEE/OSA Journal of Lightwave Technology 2011; 29(21) 3185-3195.

[14] Kim J. T. et al. Chip-to-chip optical interconnect using gold long-range surface plasmon polariton waveguides. Optics Express 2008; 16(17) 13133-13138.

[15] Raether H. Surface Plasmons on Smooth and Rough Surfaces and on Gratings. Berlin: Springer-Verlag; 1988.

[16] Gramotnev D. K., Bozhevolnyi S. I. Plasmonics beyond the diffraction limit. Nature Photonics 2010; 4(2) 83-91.

[17] Grandidier J. et al. Surface plasmon routing in dielectric-loaded surface plasmon polariton waveguides. Proceedings of SPIE 2008, 7033, 70330S-70330S-8, 10-14 August 2008, San Diego, CA, USA.

[18] Krasavin A. V., Zheludev N. I. Active plasmonics: Controlling signals in Au/Ga waveguide using nanoscale structural transformations. Applied Physics Letters 2004; 84(8) 1416-1418.

[19] Nikolajsen T., Leosson K., Bozhevolnyi S. I. Surface plasmon polariton based modulators and switches operating at telecom wavelengths. Applied Physics Letters 2004; 85(24) 5833-5835.

[20] Lereu A. L., Passian A., Goudonnet J.-P., Thundat T., Ferrell T. L. Optical modulation processes in thin films based on thermal effects of surface plasmons. Applied Physics Letters 2005; 86(15) 154101.

[21] Pacifici D., Lezec H. J., Atwater H. A. All-optical modulation by plasmonic excitation of CdSe quantum dots. Nature Photonics 2007; 1(7) 402-406.

[22] Pala R. A., Shimizu K. T., Melosh N. A., Brongersma M. L. A nonvolatile plasmonic switch employing photochromic molecules. Nano Letters 2008; 8(5) 1506-1510.

[23] MacDonald K. F., Sámson Z. L., Stockman M. I., Zheludev N. I. Ultrafast active plasmonics. Nature Photonics 2009; 3(1) 55-58.

[24] Krasavin A. V., Zayats A. V. Electro-optic switching element for dielectric-loaded surface plasmon polariton waveguides. Applied Physics Letters 2010; 97(4) 041107.

[25] Kalavrouziotis D. et al. First demonstration of active plasmonic device in true data traffic conditions: ON/OFF thermo-optic modulation using a hybrid silicon-plasmonic asymmetric MZI. Proceedings of OFC/NFOEC 2012, OW3E.3, 4-8 March 2012, Los Angeles, CA, USA.

[26] Bozhevolnyi S. I., Erland J., Leosson K., Skovgaard P. M. W., Hvam J. M. Waveguiding in Surface Plasmon Polariton Band Gap Structures. Physical Review Letters 2001; 86(14) 3008-3011.

[27] Knight M. W., Grady N. K., Bardhan R., Hao F., Nordlander P., Halas N. J. Nanoparticle-mediated coupling of light into a nanowire. Nano Letters 2007; 7(8) 2346-2350.

[28] Bozhevolnyi S. I., Volkov V. S., Devaux E., Ebbesen T. W. Channel plasmon-polariton guiding by subwavelength metal grooves. Physical Review Letters 2005; 95(4) 046802.

[29] Holmgaard T., Bozhevolnyi S. I. Theoretical analysis of dielectric-loaded surface plasmon-polariton waveguides. Physical Review B 2007; 75(24) 245405.

[30] Briggs R. M., Grandidier J., Burgos S. P., Feigenbaum E., Atwater H. A. Efficient coupling between dielectric loaded plasmonic and silicon photonic waveguides. Nano Letters 2010; 10(12) 4851-4857.

[31] Giannoulis G. et al. Data transmission and thermo-optic tuning performance of dielectric-loaded plasmonic structures hetero-integrated on a silicon chip. IEEE Photonics Technology Letters 2012; 24(5) 374-376.

[32] Gosciniak J. et al. Thermo-optic control of dielectric-loaded plasmonic waveguide components. Optics Express 2010; 18(2) 1207-1216.

[33] Hassan K., Weeber J.-C., Markey L., Dereux A. Thermo-optical control of dielectric loaded plasmonic racetrack resonators. Journal of Applied Physics 2011; 110(2) 023106.

[34] Hassan K. et al. Thermo-optic plasmo-photonic mode interference switches based on dielectric loaded waveguides. Applied Physics Letters 2011; 99(24) 241110.

[35] Hassan K. et al. Characterization of thermo-optical 2×2 switch configurations made of dielectric loaded surface plasmon polariton waveguides for telecom routing architecture. Proceedings of OFC/NFOEC 2012, OW3E.5, 4-8 March 2012, Los Angeles, CA, USA.

[36] Randhawa S. et al. Performance of electro-optical plasmonic ring resonators at telecom wavelengths. Optics Express 2012; 20(3) 2354-2362.

[37] Tsilipakos O. et al. Interfacing Dielectric-Loaded Plasmonic and Silicon Photonic Waveguides: Theoretical Analysis and Experimental Demonstration. IEEE Journal of Quantum Electronics 2012; 48(5) 678-687.

[38] Kalavrouziotis D. et al. 0.48Tb/s (12×40Gb/s) WDM transmission and high-quality thermo-optic switching in dielectric loaded plasmonics. Optics Express 2012; 20(7) 7655-7662.

[39] Vivien L., Osmond J., Fédéli J., Marris-Morini D., Crozat P., Damlencourt J., Cassan E., Lecunff Y., Laval S. 42 GHz p.i.n Germanium photodetector integrated in a silicon-on-insulator waveguide. Optics Express 2009; 17(8) 6252-6257.

[40] Sheng Z., Liu L., Brouckaert J., He S., Van Thourhout D. InGaAs PIN photodetectors integrated on silicon-on-insulator waveguides. Optics Express 2010; 18(2) 1756-1761.

[41] Bravo-Abad J., Ippen E. P., Soljačić M. Ultrafast photodetection in an all-silicon chip enabled by two-photon absorption. Applied Physics Letters 2009; 94(24) 241103.

[42] Geis M. W. et al. Silicon waveguide infrared photodiodes with >35 GHz bandwidth and phototransistors with 50 AW-1 response. Optics Express 2009; 17(7) 5193-5204.

[43] Bogaerts W. et al. Silicon-on-Insulator Spectral Filters Fabricated With CMOS Technology. IEEE Journal of Selected Topics in Quantum Electronics 2010; 16(1) 33-44.

[44] Geng M. et al. Compact four-channel reconfigurable optical add-drop multiplexer using silicon photonic wire. Optics Communications 2009; 282(17) 3477-3480.

[45] Barwicz T. et al. Reconfigurable silicon photonic circuits for telecommunication applications. Proceedings of SPIE 2008; 6872, 68720Z-1-12.

[46] Klein E. J. et al. Reconfigurable optical add-drop multiplexer using microring resonators. IEEE Photonics Technology Letters 2005; 17(11) 2358-2360.

[47] Dong P. et al. Low power and compact reconfigurable multiplexing devices based on silicon microring resonators. Optics Express 2010; 18(10) 9852-9858.

[48] Xiao S., Khan M. H., Shen H., Qi M. Multiple-channel silicon micro-resonator based filters for WDM applications. Optics Express 2007; 15(12) 7489-7498.

[49] Dahlem M. S. et al. Reconfigurable multi-channel second-order silicon microring-resonator filterbanks for on-chip WDM systems. Optics Express 2011; 19(1) 306-316.

[50] Liu A. et al. Wavelength Division Multiplexing Based Photonic Integrated Circuits on Silicon-on-Insulator Platform. Journal of Selected Topics in Quantum Electronics 2010; 16(1) 23-32.

[51] Saeung P., Yupapin P. P. Generalized analysis of multiple ring resonator filters: Modeling by using graphical approach. Optik-Int Journal for Light and Electron Optics 2008; 119(10) 465-472.

[52] Pitilakis A., Kriezis E. E. Longitudinal 2×2 switching configurations based on thermo-optically addressed dielectric-loaded plasmonic waveguides. IEEE/OSA Journal of Lightwave Technology 2011; 29(17) 2636-2646.

[53] Krasavin A. V., Zayats A. V. Three-dimensional numerical modeling of photonic integration with dielectric-loaded SPP waveguides. Physical Review B 2008; 78(4) 045425.

[54] Tsilipakos O., Pitilakis A., Tasolamprou A. C., Yioultsis T. V., Kriezis E. E. Computational techniques for the analysis and design of dielectric-loaded plasmonic circuitry. Optical and Quantum Electronics 2011; 42(8) 541-555.

[55] Al-Hetar A. M., Mohammad A. B., Supa'at A. S. M., Shamsan Z. A. MMI-MZI polymer thermo-optic switch with a high refractive index contrast. Journal of Lightwave Technology 2011; 29(2) 171-178.

[56] Fang Q. et al. Ultralow power silicon photonics thermo-optic switch with suspended phase arms. IEEE Photonics Technology Letters 2011; 23(8) 525-527.

[57] Sun P., Reano R. M. Submilliwatt thermo-optic switches using free-standing silicon-on-insulator strip waveguides. Optics Express 2010; 18(8) 8406-8411.

[58] Van Campenhout J., Green W. M., Assefa S., Vlasov Y. A. Integrated NiSi waveguide heaters for CMOS-compatible silicon thermo-optic devices. Optics Letters 2010; 35(7) 1013-1015.

[59] Xie N., Hashimoto T., Utaka K. Very low-power, polarization-independent, and high-speed polymer thermooptic switch. IEEE Photonics Technology Letters 2009; 21(24) 1861-1863.

[60] Geis M. W., Spector S. J., Williamson R. C., Lyszczarz T. M. Submicrosecond Submilliwatt Silicon-on-Insulator Thermooptic Switch. IEEE Photonics Technology Letters 2004; 16(11) 2514-2516.

[61] Espinola R. L., Tsai M.-C., Yardley J. T., Osgood R. M. Fast and low-power thermooptic switch on thin silicon-on-insulator. IEEE Photonics Technology Letters 2003; 15(10) 1366-1368.

[62] Densmore A. et al. Compact and low power thermo-optic switch using folded silicon waveguides. Optics Express 2009; 17(13) 10457-10465.

[63] Dong P. et al. Thermally tunable silicon racetrack resonators with ultralow tuning power. Optics Express 2010; 18(19) 20298-20304.

[64] http://ark.intel.com/products/64582/Intel-Xeon-Processor-E5-2687W-(20M-Cache-3_10-GHz-8_00-GTs-Intel-QPI).

[65] https://www.power.org/home/Blue_Gene_Q_Super_Efficient_Linley.pdf.

[66] Takahashi R. et al. Hybrid Optoelectronic Router for Future Energy-efficient, Large-capacity, and Flexible OPS Networks. Proceedings of IEEE Photonics Society Summer Topical Meeting Series 2011; 153-154, 18-20 July 2011, Montreal, QC, Canada.

[67] Hofmann W. et al. 44 Gb/s VCSEL for optical interconnects. Proceedings of OFC/NFOEC 2011, Paper PDPB5, 6-10 March 2011, Los Angeles, CA, USA.

[68] Geis M. W. et al. CMOS-Compatible All-Si High-Speed Waveguide Photodiodes With High Responsivity in Near-Infrared Communication Band. IEEE Photonics Technology Letters 2007; 19(3) 152-154.

Permissions

The contributors of this book come from diverse backgrounds, making this book a truly international effort. This book will bring forth new frontiers with its revolutionizing research information and detailed analysis of the nascent developments around the world.

We would like to thank Ki Young Kim, for lending his expertise to make the book truly unique. He has played a crucial role in the development of this book. Without his invaluable contribution this book wouldn't have been possible. He has made vital efforts to compile up to date information on the varied aspects of this subject to make this book a valuable addition to the collection of many professionals and students.

This book was conceptualized with the vision of imparting up-to-date information and advanced data in this field. To ensure the same, a matchless editorial board was set up. Every individual on the board went through rigorous rounds of assessment to prove their worth. After which they invested a large part of their time researching and compiling the most relevant data for our readers. Conferences and sessions were held from time to time between the editorial board and the contributing authors to present the data in the most comprehensible form. The editorial team has worked tirelessly to provide valuable and valid information to help people across the globe.

Every chapter published in this book has been scrutinized by our experts. Their significance has been extensively debated. The topics covered herein carry significant findings which will fuel the growth of the discipline. They may even be implemented as practical applications or may be referred to as a beginning point for another development. Chapters in this book were first published by InTech; hereby published with permission under the Creative Commons Attribution License or equivalent.

The editorial board has been involved in producing this book since its inception. They have spent rigorous hours researching and exploring the diverse topics which have resulted in the successful publishing of this book. They have passed on their knowledge of decades through this book. To expedite this challenging task, the publisher supported the team at every step. A small team of assistant editors was also appointed to further simplify the editing procedure and attain best results for the readers.

Our editorial team has been hand-picked from every corner of the world. Their multi-ethnicity adds dynamic inputs to the discussions which result in innovative outcomes. These outcomes are then further discussed with the researchers and contributors who give their valuable feedback and opinion regarding the same. The feedback is then collaborated with the researches and they are edited in a comprehensive manner to aid the understanding of the subject.

Apart from the editorial board, the designing team has also invested a significant amount of their time in understanding the subject and creating the most relevant covers. They scrutinized every image to scout for the most suitable representation of the subject and create an appropriate cover for the book.

The publishing team has been involved in this book since its early stages. They were actively engaged in every process, be it collecting the data, connecting with the contributors or procuring relevant information. The team has been an ardent support to the editorial, designing and production team. Their endless efforts to recruit the best for this project, has resulted in the accomplishment of this book. They are a veteran in the field of academics and their pool of knowledge is as vast as their experience in printing. Their expertise and guidance has proved useful at every step. Their uncompromising quality standards have made this book an exceptional effort. Their encouragement from time to time has been an inspiration for everyone.

The publisher and the editorial board hope that this book will prove to be a valuable piece of knowledge for researchers, students, practitioners and scholars across the globe.

List of Contributors

Amir Reza Sadrolhosseini
Center of Excellence for Wireless and Photonics Networks (WiPNet), Faculty of Engineering, Universiti Putra Malaysia, UPM Serdang, Malaysia

A. S. M. Noor
Center of Excellence for Wireless and Photonics Networks (WiPNet), Faculty of Engineering, Universiti Putra Malaysia, UPM Serdang, Malaysia
Department of Computer and Communication Systems Engineering, Faculty of Engineering, Universiti Putra Malaysia, UPM Serdang, Malaysia

Mohd. Maarof Moksin
Department of Physics, Faculty of Science, Universiti Putra Malaysia, Serdang, Malaysia

Taikei Suyama, Akira Matsushima and Yoichi Okuno
Graduate School of Science and Technology, Kumamoto University, Kurokami, Kumamoto, Japan

Toyonori Matsuda
Kumamoto National College of Technology, Suya, Nishigoshi, Japan

V.A.G. Rivera, F.A. Ferri and E. Marega Jr.
Instituto de Física de São Carlos, INOF/CEPOF, USP, São Carlos – São Paulo, Brazil

Yu-Hang Yang and Ta-Jen Yen
Department of Materials Science and Engineering, National Tsing Hua University, Hsinchu, Taiwan

K. Sathiyamoorthy and V.M. Murukeshan
Nanyang Technological University, School of Mechanical and Aerospace Engineering, Singapore

C. Vijayan
Department of Physics Indian Institute of Technology Madras Chennai, India

Young Chul Jun
Center for Integrated Nanotechnologies (CINT), Sandia National Laboratories, NM, USA

G. Ruffato, G. Zacco and F. Romanato
University of Padova, Department of Physics "G. Galilei", Padova, Italy
Laboratory for Nanofabrication of Nanodevices (LaNN - Venetonanotech), Padova, Italy
Istituto Officina dei Materiali IOM-CNR National Laboratory, Trieste, Italy

Jacak Witold
Institute of Physics, Wrocław University of Technology, Wyb. Wyspia´nskiego 27, 50-370 Wrocław, Poland

Mahnaz M. Abdi
Deprtemet of Chemistry, Faculty of Science, Universiti Putra Malaysia, Serdang, Selangor, Malaysia
Institute of Tropical Forestry and Forest Products (INTROP), Universiti Putra Malaysia, Serdang, Selangor, Malaysia

Wan Mahmood Mat Yunus
Department of Physics, Faculty of Science, Universiti Putra Malaysia, Serdang, Selangor, Malaysia

Majid Reayi
School of Chemical Sciences and Food Technology, Faculty of Science and Technology, Universiti Kebangsaan Malaysia, Bangi, Selangor D.E., Malaysia

Afarin Bahrami
Department of Physics, Faculty of Science, Universiti Putra Malaysia, Serdang, Selangor, Malaysia
Department of Physics, Faculty of Science, Islamic Azad University, Eslamshahr Branch, Iran

Qin Chen
Suzhou Institute of Nano-Tech and Nano-Bionics, Chinese Academy of Sciences, China
University of Glasgow, United Kingdom
SOC Key Laboratory, Peking University Shenzhen Research Institute, China

Xiaohua Shi
Suzhou Institute of Nano-Tech and Nano-Bionics, Chinese Academy of Sciences, China

Yong Ma
University of Glasgow, United Kingdom

Jin He
Peking University Shenzhen SOC Key Laboratory, PKU-HKUST Shenzhen-Hongkong Institution,
Shenzhen, P.R. China

Fuyi Chen, Jian Liu and Negash Alemu
State Key Laboratory of Solidification Processing, Northwestern Polytechnical University, Xian, China

Printed in the USA
CPSIA information can be obtained
at www.ICGtesting.com
JSHW011503221024
72173JS00005B/1188